Electronic Amplifiers

CHARLES H. EVANS

VAN NOSTRAND REINHOLD COMPANY
NEW YORK CINCINNATI TORONTO LONDON MELBOURNE

621.3815

E 92 e

Published in 1979 by Van Nostrand Reinhold Company
A division of Litton Educational Publishing, Inc.
135 West 50th Street, New York, NY 10020, U.S.A.

Van Nostrand Reinhold Limited
1410 Birchmount Road
Scarborough, Ontario M1P 2E7, Canada

Van Nostrand Reinhold Australia Pty. Ltd.
17 Queen Street
Mitcham, Victoria 3132, Australia

Van Nostrand Reinhold Company Limited
Molly Millars Lane
Wokingham, Berkshire, England

16 15 14 13 12 11 10 9 8 7 6 5 4 3 2 1

Library of Congress Cataloging in Publication Data

Evans, Charles H.   date
    Electronic amplifiers.

    Includes index.
    1. Amplifiers (Electronics) I. Title.
TK7871.2.E94   621.3815'35   78-12795
ISBN 0-442-22341-2

050267

Preface

ELECTRONIC AMPLIFIERS is designed primarily for the individual who understands introductory electronics and the ac and dc circuits. The mathematics level of the text requires the ability to apply the algebraic and trigonometric principles, including polar coordinates. Calculus is not required to solve the numerous problems included.

A discussion of useful equations and definitions is included in the Appendix. In addition, a brief review of semiconductor theory is included. The symbols used throughout are based on the standards of the Institute of Electronic and Electrical Engineers, Inc. for vacuum tubes and semiconductors. These standards are commonly used in industry.

This book is oriented more toward analysis of circuits than design. However, design elements are included because modern industrial practice often requires the technician to modify circuits to improve their performance. When a modification involves a change in a transistor or the addition of transistor(s), the technician must be able to bias the device properly. The technician should know how to *use* a transistor or integrated circuit to improve the performance of a circuit.

Chapter 1 presents basic amplifier theory which is common to all amplifying devices. The emphasis in this discussion is on the fact that an electronic amplifier can be made from a wide variety of devices, such as the bipolar transistor, vacuum triode, and field effect transistor, among others. The basic problem is that of developing an amplifier theory that can be applied to any of these devices, as well as others that are not described here and may not even be available yet at the present state of the technology. It is an inescapable fact that industrial technology today is growing more rapidly that curricula can be expanded to educate technicians to enter the industry. To achieve the extent of learning necessary to equip the technician with the tools needed to perform in modern society, more and more time will be required unless some common ground can be established upon which his skills can be built. So, it is not by any accident of design that this text discusses amplifier theory in detail and implies that the devices used are only ancillary to it.

Included with the discussion on the use of many of the amplifying devices are illustrative circuits. These circuits are described with sufficient detail to enable the reader to build them.

In addition to the numerous circuit examples included in the text, each chapter contains numerous review questions and problems to provide extensive practice in applying the theory presented to solve practical circuits.

Each chapter concludes with one or more laboratory experiments which are based on a "discovery/learning" philosophy rather than a "cookbook" approach. The aim in these experiments is to stress the skills of technical communication and creative analysis. These experiments have been thoroughly tested in the laboratory by the author's students over a period of several years.

Finally, each chapter contains a section "Extended Study Topics." These questions and problems further challenge and test the reader's developing skills and comprehension of amplifier theory.

ABOUT THE AUTHOR:

Charles H. Evans received his Bachelor of Electrical Engineering Degree from Clemson University, South Carolina. He has many years' experience in the field of electronics and is currently associated with the Milwaukee Area Technical College, Wisconsin. He is a Registered Professional Engineer in the state of Wisconsin.

Contents

1
Basic
Amplifier Theory

OBJECTIVES

After studying this chapter, the student will be able to

- state a theory of amplifiers that is applicable to all amplifying devices
- define linear and nonlinear amplification (gain)
- explain the construction and use of characteristic curves for a generalized amplifying device
- describe the regions of operation of a generalized amplifying device

DEFINITION OF AN AMPLIFIER

In comparison with other animals and the forces of nature, human beings are weak physically. However, human intelligence has led to the development of devices which can harness energy so that very little physical ability is needed to perform difficult tasks. Having the right equipment or tools, the average person can achieve fingertip control of an ocean liner, a large aircraft, or a speeding land vehicle. A tool called an *amplifier* is used to augment the physical ability of a human being. Such a device accepts an input signal and draws on an external power source to produce an output signal which is a desired function of the input signal. The signal may be a voltage, a current, or some nonelectrical quantity such as pressure or velocity. The amplifying device can be a transistor, gyroscope, relay, or even a large electric generator. An amplifying device is shown in block diagram form in figure 1-1, page 2.

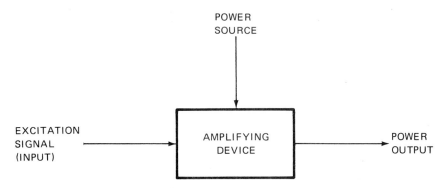

Fig. 1-1 Block diagram of a power amplifying device

Signal

The term *signal* means any driving impulse. This impulse may be an electrical potential or a mechanical movement. In communications, a signal merely contains intelligence that is to be transmitted. In a control system, the signal is a very small power or voltage value driving a small pilot device. This device, in turn, drives another device requiring a greater power input. In this case, it can be seen how the amplification of the original small signal provides control for a device requiring a larger signal.

Amplification Factor; Nonlinear and Linear Amplification

Figure 1-2 shows several familiar devices that meet the general requirements of an amplifier. Each device has an input signal (X) multiplied by an amplification factor (A) to produce an output signal (Y). The following equation shows the amplification factor expressed in terms of the output and input signals.

$$A = \frac{\text{Output Signal (Y)}}{\text{Input Signal (X)}} \qquad \text{Eq. 1.1}$$

An *efficient* (effective) amplifier has a value of A much greater than one. Ideally, the factor A is contant over a range of conditions. This means that although the input signal may be large or small, or varying rapidly or slowly, it is amplified by the same value of A.

A *practical* amplifier cannot meet these requirements exactly. The ideal device is approximated in practice by restricting the input signal to small values which can be amplified by a constant value of A without distortion. This process is known as *linear amplification.* If *nonlinear amplification* is permitted, some change in the factor A is allowed, depending on the application.

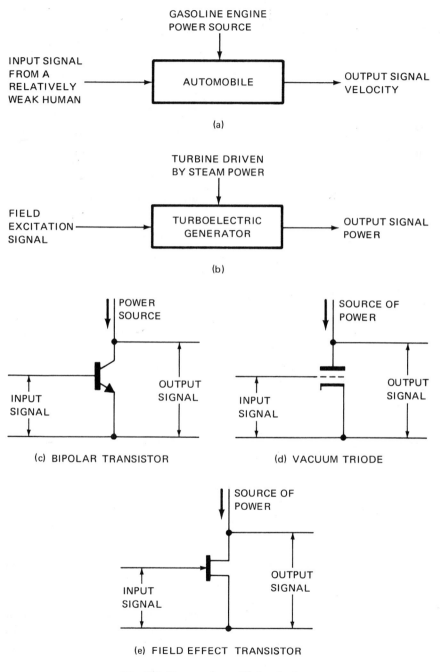

Fig. 1-2 Types of amplifying devices

The Amplifier as a Three-Terminal Electronic Device

In the field of electronics, the vacuum triode was the first three-terminal device developed for use as an amplifier. Following the invention of the solid-state transistor, it was evident that other three-terminal amplifying devices were much better suited for many applications than vacuum triodes. In recent years, developments in semiconductor electronics have increased the number of alternative devices capable of electronic amplification.

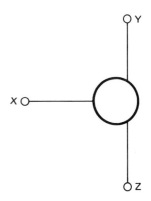

Fig. 1-3 Generalized amplifying device

For any three-terminal device, the basic theory of amplification is the same. A generalized amplifying device will be used in the following paragraphs to illustrate the principle of amplification. Figure 1-3 represents a three-terminal device which can be used to amplify power, current, voltage, or a combination of these signals.

List some devices that fit the general category of amplifier. Try to name several devices that are not mentioned in the previous paragraphs.   *(R1-1)*

CHARACTERISTIC CURVES

A single volt-ampere curve of a generalized amplifying device does not adequately describe its behavior for the following reasons: (1) the device has three terminals and (2) only a two-dimensional representation can be made of its characteristics. As a result, the volt-ampere characteristics of the generalized amplifying device are indicated by a family of curves as shown in figure 1-4.

The voltages and currents indicated on the characteristic curves are identified as follows:

I_Y = Current flowing into the Y terminal
I_X = Current flowing into the X terminal
I_Z = Current flowing into the Z terminal
V_{XZ} = Voltage drop from X to Z terminals
V_{YZ} = Voltage drop from Y to Z terminals
V_{YX} = Voltage drop from Y to X terminals

The numerical subscripts (as in I_{X1} and I_{X2}) refer to specific *constant* values of the quantities listed. When the quantities are continually changing with time, they are represented by lower case symbols, such as i_{x1}, i_{x2}, and i_{x3}.

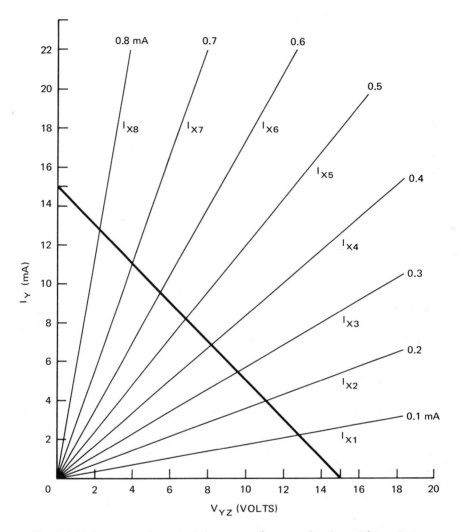

Fig. 1-4 Volt-ampere characteristic curves of a generalized amplifying device

Current Amplification

The curves in figure 1-4 are the graphical solution of the following equation for a low-impedance device such as the bipolar transistor:

$$I_Y = f (V_{YZ}, I_X) \qquad \text{Eq. 1.2}$$

Equation 1.2 reads, "I_Y is a function of V_{YZ} and I_X."

If it is assumed that a current controlled device is represented by Equation 1.2, then the amplification of the device is expressed as:

$$A_I = \frac{\triangle I_y}{\triangle I_x} = \frac{I_{y1} - I_{y2}}{I_{x1} - I_{x2}} \qquad \text{Eq. 1.3}$$

where V_{yz} has some constant value. This means the current amplification (A_I) is the ratio of change in the output current to the change in the input current. A_I is the actual amplification of an amplifier using this device, if it operates as a perfect current amplifier having zero output voltage.

Voltage Amplification

If the generalized amplifying device is a high-impedance device such as a vacuum triode or field effect transistor, the characteristic curves represent a graphical solution of:

$$I_Y = f (V_{YZ}, V_{XZ}) \qquad \text{Eq. 1.4}$$

where both V_{YZ} and V_{XZ} are independent variables having an effect on I_Y. For the high-impedance device, $I_{X1}, I_{X2}, I_{X3}, \ldots I_{Xn}$ in figure 1-4 are replaced by $V_{XZ1}, V_{XZ2}, V_{XZ3}, \ldots V_{XZn}$.

The amplification of a high-impedance device is expressed as:

$$A_v = \frac{\triangle V_{yz}}{\triangle V_{xz}} = \frac{V_{yz1} - V_{yz2}}{V_{xz1} - V_{xz2}} \qquad \text{Eq. 1.5}$$

In this equation, the current I_y has a constant value. The voltage amplification is the ratio of the output voltage to the input voltage. Thus, A_v is the actual amplification of an amplifier using this device, if it operates as a perfect voltage amplifier having zero output current.

Plotting the Characteristic Curves

The amplification, or gain, of an amplifier can be calculated from curves such as those shown in figure 1-4. These curves are produced by plotting measured currents and voltages for a circuit such as the one in figure 1-5.

In any circuit analysis problem, the current directions selected are abitrary. As a result, the calculated value may have a sign opposite to the one assumed. To simplify the following general discussion, it is assumed that the current flowing *into* the device terminals is the positive direction of current.

The following method shows how the voltage and current values for the circuit in figure 1-5 are determined and used to plot characteristic curves similar to those in figure 1-4.

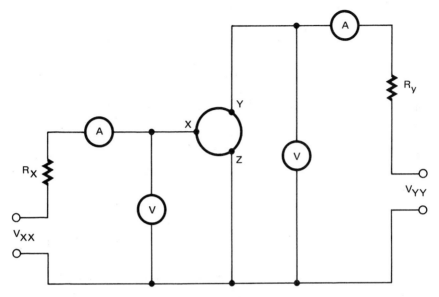

Fig. 1-5 **Circuit for measuring the volt-ampere characteristics of a generalized amplifying device**

1. Using the circuit in figure 1-5, set I_X to a specific constant value by varying R_x or V_{XX}.
2. Vary either V_{YY} or R_y to change I_Y.
3. Measure V_{YZ} for various values of I_Y.
4. Plot a curve of V_{YZ} versus I_Y for the set value of I_X.
   (Assume that the circuit being considered contains a low-impedance device that matches the characteristics shown in figure 1-4. This assumption suggests the use of the current I_X as the variable being controlled by the voltage V_{XX} and the resistance R_x. This value of I_X is called I_{X1}.)
5. Reset I_X to a new value, I_{X2}.
6. Vary I_Y and observe the effect on V_{YZ}.
7. Plot I_Y versus V_{YZ} for $I_X = I_{X2}$.
8. Continue to adjust V_{XX} and R_x to obtain new values of I_X.
9. For each value of I_X, plot I_Y versus V_{YZ}. In this manner, a family of curves such as those shown in figure 1-4 is obtained for the device.

The curves in figure 1-4 are known as common Z output curves because they show the volt-ampere characteristics of the device between terminals Y and Z. Y and Z are output terminals and X and Z are input terminals. Thus, a circuit such as the one shown in figure 1-5 is a *common Z circuit* because the input and output are common to the Z terminal.

Input Characteristics

Another set of curves, known as input characteristic curves, are useful in analyzing amplifier performance. These curves are used to represent the effect of the generalized amplifying device on a signal source.

The following method is used to obtain input characteristic curves.

1. Set I_Y to a specific value and let $V_{YZ} = V_{YZ1}$.
2. Measure the current I_X and the voltage V_{XZ} while varying I_X and holding V_{YZ} at V_{YZ1}.
3. Plot a curve of I_X versus V_{XZ} for $V_{YZ} = V_{YZ1}$.
4. Change the value of I_Y and set V_{YZ} to a new value V_{YZ2}.
5. Vary I_X and take new measurements of V_{XZ} and I_X with V_{YZ} constant at V_{YZ2}.
6. Plot a new curve of I_X versus V_{XZ} for $V_{YZ} = V_{YZ2}$. In this manner, a family of input characteristic curves for a generalized amplifying device is obtained as shown in figure 1-6.

What electronic devices can be represented by curves such as those in figures 1-4 and 1-6? No specific devices are represented because these curves reflect a hypothetical device for which there is no exact model. However, the performance of a wide variety of three-terminal devices can be described by curves similar to those in figures 1-4 and 1-6. Some of these devices are listed in Table 1-1.

Bipolar npn transistor
Bipolar pnp transistor
Junction field effect transistor (JFET or FET)
Metal oxide semiconductor field effect transistor (MOSFET)[*]
Silicon controlled rectifier (SCR)
Vacuum triode
Vacuum pentode[*]
Vacuum tetrode[*]
Unijunction transistor
Electromagnetic relay
Thyratron

[*]   These devices have more than three terminals, but only three are used for the basic amplifying action.

Table 1-1 Amplifying devices

The input signal to a high-impedance generalized amplifying device changes from 1 to 2 volts, and the output voltage changes from 100 volts to 50 volts. What is the amplification of the device?   *(R1-2)*

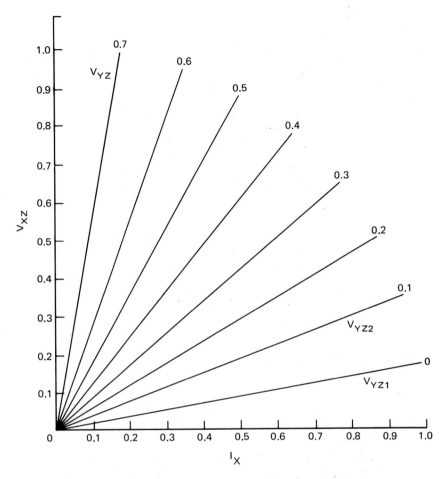

Fig. 1-6 Input characteristics of a generalized amplifying device

 The input signal to a low-impedance generalized amplifying device changes from 1 milliampere (mA) to 10 milliamperes and the output current changes from 1.2 amperes to 2 amperes. What is the amplification of the device? (R1-3)

Regions of Operation of the Generalized Amplifying Device

 There are three regions of operation of the generalized amplifying device represented by figure 1-4.

Saturation Region. This region of operation occurs at very low values of V_{YZ} and generally high values of I_Y. The region is considered to be the *on* or *saturated* state of the device. A specific extreme value of V_{XZ} or I_X is required to put the device into this region.

Cutoff Region. This region of operation occurs at very low values of I_Y and moderately high values of V_{YZ}. The region is considered to be the *off* state of the device. Cutoff is obtained by a value of I_X or V_{XZ} which is at the opposite extreme to the value required to operate the device in the saturation region.

Active Region. This region of operation is the linear amplification region for many devices. It is the region between the cutoff and saturation regions. The device is most sensitive to the V_{XZ} or I_X input signal in the active region. This region is used to amplify an input signal without distortion.

Many of the devices listed in Table 1-1 are not used in all three regions of operation. Some of the devices are designed for switching applications. In these applications, the devices are used only in the cutoff and saturation regions. The silicon controlled rectifier, the unijunction transistor, and a number of specially designed bipolar transistors are examples of devices used for switching applications.

Using figure 1-4, identify the three regions of operation for a generalized amplifying device.   *(R1-4)*

THE RELAY AS AN AMPLIFIER

An electromagnetic relay can be regarded as a three-terminal amplifying device. A schematic diagram of a relay is shown in figure 1-7.

A voltage across terminals 1 and 2 provides an electromagnetic force which closes a lever or movable contact between terminals 3 and 4. With the

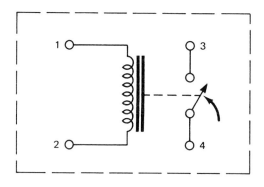

Fig. 1-7 Schematic for an electromagnetic relay

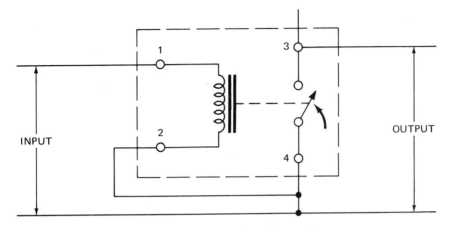

Fig. 1-8 Relay with a common connection between the input and output circuits

contact closed, there is continuity between terminals 3 and 4. Since only a very small force is required to close the contacts, the power signal from terminals 1 to 2 providing this force can also be very small. However, a very large current can flow through terminals 3 and 4 when the contacts close. With this large current, a device such as a motor or valve operator can be supplied with a large amount of electrical energy.

Refer to figure 1-8 and assume that terminals 2 and 4 of the relay are connected together so that the signal source and the power output circuit

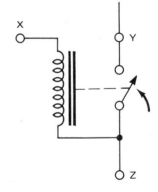

Fig. 1-9 The relay as a three-terminal device

have a common connection. In this form, the electromagnetic relay schematic resembles that of a three-terminal amplifying device. For example, compare the diagrams in figures 1-3 and 1-9. A signal applied to X produces an output at Y and the conditions for amplification are satisfied. In fact, a set of output characteristic curves can be drawn for this device, as shown in figure 1-10, page 12. A circuit using a relay as an amplifier is shown in figure 1-11, page 12.

A relay has a 500-ohm coil. The relay operates when the voltage across the coil is changed from 0 to 12 volts. The relay contacts are rated at 5 amperes maximum. What is the amplification of this relay used as a current amplifier? (R1-5)

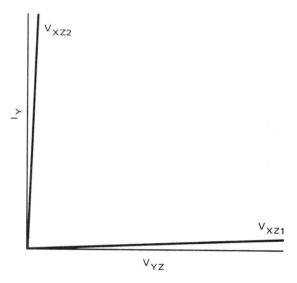

Fig. 1-10   Output characteristic of the relay considered as a three-terminal amplifying device

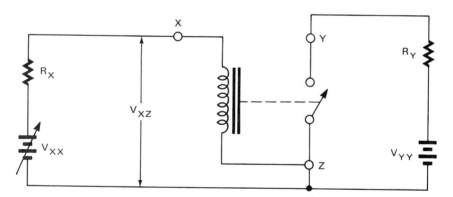

Fig. 1-11   Relay used as a three-terminal amplifying device in an amplifier circuit

LINEAR AND NONLINEAR AMPLIFICATION

An electromagnetic relay is a nonlinear amplifying device. The characteristics of figure 1-10 show that a linear operating region does not exist between V_{XZ1} and V_{XZ2}. At a value of V_{XZ1} or lower, the relay contacts do not operate; therefore, regardless of the value of V_{YZ}, I_Y can never become greater than zero. If the value of R_x or V_{XX} is adjusted so that the applied voltage across X to Z is high enough, the relay contacts operate. As a result, a high current capability exists through Y to Z and a high value of I_Y can exist at a low value of V_{YZ}. In this manner it is possible to obtain a considerable current, voltage, or power amplification. However, this amplification is not linear, severely limiting the relay as an amplifying device. Vacuum triodes and various types of transistors are not limited because they have a controllable active region.

Depending upon the requirements, an amplifier can be either linear or nonlinear. For a linear amplifier, the actual shape of the input signal as it varies with time is reproduced at the output. For a nonlinear amplifier, only gross amplification is possible. A relay is an example of the latter case, where a relatively weak signal is used to turn a large power device completely on or completely off.

Many electronic amplifying devices are used for switching purposes, just as relays are. Although these devices have linear operating regions, they are rarely used as linear amplifiers. Rather, these devices are operated either in the saturation region (*on* state) or in the cutoff region (*off* state). Many of these devices can have more or less linear operation depending on the specific setting of the operating point and the relative magnitude of the input signal.

Explain the difference between linear and nonlinear amplification. (R1-6)

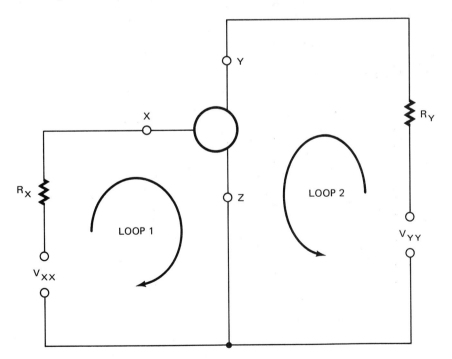

Fig. 1-12 Basic circuit of a generalized amplifying device used as an amplifier.

AMPLIFIER CIRCUITS

A basic amplifier circuit is shown in figure 1-12. It is assumed that current flows *into* the X and Y terminals. The polarity of the voltages V_{YY} and V_{XX} depends upon the amplifying device used. In this circuit, V_{YY} and V_{XX} are assumed to have the polarity necessary to drive the current into the X and Y terminals.

Apply Kirchhoff's Voltage Law* around Loop 2 to obtain

$$V_{YY} - I_Y R_y - V_{YZ} = 0 \qquad \text{Eq. 1.6}$$

Draw a *load line* on the characteristic curves for the device by finding the end points and drawing a straight line through them. This straight line has a slope equal to I/R_y. Figure 1-13 shows a load line constructed on the characteristic curves.

The I_Y intercept used to determine one end of the load line is found by setting $V_{YZ} = 0$ and solving Equation 1.6 for I_Y.

$$I_Y = \frac{V_{YY}}{R_y} \qquad \text{Eq. 1.7}$$

The V_{YZ} intercept for the other end of the load line is determined by setting $I_Y = 0$ in Equation 1.6 and solving for V_{YZ}.

$$V_{YZ} = V_{YY} \qquad \text{Eq. 1.8}$$

Quiescent Point

A satisfactory operating point for linear operation is in the approximate center of the characteristic curves, along the load line. This operating point is called the *Q-point* or *quiescent point.*

For any input signal, the resulting output is traced along the load line above and below the quiescent point. In figure 1-13, note that the output voltage decreases as the input signal increases. This behavior is characteristic of many amplifying devices.

The gain of the generalized amplifying device is modified when it is used in an amplifier circuit. The actual gain can be calculated from the load line. The current amplification is given by

$$A_I = \frac{I_{Y1} - I_{Y2}}{I_{X1} - I_{X2}} \qquad \text{Eq. 1.9}$$

* See the Appendix for a discussion of Kirchhoff's Laws.

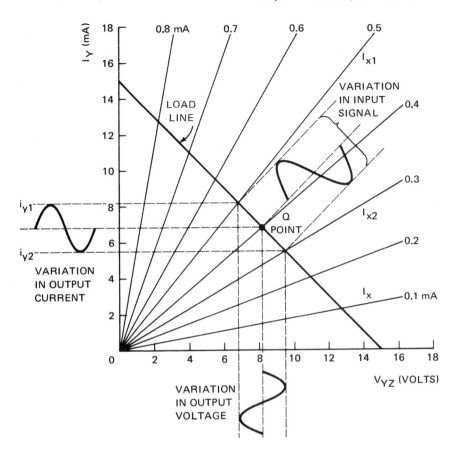

Fig. 1-13 Result of applying an input signal to a generalized amplifying device, showing the resulting output signal

where changes in I_Y and I_X are taken along the load line. By selecting arbitrary locations along the load line about the Q-point in figure 1-13, the gain A_I for the generalized amplifying device is determined as follows:

$$A_I = \frac{(8) - (5.3)}{(0.5) - (0.3)} = 13.5$$

Figure 1-14, page 16, shows that three circuit configurations are possible with three-terminal generalized amplifying devices. Each configuration has certain relative advantages and disadvantages, such as high or low output impedance, high or low input impedance, or high or low voltage gain. Each configuration has its place in a complex system. Most of the devices listed in Table 1-1 can be used in more than one circuit configuration.

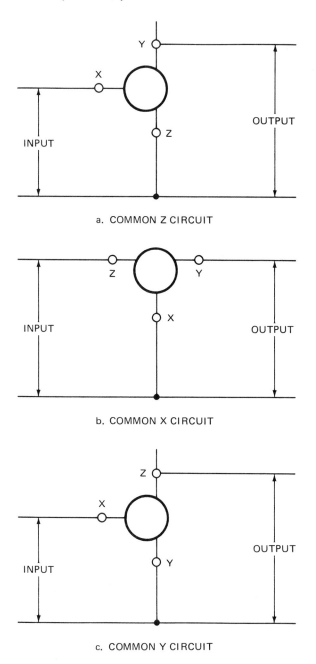

a. COMMON Z CIRCUIT

b. COMMON X CIRCUIT

c. COMMON Y CIRCUIT

Fig. 1-14   Three circuit configurations in which the generalized amplifying device is used as an amplifier

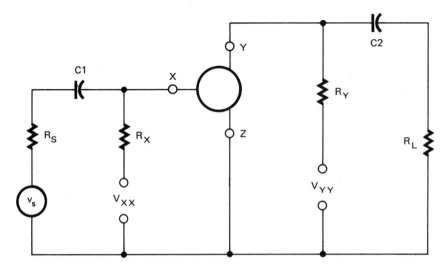

Fig. 1-15 RC coupled amplifier circuit

The device described by figure 1-13 has its Q-point located as described in the previous paragraphs and as defined by Equations 1.7 and 1.8. The input signal I_X changes from 0.2 milliamperes below the Q-point to 0.6 milliamperes above the Q-point. What is the change in output voltage V_{YZ}? What is the change in output current I_Y? What is the current amplification, using this change in input signal? (R1-7)

RC COUPLED AMPLIFIER

A common amplifier circuit is the RC coupled amplifier shown in figure 1-15. The input signal is coupled to the amplifier through a series capacitor. The amplifier is coupled to the load through a series capacitor.

The capacitors are used to prevent the flow of dc bias current through the load or the signal source. Thus, the signal source and load are not affected by quiescent currents and voltages. In addition, any variations in the source and load circuits do not interfere with the proper biasing conditions.

It is assumed that the signal v_s varies sinusoidally with time so that it has the waveform shown in figure 1-16, page 18. It is also assumed that at the input signal frequency the capacitors behave as though a wire is shorting them. (Refer to the Appendix for a discussion of how capacitors behave in the presence of sinusoidally varying electrical signals.) For an ac signal only, the circuit can be represented by figure 1-17, page 18.

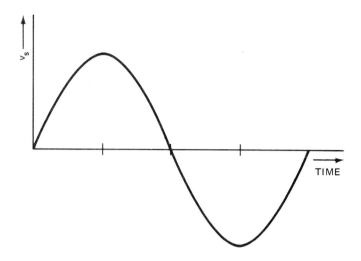

Fig. 1-16 Sinusoidal variation of amplifier input signal

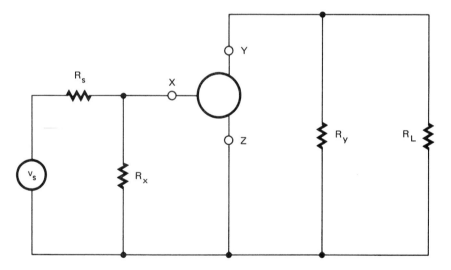

Fig. 1-17 Ac equivalent circuit of RC coupled amplifier with capacitors shorted

NOTE: lower case letters are used in the following analysis to represent the subscripts of voltage and current signals when dealing with ac rms values. Since the meters used in ac circuits are usually calibrated in rms quantities, these quantities are used occasionally in circuit analyses. Instantaneous values of the ac signals are indicated by the use of lower case letters for current and voltage symbols as well as their subscripts. Instantaneous values of the total voltages and currents (ac and dc) are indicated by the use of lower case letters for the symbols and upper case letters for the subscripts. (See the Appendix.)

Additional assumptions are made as follows for the circuit in figure 1-17:

1. the amplifying device has a negligibly low impedance between X and Z
2. the characteristic curves of the device match those of figure 1-4
3. R_X is so large by comparison that it is considered to be infinite.

The following equation expresses the approximate value of i_X.

$$i_x = \frac{v_s}{R_s}$$

Eq. 1.10

Using an Ac Load Line for Analysis

The performance of the circuit can be determined by referring to the curves in figure 1-13. The value of i_Y varies along a load line established by the parallel combination of R_y and R_L. A new load line is drawn through the quiescent point as shown in figure 1-18, page 20. This line is called the ac load line and has the slope

$$\frac{R_L + R_y}{R_L R_y}$$

The voltage gain of the amplifier circuit is given as follows:

$$A_v = \frac{v_{YZ1} - v_{YZ2}}{v_{S1} - v_{S2}} = \frac{v_{yz}}{v_s}$$

Eq. 1.11

To determine the voltage gain, it is assumed that i_X varies between 0.2 milliampere and 0.6 milliampere with the quiescent value (I_{XQ}) equal to 0.4 milliampere. The value of v_{YZ} varies between 6.6 volts and 9.8 volts for the values of i_X (0.2 milliampere to 0.6 milliampere) along the ac load line. Combining Equations 1.10 and 1.11 gives a new expression for the voltage gain.

$$A_v = \frac{v_{YZ1} - v_{YZ2}}{(i_{X1} - i_{X2})R_s}$$

Eq. 1.12

$$A_v = \frac{9.8 - 6.6}{0.2 - 0.6}\left(\frac{1}{R_s}\right) = (-8000)\left(\frac{1}{R_s}\right)$$

If R_s is 100 ohms, the voltage gain or amplification of the circuit is -80. The negative sign means that as v_S increases (becomes more positive), v_{YZ} decreases (becomes more negative).

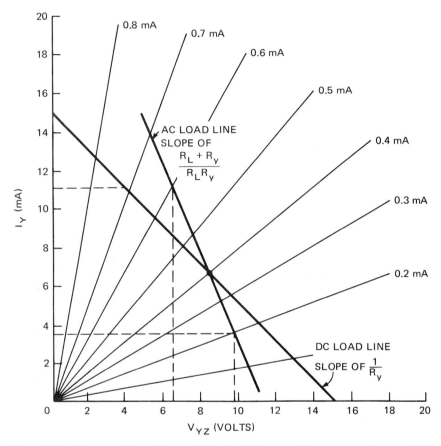

Fig. 1-18   Construction of ac and dc load lines on the output characteristic curves of a generalized amplifying device for an RC coupled amplifier

The current gain (A_I) of the amplifying device is determined as follows:

$$A_I = \frac{i_{Y1} - i_{Y2}}{i_{X1} - i_{X2}} = \frac{3.6 - 11.3}{0.2 - 0.6} = 19.3$$

NOTE: if the current gain is defined as the ratio of the load current through R_L to $\triangle i_X$, then the gain value is reduced by the shunting effect of R_X.

Power Gain

The power amplification is expressed as

$$A_p = \frac{\text{Output power}}{\text{Input power}} = \frac{(V)\,(I)\text{ out}}{(V)\,(I)\text{ in}} \qquad \text{Eq. 1.13}$$

$$A_p = \frac{(v_{YZ1} - v_{YZ2})}{(v_{S1} - v_{S2})} \frac{(i_{Y1} - i_{Y2})}{(i_{X1} - i_{X2})} \qquad \text{Eq.1.14}$$

Substituting Equations 1.9 and 1.11 into Equation 1.14 and solving for A_p yields

$$A_p = -A_v A_I \qquad \text{Eq. 1.15}$$

The minus sign is required to insure that the power value is positive . In other words, Equation 1.15 must have a minus sign to cancel out the minus sign of the voltage amplification (A_v). Numerical values taken from figure 1-18 are substituted in Equation 1.15 to yield a value for the power amplification.

$$A_p = - (-80) \; (19.3) = 1\,540$$

Thus, it can be seen that the power of the incoming signal is amplified by a factor of 1 540. The actual power to provide this power increase is supplied by the dc bias supply which sets the quiescent point for the generalized amplifying device.

If the generalized amplifying device used in the circuit in figure 1-17 has a high-input impedance, i_X may be set equal to zero. Equation 1.4 then describes the device and the characteristic curves used are those for constant values of v_{XZ}. The value of v_{XZ} as it varies with time is determined by the voltage divider formed by R_x and R_s in the circuit of figure 1-17. The equation is:

$$v_{XZ} = \frac{R_x v_S}{(R_s + R_x)} \qquad \text{Eq. 1.16}$$

Such a high-impedance device generally is used as a voltage amplifier. As a result, the gain is expressed as a ratio of voltages using Equation 1.11.

EXAMPLE OF AN RC COUPLED AMPLIFIER CIRCUIT

The output characteristics of a high-impedance amplifying device are measured and plotted for various values of v_{XZ}. The characteristics are shown in figure 1-19, page 22. It is desired to use the device in a circuit such as the one shown in figure 1-15. The circuit is to be operated as a linear amplifier.

a. Find a quiescent point for a 100,000-ohm load.

The device characteristics indicate that a value of $V_{YY} = 300$ volts should be satisfactory if this value is below the maximum voltage recommended by the manufacturer. Thus, one end of the load line is established using Equation 1.8.

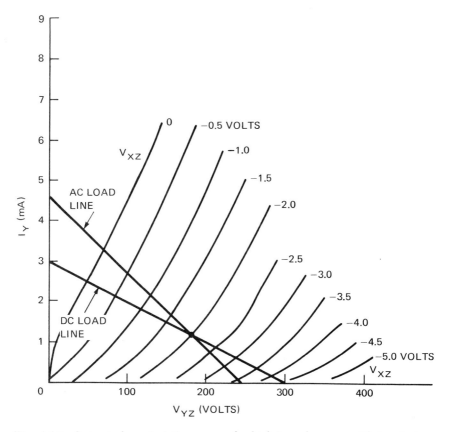

Fig. 1-19 Output characteristic curves of a high-impedance amplifying device. The figure shows the construction of the ac and dc load lines when the device is used in an RC coupled amplifier circuit.

When $V_{YZ} = 0$,

$$I_Y = \frac{V_{YY}}{R_y} = \frac{300}{100,000} = 3 \text{ milliamperes}$$

This value establishes the other end of the load line. Now the dc load line can be drawn on the characteristic curves. A desirable operating point of about -2 volts can be selected for V_{XZ}. Because a high-impedance device is used in this problem, R_x is considered to have a negligible effect. V_{XX} is set at -2 volts to obtain the bias voltage.

b. Assume the following conditions:

1. the load impedance R_L also has a value of 100,000 ohms,
2. the input signal v_s is 2 volts peak-to-peak
3. the capacitors are selected to have negligible resistance

What is the voltage amplification of the circuit?

The ac load line must be used to find the voltage amplification. The slope of the line is I_y/V_{yz}. From the discussion on page 19, the slope is also equal to:

$$\frac{I_y}{V_{yz}} = \frac{1}{\dfrac{R_L \, R_y}{R_L + R_y}}$$

Since R_y and R_L both equal 100,000 ohms, the slope is exactly 1/50,000 or twice that of the dc load line. The ac load line is drawn through the operating point Q at the new slope.

The amplification (A_v) can be read from the characteristic curves in figure 1-19.

$$A_v = \frac{220 - 135}{(-3) - (-1)} = -42.5$$

Figure 1-20, page 24, represents the output characteristics of a low-impedance amplifying device. The manufacturer states that the maximum value of V_{YZ} is 16 volts. (This requirement also limits V_{YY} to 16 volts.) Select an appropriate quiescent point for a bias resistance of 2000 ohms. (R1-8)

Use the device described in question (R1-8) in an RC coupled amplifier circuit such as the one shown in figure 1-15. Assume that the load resistance R_L is 2000 ohms. Assume that (1) i_x varies sinusoidally by 0.2 milliampere peak-to-peak and (2) the capacitors have negligible reactance at the frequency of the signal input current. What is the amplification of this circuit when it is used as a current amplifier? (R1-9)

Assume that the device described in question (R1-8) has the input characteristics shown in figure 1-21, page 24. The device is used in the circuit described in question (R1-9) with $R_s = 0$. For a peak to peak current input signal of 0.2 milliampere, what is the amplification of this circuit when it is used as a voltage amplifier? (R1-10)

Using the results of questions (R1-9) and (R1-10), what is the power amplification of the circuit? (R1-11)

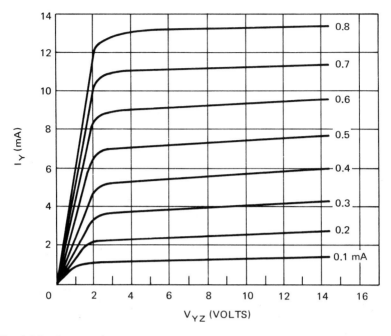

Fig. 1-20 Output characteristic curves for a low-impedance amplifying device described in question (R1-8)

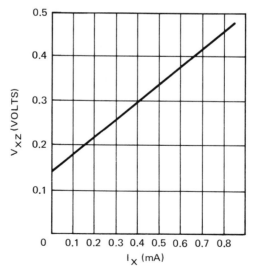

Fig. 1-21 Input characteristics for the low-impedance device described in question (R1-9)

EXTENDED STUDY TOPICS

1. List as many three-terminal electrical devices as possible that can be used as the major component of an amplifying system.

2. What relationship is there between the input and output for a nonlinear amplifier?

3. Why is the quiescent point selected at the approximate center of the active region of an amplifying device if it is to be used as a linear amplifier?

4. Assume that a quiescent point is selected such that it is at cutoff ($I_Y = 0$). If a sinusoidal waveform is used as the input signal, what is the approximate shape of the output waveform?

5. Discuss the possible uses of the input characteristics of an amplifying device in the analysis of an amplifier circuit.

6. Referring to figures 1-19 and 1-20, identify the three operating regions for the devices concerned.

7. Is the current gain calculated from figure 1-18 always the same regardless of the changes in I_X and I_Y? Explain.

2

Bipolar
Transistor
as an Amplifier

OBJECTIVES

After studying this chapter, the student will be able to:

- discuss semiconductor and transistor theory
- explain the function of a transistor amplifier
- analyze and modify a basic transistor amplifier
- make circuit changes to maintain the stable operation of a transistor amplifier

REVIEW OF TRANSISTOR THEORY

Transistors are made from two types of semiconductor crystals: germanium and silicon. Circuit requirements will dictate whether silicon or germanium transistors are to be used for particular applications. A minute amount of an impurity substance is added to each germanium or silicon crystal. As a result, the crystal will have either an abundance of free electrons or an excess of spaces for free electrons *(holes)*. The type of impurity added determines whether the electrons or holes are in excessive supply. The holes are considered to be carriers of electric current (just as electrons are considered to be current carriers). The difference between the two is that *hole current* is opposite in direction to electron current. Semiconductor materials with impurities are classified as follows:

- the semiconductor crystal containing impurities that result in an abundance of free electrons is an *n*-type material.
- the semiconductor crystal containing impurities that result in an abundance of holes is a *p*-type material.

Explain how a hole can be considered to be a carrier of electric current when it has no mass or other distinguishing physical properties.   *(R2-1)*

During the production of semiconductor crystals, it is possible to produce *n*-type regions and *p*-type regions in the same crystal. Various combinations of materials can be maufactured by the careful control of the kinds of impurities added during the process. Transistors and diodes are made of single crystals in which both types of impurities are injected. For example, diodes have one *n*-type region and one *p*- type region. Bipolar transistors have either two *n*-type regions and one *p*-type region, or two *p*-type regions and one *n*-type region.

Forward and Reverse Biased Diodes

The ideal diode allows current to flow in only one direction. If a battery is connected across the terminals of a semiconductor diode and the polarity is as shown in figure 2-1, there is no current. This no-current condition (reverse bias) occurs because the free holes in the *p*-type material and the free electrons in the *n*-type material are driven further away from the junction between the *p*- and *n*-type sections of the crystal by the applied voltage. As a result, there are no free current carriers near the junction between the *n*- and *p*-type materials. If there are no current carriers, then there can be no flow of current through the semiconductor.

If a direct-current voltage is applied across the terminals of the diode and the polarity is as shown in figure 2-2, page 28, then current flows through the diode. The diode is said to be forward biased. In this case, the applied voltage drives the free electrons in the *n*-type material and the free holes in the *p*-type material toward the junction. The free electrons and the free holes at the

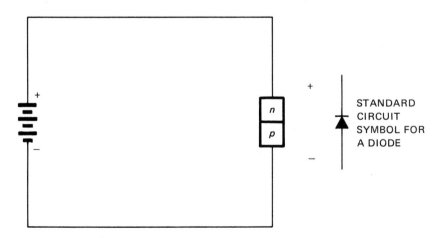

Fig. 2-1 Reverse biased semiconductor diode

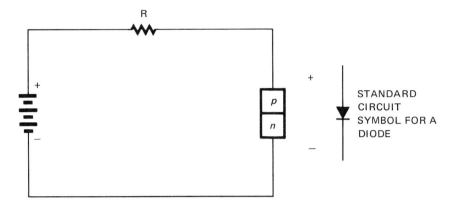

Fig. 2-2 Forward biased semiconductor diode

junction then combine. The applied voltage continuously supplies additional free electrons to the *n*-type material to replace those which have combined with the free holes in the *p*-type material. Since the moving charges are drawn continuously from the voltage supply, there is no absence of free current carriers anywhere in the diode. In the ideal diode, the amount of current carriers is limited by the external resistance (R) in the circuit. As a result, when a diode is used in a circuit, a current limiting resistance must always be present. (Such a current limiting resistance is mandatory for most semiconductor devices. This requirement is due to the fact that as the current increases, the internal resistance of the device decreases. This condition creates an even greater current flow which can destroy the device.) A typical volt-ampere characteristic of a forward biased diode is shown in figure 2-3.

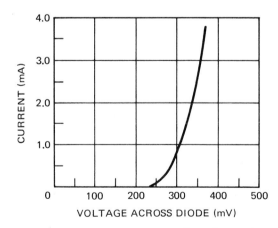

Fig. 2-3 Volt-ampere characteristic curve for a forward biased diode

What is the electrical resistance of an ideal diode which is forward biased?
(R2-2)

What is the electrical resistance of an ideal diode that is reverse biased?
(R2-3)

Definition of a Transistor

A transistor appears to consist of two diodes wich have either their *p*-terminals or their *n*-terminals connected together. Figure 2-4 shows the two possible types of transistors: the pnp transistor in part (a) and the npn transistor in part (b).

In practice, a bipolar transistor does not consist of two separate diodes connected together, but rather is constructed as one single npn or pnp semi-conductor crystal. The terminals of the transistor are called the *emitter,* the

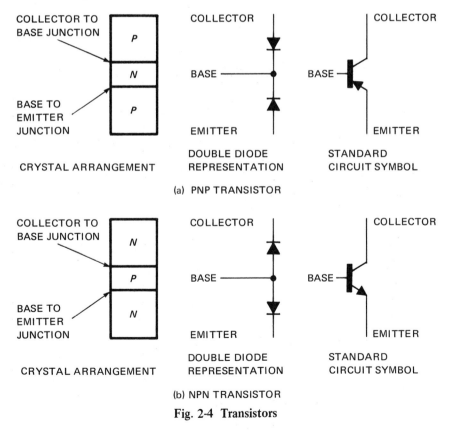

Fig. 2-4 Transistors

base, and the *collector,* as shown in figure 2-4. Note that the base is the terminal which represents the connection between the diodes.

The base is a thin section containing only a small amount of impurities (it is referred to as being *lightly doped*). The emitter and collector sections are larger and more heavily doped. If desired, the bipolar transistor can be used as two separate diodes in an electronic circuit. However, the bipolar transistor has properties that are unlike those of two diodes connected together.

Sketch an npn transistor with its emitter diode forward biased. Show the voltage polarity and a current limiting resistor.   *(R2-4)*

Sketch a pnp transistor with its collector diode reverse biased. Show the proper voltage polarity.   *(R2-5)*

Transistor Circuit Connections and Transistor Currents Defined

Examine the npn and pnp transistors connected in the circuits shown in figure 2-5. By forward biasing one diode with a dc voltage (the base-emitter junction) and by reverse biasing the other diode with another dc voltage (the collector-base junction), a device with amplifying properties is produced.

As a result of the forward biased base-emitter junction, current carriers are introduced into the base. At the same time, reverse biasing of the collector-base junction causes the same type of current carriers to move toward the collector-base junction. The overall flow of current carriers in the base is in the same direction across both junctions (although one junction is forward biased and the other junction is reverse biased). The current through the base terminal is very small when compared to the current flowing between the collector and emitter terminals. An increase in the base current causes a greater number of recombinations of electrons and holes in the base section. As a result, the apparent width of the base section seems thinner and a larger collector to emitter current flows.

Current Gain of the Common Emitter Circuit

In the circuits in figure 2-5, the base-emitter junction is forward biased and a small current is allowed to flow around Loop 1, giving rise to a much larger current, driven by V_{cc}, which flows around Loop 2. In other words, a large change in collector current is controlled by a small change in the base current. Thus, the bipolar transistor can be considered as a current amplifier. The current gain of the bipolar transistor connected in a common emitter configuration is given by:

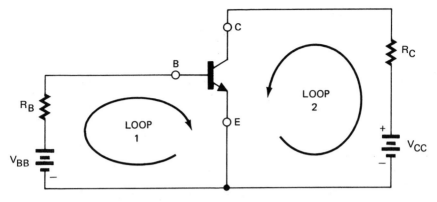

A. NPN TRANSISTOR CONNECTED AS A CURRENT AMPLIFIER

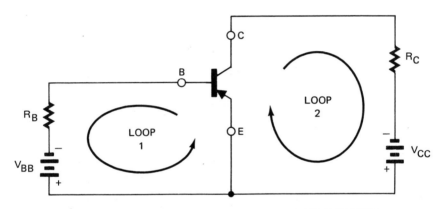

B. PNP TRANSISTOR CONNECTED AS A CURRENT AMPLIFIER

Fig. 2-5

$$A_I = \frac{I_{C1} - I_{C2}}{I_{B1} - I_{B2}} \qquad \text{Eq. 2.1}$$

Compare Equation 2.1 with Equation 1.9. I_{C1} and I_{C2} are the initial and final values of the collector current. I_{B1} and I_{B2} are the initial and final values of the base current.

By examining the diode representation of the transistor as shown in figure 2-4, it can be seen that if the diodes are to be properly biased, the bias voltages for the npn transistor must be opposite in polarity to the bias voltages for the pnp transistor. This is the only difference between the use of an npn or a pnp transistor in a circuit.

If the initial or final values of current are zero in Equation 2.1 and the remaining values are those for the transistor near or in the saturation region, then the current gain A_I is the dc or large signal current gain. This value is commonly known in technical literature and transistor specifications as β (beta) or h_{FE}. If the initial and final current values in the equation are in the active region for a constant value of collector to emitter voltage (V_{CE}) and there is only a small difference between the initial and final values, then the current gain A_I is the ac or small signal current gain. This value is commonly known as β' or h_{fe}. Note that the symbol h_{fe} for the small signal current gain has lower case letters as subscripts and the symbol h_{FE} for the large signal current gain has upper case letters as subscripts. The symbol h_{fe} is one of a large number of so-called h-parameters for a transistor. The h-parameters are used in small signal transistor circuit analysis and are discussed in detail in Chapter 4.

Define the symbols β and h_{FE} with regard to transistor performance.
(R2-6)

An npn transistor is connected into the circuit of figure 2-5 (a). A change in the base current of 10 microamperes (μA) results in a change in the collector current of 0.5 milliampere. What is the current gain of the circuit?   *(R2-7)*

A pnp transistor having a β of 100 is used in the circuit in figure 2-5 (b). The collector current is measured on an ammeter connected in series with resistor R_c. The base bias voltage initially is disconnected and there is no reading on the collector current ammeter. When the base bias voltage is applied so that the base current flows, the collector current changes to 10 mA. What is the base current?   *(R2-8)*

FIXED BIAS CIRCUIT, COMMON EMITTER CONFIGURATION

A common transistor circuit is the *common emitter* (CE) circuit as shown in figure 2-6. In this circuit, the emitter is the common terminal to both the input and output connections. Note that the circuit is similar to the common Z circuit shown in figure 1-12. The performance of this transistor can be expressed by a family of input and output characteristic curves, just as the performance of the generalized amplifying device in Chapter 1 was expressed by a family of curves. Typical common emitter output characteristic curves for a bipolar transistor are given in figure 2-7. A load line can be drawn on these curves for the particular circuit being analyzed. Using the circuit in figure 2-6, apply

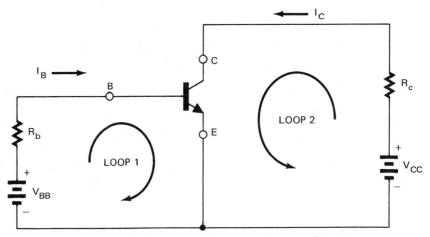

Fig. 2-6   Common emitter transistor amplifier circuit

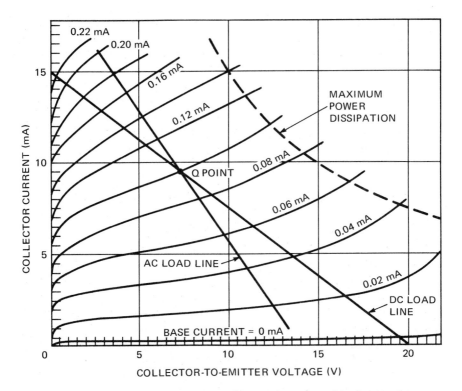

Fig. 2-7 Common emitter output characteristics for a bipolar transistor

Kirchhoff's Voltage Law around the output circuit, Loop 2. (The symbols and subscripts are specified on the circuit.*)

$$V_{CC} - I_C R_c - V_{CE} = 0 \qquad \text{Eq. 2.2}$$

When I_C is zero, $V_{CC} = V_{CE}$ and one end of the load line is established. When $V_{CE} = 0$, $I_C = V_{CC}/R_c$ and the other end of the load line is established. Assume that $V_{CC} = 20$ volts and $R_c = 1340$ ohms for the transistor whose characteristics are given in figure 2-7. The intersection of the load line on the collector current axis is at $I_C = V_{CC}/R_c = 20/1340 = 15$ milliamperes. The intersection of the load line on the V_{CE} axis is at 20 volts ($V_{CC} = V_{CE}$). This load line is drawn on figure 2-7.

Quiescent Point Established

It is necessary to establish the quiescent value of the base current. The least distortion of the signal is obtained when the base bias is selected in the center of the active region. The value of the base bias current establishes the quiescent point for the transistor. V_{BB} in figure 2-6 represents the base bias voltage necessary to establish the exact location of the Q point.

For the input circuit of figure 2-6, the transistor base to emitter terminals are representative of a forward biased diode. As a result, the shape of the input characteristic is essentially the same as that of the volt-ampere curve for a semiconductor diode (figure 2-3). However, the input characteristics of the transistor are influenced slightly by the output circuit operating conditions, resulting in a family of curves as shown in figure 2-8. Note that the transistor is basically a current operated device. That is, the base to emitter voltage V_{BE} changes only a little with large changes of base current. At the small level of base current needed to establish bias conditions, V_{BE} has a maximum value of about 0.250 volt (for a germanium transistor). The corresponding value of V_{BE} for a silicon transistor is about 0.7 volt.

Apply Kirchhoff's Voltage Law to the input circuit (Loop 1 in figure 2-6) and assume that there is no signal input voltage source. The resulting expression is given by

$$V_{BB} - I_B R_b - V_{BE} = 0 \qquad \text{Eq. 2.3}$$

To determine R_b using Equation 2.3, select a value of V_{BE} from figure 2-8.

For the circuit in figure 2-6, select a transistor having the common emitter characteristics shown in figure 2-9, page 36. Assuming values of $V_{CC} = 40$ volts and $R_c = 5000$ ohms, draw a load line indicating the operating conditions. (R2-9)

---

* All symbols are identified and the meanings of their subscripts are defined in the Appendix.

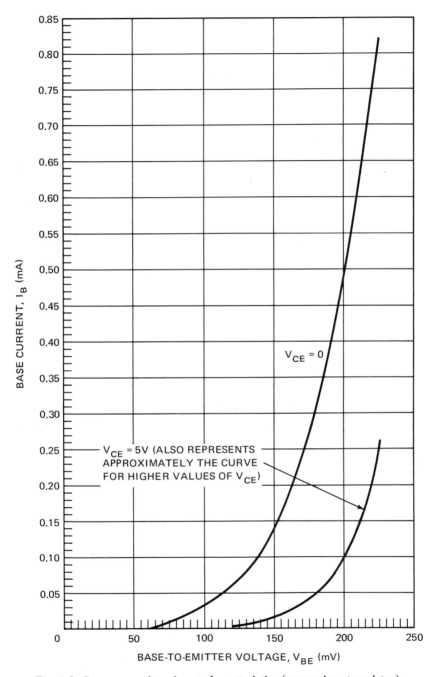

Fig. 2-8 Common emitter input characteristics (germanium transistor)

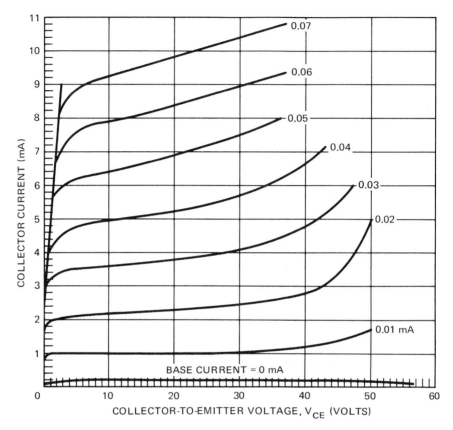

Fig. 2-9   Common emitter characteristics used in the solution of questions (R2-9) and (R2-10)

Using the circuit of figure 2-6 and the characteristics shown in figure 2-9, find the Q-point and specify the value for V_{BB} such that $V_{BE} = 0.7$ volt and $R_b = 100$ kilohms (k Ω). Use the results of (R2-9).   (R2-10)

Draw the circuit using this transistor. Show the proper voltage polarities and identify all of the values from questions (R2-9) and (R2-10).   (R2-11)

The values of the collector-to-emitter voltage V_{CE}, the collector current I_C, and the base current I_B at the quiescent or Q-point will be identified throughout the text as V_{CEQ}, I_{CQ}, and I_{BQ}, respectively.

Analysis of a Transistor Circuit with a Single Power Supply

The circuit shown in figure 2-5 requires two power supplies and is seldom used in practice. A more common circuit arrangement provides the base current from the same power supply that furnishes the collector bias, such as the circuit of figure 2-10. The same basic circuit is used but is rearranged slightly to permit the use of a single power supply.

Kirchhoff's Voltage Law can be applied to the circuit in figure 2-10 to obtain the following expression:

$$V_{CC} - I_B R_b - V_{BE} = 0 \qquad \text{Eq. 2.3}$$

Solving for R_b yields:

$$R_b = \frac{V_{CC} - V_{BE}}{I_B} \qquad \text{Eq. 2.4}$$

For the value of I_B at the desired Q-point, a value of V_{BE} can be selected from figure 2-8.

For example, assume that $V_{CC} = 20$ volts, $I_B = 0.10$ milliampere, and $V_{BE} = 0.250$ volt (for a germanium transistor).

$$R_b = \frac{20 - 0.250}{0.1} = 197 \text{ kilohms}$$

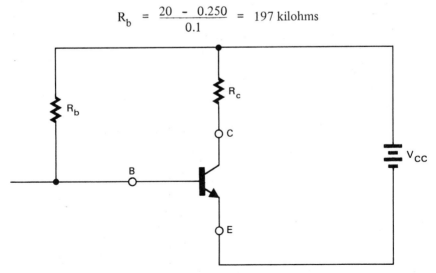

Fig. 2-10 Common emitter transistor amplifier circuit with a single power supply

For an approximate calculation, it is sufficient to assume that $V_{BE} = 0$. R_b is then calculated from the following expression:

$$R_b = \frac{V_{CC}}{I_B}$$

Eq. 2.5

The calculated value of R_b fixes I_B for the Q-point. An applied value of the input signal voltage (v_s) causes the transistor to operate above and below this point along the load line. The device operates in a manner similar to that of the generalized amplifying device described in Chapter 1.

Using the transistor characteristics shown in figure 2-9, determine a value of R_b for a satisfactory Q-point when the transistor is being used in the circuit given in figure 2-10. Use $V_{CC} = 40$ volts, $R_c = 5000$ ohms, and $V_{BE} = 0.7$ volt. (R2-12)

Ac Coupling

In general, amplifiers are not directly coupled to their input and output circuits. The signal source and the load can be affected by quiescent currents and voltages. Variations in the input and output signals can affect the biasing conditions for the transistor. For this reason, the transistor amplifier is usually ac coupled to input and output circuits by means of capacitors or transformers. The circuit of a typical capacitively coupled common emitter amplifier is shown in figure 2-11.

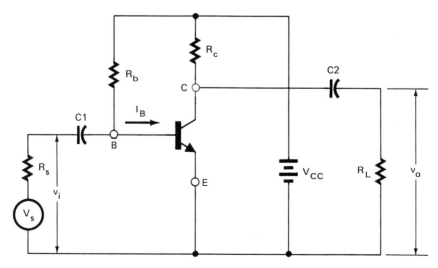

Fig. 2-11 Capacitively coupled common emitter amplifier

What effect do the coupling capacitors have on the bias conditions and quiescent point of a transistor that is being used in an RC coupled amplifier?
(R2-13)

What effect does the input capacitor have on the signal generator? (R2-14)

Transformer Coupled Amplifier

Another type of ac coupled amplifier is the transformer coupled amplifier circuit shown in figure 2-12. The bias conditions of the ac coupled amplifier circuits given in figures 2-11 and 2-12 are the same as those for the circuits in figures 2-10 or 2-6. These conditions arise from the fact that the dc current is limited to the transistor bias circuit itself. However, the response of the ac coupled amplifier is along the ac load line. (Refer to figure 2-7.) The load resistance of the capacitively coupled amplifier (figure 2-11) is the resistor R_L in parallel with the resistance R_c. For ac signals within the normal frequency range of the amplifier, the capacitor acts like an ordinary conductor.

The ac coupled amplifier can also take the form of a transformer coupled output and a capacitively coupled input, or a capacitively coupled output and a transformer coupled input.

It is assumed that the circuit in figure 2-11 uses a transistor having the output characteristics shown in figure 2-9. R_L is infinite, R_c = 5 kilohms, and V_{CC} = 50 volts. Calculate and draw the dc load line.   *(R2-15)*

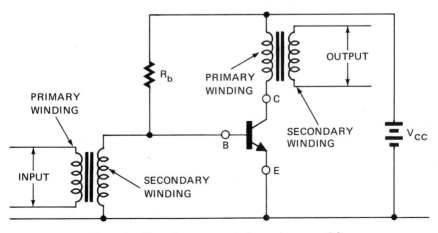

Fig. 2-12 Transformer coupled transistor amplifier

Calculation of the Ac Load Line

The slope of the ac load line is determined by the parallel arrangement of R_c and R_L as shown in the following equation:

$$\text{Slope} = \frac{1}{\dfrac{R_c\,R_L}{R_c + R_L}} = \frac{R_c + R_L}{R_c\,R_L} \qquad \text{Eq. 2.6}$$

As an example, the same transistor and bias conditions that were used in the previous example are assumed, and a load is applied to the circuit of figure 2-11. Let $R_c = 1340$ ohms and $R_L = 1500$ ohms.

$$\text{Slope} = \frac{1340 + 1500}{(1340)(1500)} = 0.00141$$

Thus, for a 5-volt change in V_{CE}, there is a change in I_C of $(5)(0.00141) = 0.00705$ ampere or 7.05 milliamperes.

The following discussion of the operation of an ac coupled amplifier is presented in conjunction with figure 2-7. Use the value of the slope just calculated. Starting at the quiescent level of I_B, a change in V_{CE} from 7.2 volts to 5 volts results in a change in the collector current from 9.6 milliamperes to 12.7 milliamperes. The ac load line can be drawn through the Q-point at the proper slope by drawing a straight line through the indicated points. The operation of the amplifier is along the ac load line. If the signal wave shape is to be traced from the input to the output, then the ac load line must be used.

For the transformer coupled amplifier circuit in figure 2-12, the total value of the bias resistance is the dc resistance of the transformer. To determine the slope of the dc load line, the dc resistance of the transformer is used. The ac resistance of the primary transformer winding is higher than the dc resistance. Therefore, the ac load line for a transformer coupled amplifier must be calculated from values given in the transformer specifications. Chapter 12 continues the discussion on transformer coupled amplifiers.

For ac coupled amplifier circuits, the choice of the capacitor values or the frequency range of the transformers used is determined by the frequencies of the signals to which the amplifier is expected to respond. For the circuits of this chapter and all of the circuits in Chapter 3, the calculation of the ac load line is based on the assumption that the capacitors are selected so that the capacitive reactance is negligible at signal frequencies. For a transformer coupled amplifier, the transformers are selected so that their impedance is predictable and available in the transformer specifications for the operating frequencies that are used.

To prevent excessive signal distortion, the input and output capacitors are selected to insure that the capacitive reactances at the lowest expected level of signal frequency is less than 10% of the lowest value of load or source resistance to be expected.

A problem is to be solved using these assumptions: the circuit is an RC coupled amplifier such as the one shown in figure 2-11; the transistor used is described by figure 2-9; the component values and operating point are those determined in review questions (R2-9) and (R2-10); and the ac load resistance (R_L) is 3 kilohms. Compute the slope of the ac load line. Draw the load line correctly on a graph of the transistor common emitter characteristics.   (R2-16)

The values of the coupling capacitors used in an RC coupled amplifier are 10 microfarads for both the input and output circuits. Can the amplifier be used for a signal frequency as low as 1500 hertz?   (R2-17)

For an RC coupled amplifier circuit, such as figure 2-11, using a transistor described by figure 2-9, V_{CC} = 30 volts, R_c = 4 kilohms, R_b = 1 megohm, V_{BE} = 0, and R_L = 3 kilohms. Draw the dc and ac load lines and locate the Q-point on a graph of the common emitter curves. (Note: It is not necessary to copy the common emitter curves completely but the scale of the graph is necessary.)   (R2-18)

NEED FOR STABILIZATION OR COMPENSATION

Transistors of the same type vary widely in their characteristics. Figure 2-13 represents the common emitter characteristics for two transistors of the same type and having the same technical specifications. For some applications, it may be necessary to construct a large number of identical circuits, each of which uses the same type of transistor. Because of the wide variation in performance of the transistors, each of these circuits will behave differently. Some circuits may not perform properly and others may not perform at all.

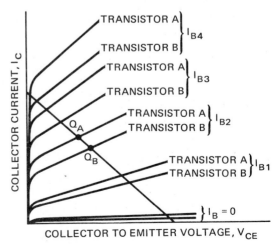

Fig. 2-13 Common emitter characteristics for two transistors having the same type number

Transistors are also very sensitive to temperature changes. Figure 2-14, page 42, shows the common emitter characteristics of a typical transistor for two different temperatures. The dashed lines represent operation at 25° C and the solid lines represent operation at 100°C. It is obvious that an amplifier circuit using this transistor will behave differently at one temperature than at another. The temperature dependence of the transistor is due to thermally generated current carriers which make up part of the collector current. When the base-to-emitter junction is reverse biased, the only current that exists is the thermally generated current in the transistor. This current is known in technical literature as I_{CO} or I_{CBO}. The value of this current doubles for every 10°C

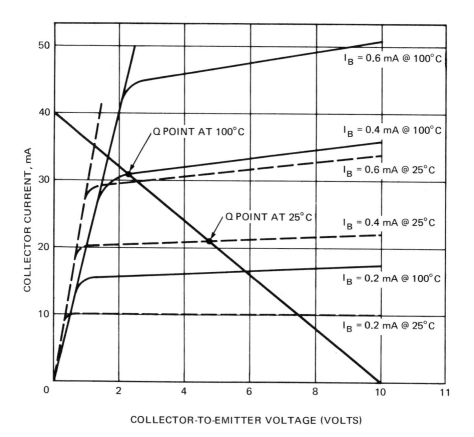

Fig. 2-14 Common emitter characteristics for two different temperatures

rise in the temperature. I_{CO} is much larger for germanium transistors than it is for silicon transistors. Thus, germanium transistors appear to be more temperature sensitive than silicon transistors.

A transistor amplifier is built, tested, and then is accidently destroyed in a fire. Another amplifier is built according to the same specifications to replace the first amplifier. The new amplifier does not have the same gain as the first device. Why? (R2-19)

Thermal Runaway

A different operating temperature level can not only cause a change in the transistor behavior, but also can make the circuit completely inoperative because of the change in the transistor operating point. It is possible that the transistor may be destroyed.

The resistance of semiconductors tends to decrease as the temperature increases. As a result, there is an increase in collector current that leads to increased I^2R losses. This increased power dissipation creates a higher temperature, resulting in an additional decrease in the resistance and a further increase in the collector current. This effect is cumulative and results in a condition known as *thermal runaway*. The selection of the transistor operating point is made not only to achieve operation in the active region, but also to avoid the condition of thermal runaway. Figure 2-15 illustrates how the selection of the operating point is made. The power curves (P) are drawn through a locus of points representing a constant power loss which is the maximum amount of power the

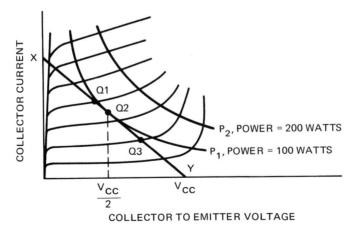

Fig. 2-15 Selection of the Q-point to insure that the maximum power dissipation cannot be reached

transistor can safely dissipate. Curve P_1 is for one operating temperature and curve P_2 is for another temperature. The power curves are drawn as an overlay over the common emitter output characteristics of a transistor. In addition, a negatively sloping straight line XY is drawn to represent a typical load line. Note that there is a selection of operating points: Q_1, Q_2, and Q_3. The choice of Q_1 as the operating point insures that the maximum power dissipation cannot be exceeded by an increase in the collector current. Thus, thermal runaway cannot take place. An operating point of Q_3 should not be selected because an increase of collector current leads to a higher power dissipation. As a result, there is a further increase in the collector current and thermal runaway is more likely to occur. The V_{CE} scale indicates that the Q-point should not be allowed to occur at a V_{CE} value greater than $V_{CC}/2$.

Careful circuit design and the judicious choice of circuit components can minimize the effects of temperature changes and the variability in the characteristics of like devices.

The manufacturer lists the maximum power dissipation for the transistor whose characteristics are given in figure 2-9 as 150 milliwatts at 25°C. On a graph of the common emitter characteristics for this device, plot a curve of I_C versus V_{CE} for a constant power dissipation of 150 milliwatts.   *(R2-20)*

In figure 2-14, it is desired to keep the Q-point at approximately the center of the linear region when the device is operating at 100°C. I_{BQ} is 0.4 milliampere for operation at 25°C. Extrapolate between the curves and estimate the value that I_{BQ} must be lowered to so that operation occurs in the center of the linear region at 100°C.   *(R2-21)*

Assume V_{CC} is 50 volts for the amplifier circuit shown in figure 2-11. Determine a maximum value of V_{CEQ} that will insure that thermal runaway does not occur at heavy loads (R_L small).   *(R2-22)*

CIRCUITS FOR STABILIZATION

Two circuits will be discussed that are commonly used to stabilize the operating point of common emitter transistor amplifiers. These circuits are the collector-to-base bias circuit shown in figure 2-16 and the emitter self-bias circuit shown in figure 2-17.

Figures 2-13 and 2-14 show that the collector current (I_C) is the transistor variable most sensitive to changes in temperature. It is also the transistor variable that is most sensitive to the differences between transistors due to the

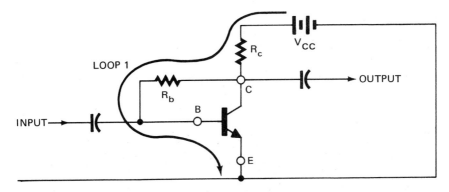

Fig. 2-16 Collector-to-base bias circuit

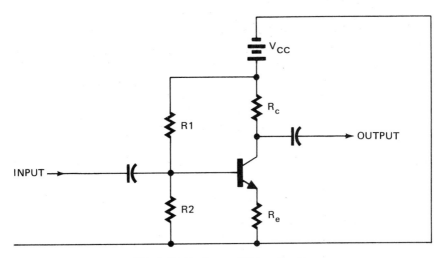

Fig. 2-17 Emitter self-bias circuit

difficulty of making two or more transistors exactly alike. Both the emitter
self-bias circuit and the collector-to-base bias circuit compensate for changes in
I_C. For the circuit of figure 2-16, Kirchhoff's Voltage Law is applied around
Loop 1 to yield:

$$V_{CC} - (I_C + I_B)R_c - I_B R_b - V_{BE} = 0 \qquad \text{Eq. 2.7}$$

Equation 2.7 can be solved for the base current at the Q-point.

$$I_B = \frac{V_{CC} - V_{BE} - I_C R_c}{R_c + R_b} \qquad \text{Eq. 2.8}$$

It can be seen that if I_C increases, the base current I_B decreases. A lower I_B results in a smaller increase in I_C and compensates for the change in temperature. The circuit of figure 2-17 is also used to stabilize the transistor operating point. A Thévenin equivalent circuit for the base circuit in figure 2-17 is made by using terminals B and E of the transistor as the output terminals of the equivalent circuit. Figure 2-18 shows a circuit from which a Thévenin equivalent circuit is to be obtained. Refer to the Appendix for a review of Thévenin's theorem and how it is applied.

The Thévenin equivalent circuit for the base circuit of figure 2-17 is given in figure 2-19. The Thévenin equivalent voltage is expressed by the following equation:

$$V_{th} = \frac{R_2 V_{CC}}{R_1 + R_2}$$

Eq. 2.9

The Thévenin equivalent resistance is the equivalent series resistance of the base circuit. The Thévenin resistance is given as:

$$R_b = \frac{R_1 R_2}{R_1 + R_2}$$

Eq. 2.10

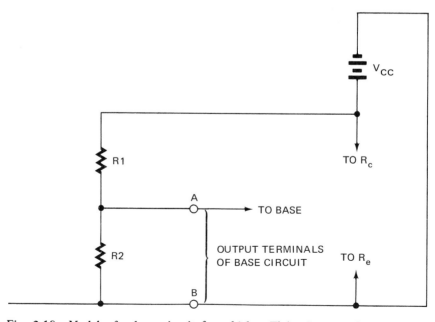

Fig. 2-18   Model of a base circuit for which a Thévenin equivalent is to be obtained

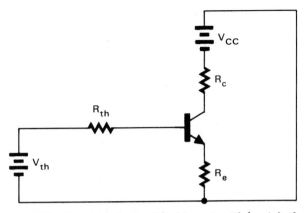

Fig. 2-19 Base circuit simplified by using Thévenin's theorum

Apply Kirchhoff's Voltage Law around the base circuit to obtain the following:

$$V_{th} - I_B R_b - V_{BE} - (I_B + I_C)R_e = 0 \qquad \text{Eq. 2.11}$$

Equation 2.11 can be solved to obtain I_B.

$$I_B = \frac{V_{th} - V_{BE} - I_C R_e}{R_b + R_e} \qquad \text{Eq. 2.12}$$

An increase in I_C results in a decrease of I_B. Such a decrease of I_B tends to reduce the amount by which I_C tends to increase. As a result, as I_C becomes smaller, I_B becomes larger, and I_C tends to decrease and compensate for the change that produced it.

The setting for the collector circuit bias (refer to the circuit of figure 2-16) can be determined approximately by the method which begins on page 32, using Equation 2.2. The base current is so small when compared to the value of I_C that it can be neglected in the calculations. The base bias for the circuit of figure 2-16 is determined by selecting the approximate location for the Q point along the load line for the device. The values of I_B and I_C are read at the Q-point. A value of V_{BE} is selected from the input characteristics.* (See figure 2-8.) Then solve Equation 2.7 to obtain the value of resistor R_b.

$$R_b = \frac{V_{CC} - V_{BE} - I_C R_c}{I_B} - R_c \qquad \text{Eq. 2.13}$$

* If the input characteristics for the device are not available, assume that V_{BE} = 0.25 volt for a germanium transistor and 0.6 volt for a silicon transistor.

R_b fixes the value of the base current at the desired Q point.

The circuit in figure 2-16 uses a transistor whose characteristics are given in figure 2-7. It is assumed that the operation of the transistor is about the Q-point shown in figure 2-7. Assume that $V_{BE} = 0.2$ volt. Determine the values of R_c and R_b necessary for the operation of this transistor.   (R2-23)

It is assumed that the collector current of the transistor in the circuit considered in question (R2-23) changes by 1 milliampere. Using the component values determined in question (R2-23), what change in base current results from the use of the transistor in the circuit of figure 2-16?   (R2-24)

For a circuit such as the one shown in figure 2-16, assume that $R_c = 1500$ ohms and $R_b = 100$ kilohms. Use a transistor which behaves according to the curves in figure 2-7. $V_{CC} = 20$ volts, $V_{BE} = 0.2$ volt, and $\beta = 100$. Find the value of I_B at the Q point.   (R2-25)

Solve question (R2-18) using the circuit of figure 2-16. Neglect the base current during the calculation of the collector current. Assume $\beta = 127$. (R2-26)

An approximate value of the collector bias for the circuit of figure 2-17 can be determined by assuming that the base current is small when compared to the collector current. Kirchhoff's Voltage Law is then taken around the collector circuit to yield:

$$V_{CC} - I_C (R_c + R_e) - V_{CE} = 0 \qquad \text{Eq. 2.14}$$

When $I_C = 0$, $V_{CC} = V_{CE}$, and one end of the load line is established. When $V_{CE} = 0$, Equation 2.14 is solved for I_C.

$$I_C = \frac{V_{CC}}{R_c + R_e} \qquad \text{Eq. 2.15}$$

This value of I_C establishes the other end of the load line. To determine the base bias, use Equations 2.9, 2.10, and 2.12 together to find the Thévenin voltage and the base current at the Q-point.

As an example of the use of the circuit in figure 2-17, assume that the transistor shown in figure 2-14 is to be used such that the junction temperature of the transistor is allowed to be at $100^\circ C$ while the equipment is in operation and at 25°C when most of the equipment in the room is shut down. It is assumed that the transistor is used in a fixed bias circuit such as the one shown in figure 2-10, with typical values of R_c = 250 ohms, R_b = 25 kilohms, V_{CC} = 10 volts, and $V_{BE} \approx 0$. These conditions establish the 25°C Q-point shown in figure 2-14. At 100°C, the Q-point is shifted to I_{CQ} = 32 milliamperes and V_{CEQ} = 2.3 volts. This Q-point is obviously undesirable for linear amplification. A better circuit is the one shown in figure 2-17. To insure that circuit operation remains on the same load line, these typical bias resistors are used: R1 = 25 kilohms, R2 = 100 kilohms, R_c = 150 ohms, and R_e = 110 ohms. I_B is adjusted automatically so that the Q-point remains in approximately the same place. If I_C changes to 32 milliamperes as a result of the temperature conditions, then the typical resistance values given can be used as follows to find I_B:

$$V_{th} = \frac{V_{CC}\ R2}{R1 + R2} = \frac{(10)\ (100)}{25 + 100} = 8 \text{ volts}$$

$$R_b = \frac{(R1)\ (R2)}{R1 + R2} = \frac{(25)\ (100)}{25 + 100} = 20 \text{ kilohms}$$

$$I_B = \frac{V_{th} - V_{BE} - I_C\ R_e}{R_b + R_e} = \frac{8 - 0 - (32)\ (0.11)}{20 + 0.11} = 0.225 \text{ milliampere}$$

This means that the new value of the quiescent base current I_{BQ} is 0.225 milliampere. The circuit allows I_{CQ} to change due to the change in the temperature. However, it still maintains the Q-point in the linear region.

The dc load line is calculated using the series combination of R_c and R_e. According to Equation 2.15, if R_c is lowered to 150 ohms, the load line intercepts the I_C axis at the same point as in the fixed bias circuit.

Verify the previous statement by calculation.   *(R2-27)*

Verify the Q-point at 25°C for the fixed bias circuit used to draw the dc load line on figure 2-14.   *(R2-28)*

For the circuit in figure 2-17, note that an increasing I_C results in a decreasing I_B. This condition can be objectionable, because the overall current gain, I_C/I_B, is reduced. To eliminate this effect a large capacitance (in the order of 10 microfarads or larger) is connected across R_e. With or without the capacitance, slow changes in I_C still provide temperature stabilization. However, rapid changes in I_C, such as the changes caused by the input signal, are conducted around R_e to ground. The result of this configuration is a high current gain, A_I, for signal amplification and a low current gain for the slowly changing signal which occurs as a result of temperature changes.

Assume R2 = 1 megohm and V_{BE} = 0 for the circuit of figure 2-17 and calculate the operating point for the transistor of figure 2-7 when R_c = 890 ohms, R_e = 450 ohms, V_{CC} = 20 volts, R1 = 145 kilohms, and β = 100.   *(R2-29)*

CIRCUITS USING COMPENSATION

Figures 2-20, 2-21, and 2-22 (page 52) show circuits in which other semiconductor devices function to stabilize the transistor operating point against changes in temperature.

In figure 2-20, the diode is maintained in the forward biased condition by the power supply V_{DD}. A voltage drop is created which is opposite in polarity to the voltage V_{BE} of the transistor. Recall that the base-emitter junction of the transistor is simply forward biased diode in normal operation. If the compensating diode is made of the same material as the transistor, it has the same temperature coefficient. Thus, the transistor voltage V_{BE} and the diode voltage V_D track each other. When Kirchhoff's Voltage Law is applied around Loop 1 of the circuit, it is evident that $(V_{BE} - V_D)$ is one of the voltages in the circuit. Since these voltages tend to cancel each other, the circuit is less sensitive to temperature changes.

In figure 2-21, the diode is reverse biased so the base current is given by the following equation:

$$I_B = I_1 - I_D \qquad \text{Eq. 2.16}$$

As the temperature rises, the resistivity of a semiconductor decreases. Therefore, for constant bias voltages in the circuit, I_D increases as the temperature increases. At the same time, the collector-to-base junction (also reverse biased) must endure an increased current arising from the temperature. The base current, however, drops as the temperature increases (according to Equation 2.17). This effect tends to compensate for the increase in the collector current because a decrease in the base current results in a decrease of the collector current. The characteristic curves of the transistor illustrate this action.

A diode added to the circuit of figure 2-17 changes the circuit to that of figure 2-20. Assume that V_{BE} and V_D = 0.7 volt. Apply Kirchhoff's Voltage Law and write the equation for the base circuit. Describe how V_{BE} and V_D cancel each other.   *(R2-30)*

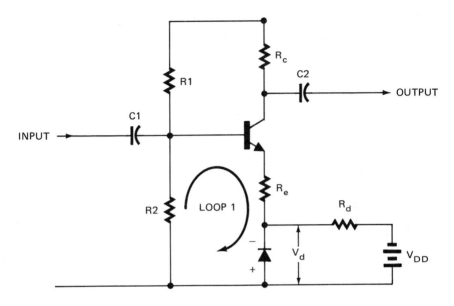

Fig. 2-20 Emitter self-biasing circuit with forward biased diode for temperature compensation

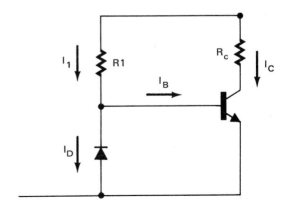

Fig. 2-21 Common emitter circuit with reverse biased diode for temperature compensation

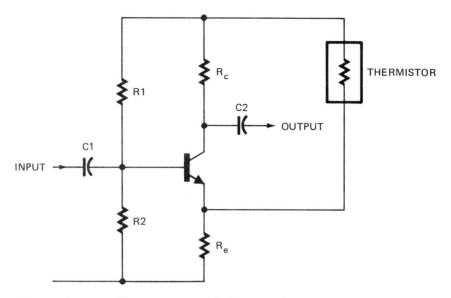

Fig. 2-22 Emitter self-biased circuit with thermistor for temperature compensation

Semiconductor devices are available which are particularly sensitive to temperature changes. These devices are frequently used in transistor amplifier circuits to compensate for temperature changes. Figure 2-22 shows one such device, the thermistor, being used in the collector circuit of a common emitter amplifier. Because of the thermistor, there is less voltage available for the base bias as the temperature increases. The resistivity of the thermistor drops drastically as the temperature increases. Thus, it is considered to have a *negative* temperature coefficient of resistance. Another device that can be used for temperature compensation is a heavily doped semiconductor known as a *sensistor*. The temperature coefficient of the sensistor is changed to *positive* by the addition of impurities rich in current carriers. The resistor R_e in figure 2-17 can be replaced by a sensistor having the proper resistance to insure temperature compensation. As the temperature increases, the sensistor resistance also increases. As a result, the collector current remains low although the temperature of the transistor collector tends to cause it to increase.

LABORATORY EXPERIMENTS

The following experiments have been tested and found to be useful for illustrating the theory in this chapter.

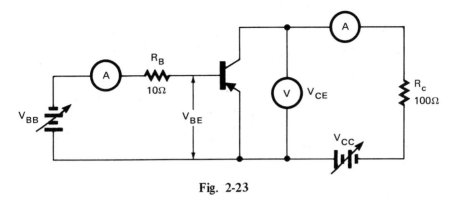

Fig. 2-23

LABORATORY EXPERIMENT 2-1

OBJECTIVE

To develop transistor common emitter output characteristic curves.

EQUIPMENT AND MATERIALS REQUIRED

1	Transistor, 2N1541
2	Power supplies, dc, variable 0 to 40 volts
3	Volt-ohm-milliammeters (VOM)
1	Resistor, 10 ohm
1	Resistor, 100 ohm, 2 watts

PROCEDURE

1. Connect the transistor in the circuit shown in figure 2-23. The power supplies are to be in the *off* position with the voltage adjusters set at the minimum voltage value.

2. Switch on the supplies. With I_B set at zero, vary V_{CC} such that V_{CE} is varied from 0 to 20 volts in about five steps. Measure V_{CE} and I_C at each step.

3. Plot a curve of V_{CE} versus I_C.

4. Increase V_{BB} until I_B = 10 milliamperes.

5. Vary V_{CE} from 0 to 20 volts in about eight steps. Measure V_{CE} and I_C at each step. Take data at about three places in the region where the curve bends.

6. Plot a second curve of V_{CE} versus I_C on the graph plotted in step 3.

7. Repeat steps 3 through 6 for values of I_B = 20 milliamperes, 30 milliamperes, and 50 milliamperes.

8. The resulting curves are the common emitter characteristics of the 2N1541 transistor. Draw a dc load line for the circuit used following the procedure which begins on page 32.

LABORATORY EXPERIMENT 2-2 (Alternate form for experiment 2-1)

OBJECTIVE

To obtain transistor common emitter output characteristic curves.

EQUIPMENT AND MATERIALS REQUIRED

1   Transistor, 2N1808
1   Curve tracer, Tektronix Model 575
Note: Other curve tracers can be used, but the following experiment is based on the Tektronix instrument.

PROCEDURE

1. Connect the 2N1808 transistor in the appropriately marked sockets of the curve tracer. The socket switch is to be left at the disconnection position while the curve tracer is turned on.

2. Turn on the curve tracer. Adjust the collector sweep controls to the following positions:

   Collector Sweep Voltage Switch: 0-20 V.
   Collector Sweep Voltage Fine Adjustment: Fully clockwise.
   Dissipation Limiting Resistor: 500 ohms
   Polarity Switch: +

3. Set the base step generator controls to the following positions:

   Step Selector: 0.01 mA/step
   Step Zero: Fully CCW
   Polarity: +
   Steps/family: Fully CW
   Repetitive/Off/Single Family Switch: Repetitive
   Step/Sec Switch: 240

4. Set the other controls as follows:

   Vertical Axis Scale Factor: 2 mA/div.
   Horizontal Axis Scale Factor: 2 volts/div.
   Emitter Grounded/Base Grounded Switch: Emitter Grounded

5. Switch the socket switch on to view the trace.

The common emitter output characteristic curves can now be traced for this transistor. Care must be taken to insure that (1) the base and collector voltage polarity switches are both on the + setting, (2) the collector sweep voltage is not set beyond 20 volts, and (3) the collector dissipation resistor is not set lower than 500 ohms. The transistor socket must be disconnected from the curve tracer while the tracer controls are set. To obtain accurate readings, the curve tracer must be warmed up for 15 minutes before any readings can be taken.

LABORATORY EXPERIMENT 2-3

OBJECTIVE

To build and operate a common emitter amplifier.

EQUIPMENT AND MATERIALS REQUIRED

2   Transistors, 2N1808 (tested and determined to be acceptable)
2   Resistance substitution boxes, Heathkit 1N-37 or equivalent
1   Power supply, dc, variable 0-20 volts
1   Volt-ohm-milliammeter
1   Signal generator, 50-ohm source impedance, sinusoidal signal of 1 volt at 1000 to 2000 hertz
2   Capacitors, 10 microfarads, rated to 20 volts at least

PROCEDURE

1. Connect the transistor in the circuit shown in figure 2-24. Omit the load resistance at this time.

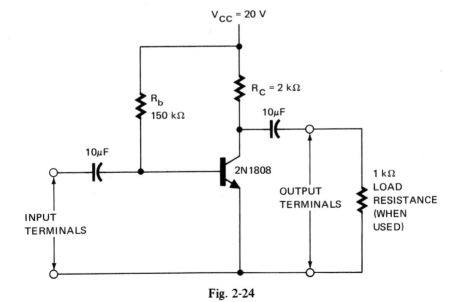

Fig. 2-24

2. Operate the transistor as an amplifier using a 1000-hertz sinusoidal signal. If necessary, readjust the bias resistors and the dc bias voltage to achieve a maximum voltage gain without distortion. The ideal location of the Q-point is in the approximate center of the active region of the characteristic curves for the transistor.

To find the Q-point, increase the input signal until some distortion takes place in the output signal. If distortion occurs for both positive and negative peaks of the signal, the Q-point is located in the center of the linear region. If the distortion is predominantly at either peak and there is very little or none at all at the other peak, the Q-point is not located properly. Adjust the bias resistors and voltage until the Q-point is located correctly. The main adjustment should be R_b. For operation within the linear region, the signal should then be reduced until the distortion disappears.

3. Draw the dc load line and locate the Q-point on the characteristic curves for the transistor. A copy of the characteristics for the 2N1808 transistor tested should be used. If these characteristics are not available, use the curves shown in figure 2-25.

4. Measure the input voltage and the output voltage. Calculate the voltage gain.

5. Sketch the waveforms for the input and output signals for the undistorted signal.

6. Apply a load of 1 kilohm to the amplifier.

7. Repeat steps 4 and 5.

8. Draw the ac load line on the transistor characteristic curves.

9. Using the characterisitic curves (with the ac and dc load lines drawn), calculate the voltage gain. Compare this value with the voltage gain value obtained by measurement.

10. Remove power from the amplifier. Replace the 2N1808 transistor with a second 2N1808 transistor.

11. Again apply power to the amplifier. There should be a change in the performance of the amplifier. If no change takes place, remove the second transistor and insert a transistor that does cause at least a slight change in the amplifier performance. Do *not* change any circuit values; change only the transistor.

12. Repeat steps 4 and 5. The data obtained for this step is to be kept for use in the next Laboratory Exercise. Label the transistors used in the circuit No. 1 and No. 2 and keep them for use in the next Laboratory Exercise.

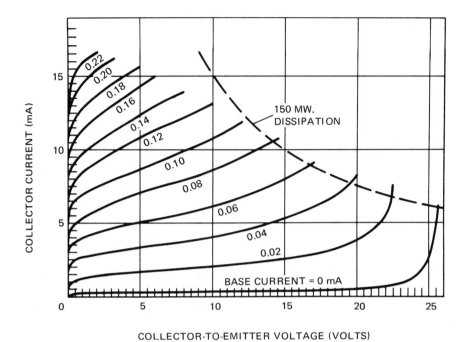

Fig. 2-25 Characteristic curves for 2N1808 transistor

LABORATORY EXPERIMENT 2-4

OBJECTIVE

To build and operate a common emitter amplifier circuit with compensation; to compare the performance of this circuit with the performance of a fixed bias circuit.

EQUIPMENT AND MATERIALS REQUIRED

2	Transistors, 2N1808 (tested and determined to be acceptable)
2	Resistance substitution boxes, Heathkit 1N-37 or equivalent
1	Power supply, dc, variable 0-20 volts
1	Volt-ohm-milliammeter
1	Signal generator, 50-ohm source impedance, sinusoidal signal of one volt at 1000 to 2000 hertz
2	Capacitors, 10 microfarads, rated to 20 volts at least

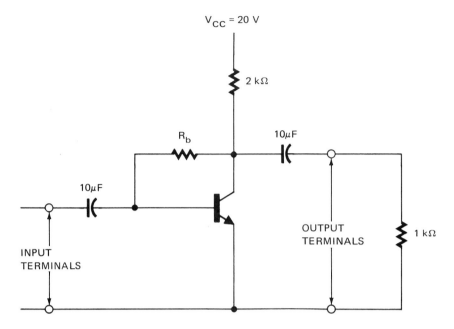

Fig. 2-26 Circuit with collector-to-base bias

PROCEDURE

1. Connect the circuit shown in figure 2-26, using transistor No.1 from Laboratory Experiment 2-3.

2. Adjust R_b to achieve approximately the same Q-point determined in the previous exercise.

3. Adjust a 1000-hertz input signal to a value that is the maximum input level that can be tolerated and still maintain linear operation (minimum distortion).

4. Calculate the voltage gain. Is the gain the same as the gain obtained in Laboratory Experiment 2-3? Why?

5. Replace transistor No. 1 with transistor No. 2. Does the amplifier still operate satisfactorily? Is the gain different? What is the gain using transistor No. 2? Compare the gain calculations for the two transistors with the gain calculations of Laboratory Experiment 2-3. For the circuit in figure 2-26, is there less change in gain when the transistors are changed? Explain.

6. Connect a transistor in the self-bias amplifier circuit shown in figure 2-27.

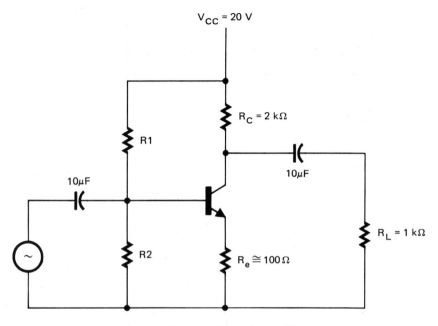

Fig. 2-27 Circuit with emitter self-bias

To determine values of R1 and R2, recall the Thévenin equivalent circuit covered on pages 46 and 47. The values R1 and R2 should be determined so that approximately the same base current is obtained. (Note: To achieve the best results for this circuit, R1 and R2 should be as small as possible and still limit the base current to acceptable values. R_e should be in the order of 100 ohms.)

7. Using a signal at about 1000 hertz, adjust the input signal to a value that is the maximum input level that maintains linear operation.

8. Calculate the gain of the amplifier circuit. Compare this gain to the values obtained for the other circuits tested previously.

9. Vary R_e and note the effect on the gain. Explain why R_e has the effect it does.

10. Replace transistor No. 1 with transistor No. 2. Operate the circuit using the conditions given in steps 6 and 7. Recheck the gain. Has the value of the gain changed from the gain value for transistor No. 1? Compare the gain calculations for the two transistors with the gain calculations completed for Laboratory Experiment 2-3. For the circuit given in figure 2-27, is there less change in gain when the transistors are changed? Explain.

EXTENDED STUDY TOPICS

1. Explain the concept of holes and electrons as carriers of electric current.

2. Name some materials that comprise semiconductor devices.

3. Define the difference between n-type and p-type impurities.

4. What are the relative magnitudes of the forward biased resistances and the reverse biased resistances of a semiconductor diode?

5. Explain the physical mechanism involved in the reverse biased semiconductor diode.

6. Explain the physical mechanism involved in the forward biased semiconductor diode.

7. Draw the two-diode representation of a pnp transistor.

8. If a collector to base bias voltage is applied, but there is no base current, the transistor is operating in its (a) saturation region, (b) active region, (c) cutoff region.

9. Why must there be some resistance in series with a forward biased diode?

10. Discuss the reasons for the temperature sensitivity of transistors. Does the resistivity of a semiconductor increase or decrease with temperature? What effect does a change in resistivity have on the performance of the transistor?

11. Does the current gain of a transistor amplifier change with temperature? Explain.

12. Name several methods of compensating for the temperature dependence of transistors in an amplifier circuit.

13. How can the condition known as thermal runaway be avoided in biasing a transistor?

14. What property of a semiconductor diode permits it to be used in temperature compensation for a transistor amplifier?

3
Amplifier
Techniques Using
Bipolar Transistors

OBJECTIVES

After studying this chapter, the student will be able to:

- analyze the performance of common emitter amplifiers under large signal conditions

- define common base and common collector circuit configurations, including an analysis of their performance

- specify multistage amplifiers and specialized transistor amplifier circuits

AMPLIFIER BEHAVIOR FOR LARGE SIGNALS

The bipolar transistor is a nonlinear device. Therefore, the output signal is not a perfect reproduction of the input signal. Assume the common emitter amplifier circuit shown in figure 2-10. Refer to the common emitter or CE characteristic curves of figure 2-7. The dc and ac load lines are drawn on these curves. If the curves for constant values of I_B in figure 2-7 are equal distances apart for equal increments of I_B at all values of V_{CE}, then the device is linear. As a result, the transistor operates as a perfectly linear device between the cutoff and saturation regions. However, the transistor is not linear; therefore, the calculations for large signals must be based on the actual operating characteristic curves of the transistor. The solution must be entirely graphical or an equivalent circuit must be used which represents the extremes of the signals.

To trace the operation of a common emitter amplifier, assume that the base current varies sinusoidally and causes the collector current to vary from a value close to saturation to a value close to cutoff. Figure 3-1, page 62, illustrates

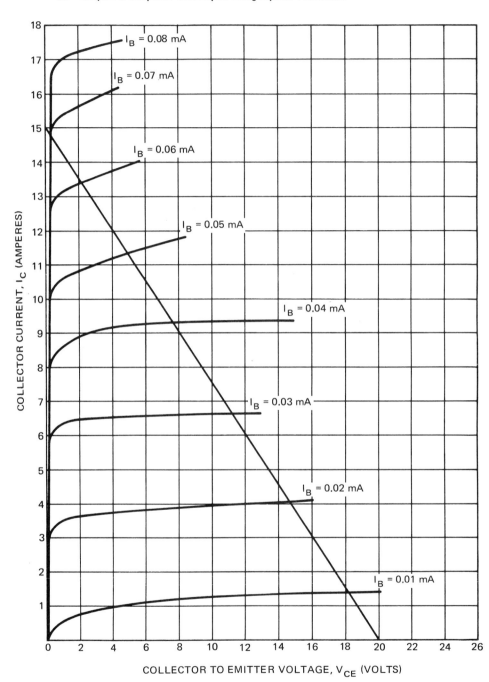

Fig. 3-1 Common emitter characteristics for silicon transistor

the common emitter characteristics of a transistor. The dc load line is drawn to indicate the biasing for the collector-to-emitter circuit. It is assumed that the quiescent base current is 0.04 milliampere. Assume also that the base current varies sinusoidally by 0.08 milliampere peak-to-peak. A *dynamic transfer characteristic* for the circuit is made by plotting the collector current against the base current where the base current is the independent variable. Such a plot is shown in figure 3-2, page 64, for the transistor described by the characteristics in figure 3-1. Note in figure 3-2 that the collector current is somewhat distorted for a sinusoidal variation of the base current.

Some amplifiers may be acceptable with this amount of distortion present, depending upon the intended use of the amplifier. This distortion is known as *amplitude distortion*.

Plot the dynamic transfer characteristic for the transistor whose curves are given in figure 2-9. Make the plot small enough so that sketches of the sinusoidal input and output can be made on the same sheet (refer to figure 3-2). Assume V_{CC} = 40 volts and R_c = 5 kilohms in a circuit such as the one given in figure 2-11. I_{BQ} is set at 0.04 milliampere. (R3-1)

Assume a quiescent base current of 0.04 milliampere and an applied sinusoidal input signal current of 0.07 milliampere, peak-to-peak. Sketch the input signal and the collector current signal as the output signal on the graph prepared for problem (R3-1). (Refer to figure 3-2.) (R3-2)

The variation in collector to emitter voltage, v_{ce},* also displays some amplitude distortion. Figure 3-3, page 65, shows how v_{ce} varies with the sinusoidal base current input signal. Note that the waveform of v_{ce} is a distorted sinusoid. Also note that the collector-to-emitter voltage increases from the Q-point value as the collector current decreases from the same value. In general, the base current and the collector current are in phase and the output voltage (v_{ce}) is approximately 180° out of phase with the input signal current. The signal input current is approximately in phase with the base-to-emitter voltage v_{be}. It is typical of a common emitter (CE) amplifier used to amplify voltage that there is a 180 degree phase shift between the input signal voltage and the output signal voltage.

---

* In this chapter, lower case symbols and subscripts (such as v_{ce}) are used frequently to indicate instantaneous values. A full explanation of the use of symbols and subscripts is given in the Appendix.

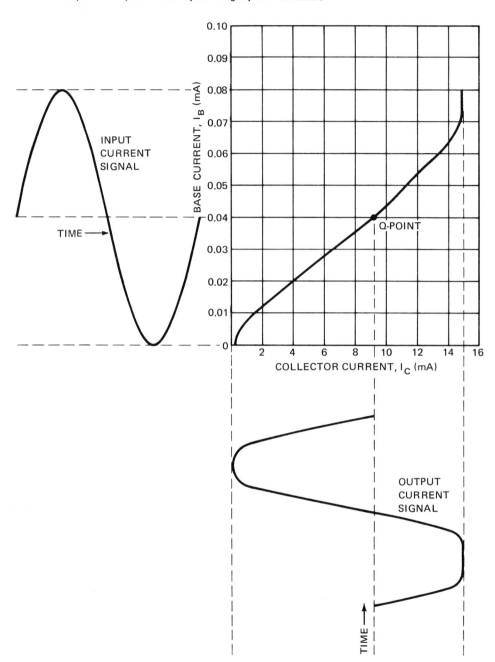

Fig. 3-2   Dynamic transfer characteristic for the transistor described by the curves shown in figure 3-1

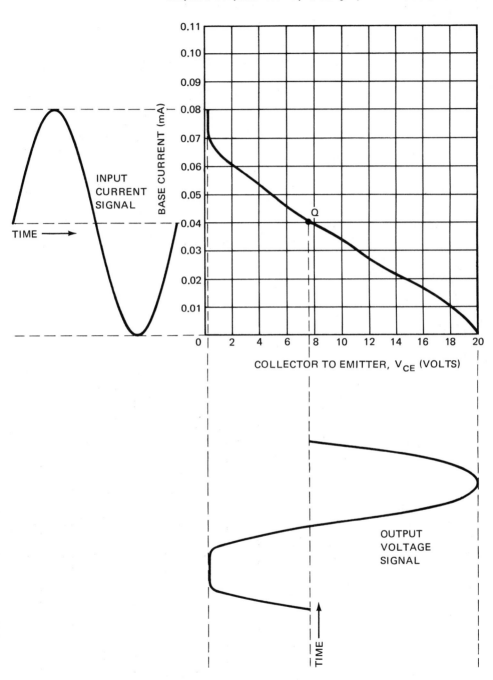

Fig. 3-3 Variation in V_{ce} with base current

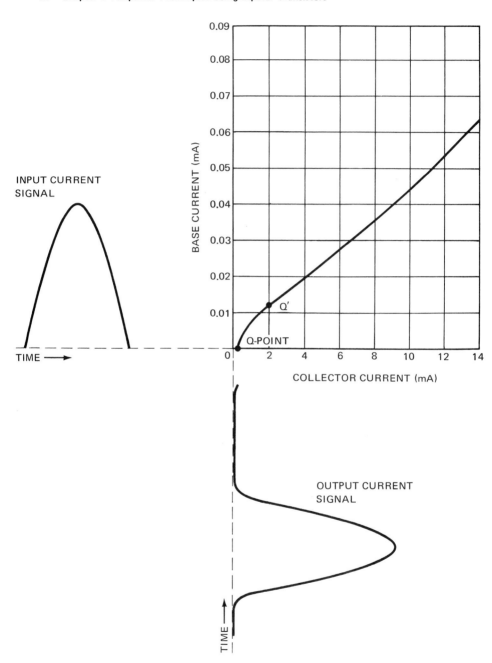

Fig. 3-4 Dynamic characteristic of the transistor characteristics shown in figure 3-1 for $I_{BQ} = 0$

Plot the dynamic characteristic of the base current versus the collector to emitter voltage v_{ce} for the transistor whose characteristics are shown in figure 2-9. Use the circuit conditions of question (R3-1). Assume a quiescent base current of 0.04 milliampere and a sinusoidal applied signal of 0.07 milliampere peak-to-peak. Sketch the input signal and v_{ce} versus time (refer to the graph in figure 3-3). (R3-3)

It is interesting to examine what happens to the output signal waveform if the Q-point is selected at zero base current. The base current and the output current appear clipped and are approximately one-half cycle of a nearly sinusoidal waveform. A sinusoidal base current cannot be obtained because the base to emitter junction becomes reverse biased. Refer to figure 3-4 and note that the shape of the output signal below the zero value of the base current is only a rough approximation. Any signal that exists at that level represents the movement of thermally generated carriers alone. The peak-to-peak value of the collector to emitter voltage also has its amplitude clipped and reduced by half.

If the Q-point in figure 3-4 is moved up to a new value Q', then some of the negative values of the input signal are reproduced. Figure 3-5 shows the approximate output current waveform for the Q' condition. This type of waveform is known as a *clipped signal*.

Assume that the Q-point is located high on the load line near $V_{CE} = 0$. Refer to figure 3-3 and note that V_{CE} does not reach a value of zero. This condition results from the fact that there is always some voltage drop from the collector to the emitter, even if the transistor is operated in the saturation region. Changes in the base current cause little change in the collector-to-emitter voltage at low values of V_{CE}. However, it can be seen that clipping of the signal occurs. Clipping results because the transistor is relatively insensitive to the base current in this region of the characteristic curves.

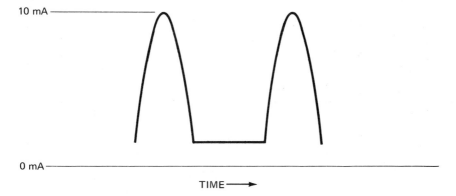

Fig. 3-5 Clipped sinusoidal waveform

From the previous discussion, the Q-point should be located approximately in the center of the load line to realize the full capability of the transistor. Then, if an input signal is too large to be amplified without distortion, the signal is clipped an equal amount at both the positive and negative peaks. This is one method of obtaining a square wave signal from a sinusoidal signal, as shown in figure 3-6. The transistor circuit used to obtain a square wave in this manner is called a *clipper circuit*. Such a circuit can be used to *square off* a particular waveform so that undesirable high voltage pulses are eliminated.

OPERATION IN CUTOFF AND SATURATION REGIONS; THE TRANSISTOR AS A SWITCH

Many circuits use the bipolar transistor as a simple switch. In this mode, the active region of the transistor is not used at all in circuit operation. When

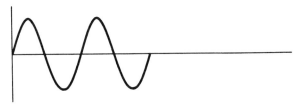

(a) TRANSISTOR INPUT SIGNAL

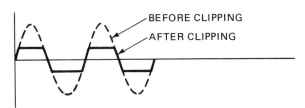

(b) SIGNAL AFTER BEING CLIPPED BY TRANSISTOR

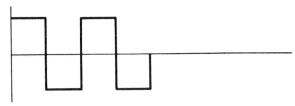

(c) SQUARE WAVE FORMED BY PASSING THE CLIPPED
SIGNAL THROUGH A LINEAR AMPLIFIER SECTION

Fig. 3-6 Changing a sinusoidal signal to a square wave signal

the transistor is operating in its saturation region, V_{CE} is a very low value (usually about 0.3 volt for silicon and 0.1 volt for germanium). When the transistor is saturated, the voltage drop across the collector-to-emitter terminals is almost zero ($V_{CE} \approx 0$) when compared with the usual levels of V_{CC}. This observation is true regardless of the value of the collector current. The effect is that of a closed switch. As a result, when the transistor is in the saturation region, it is said to be *on*. For a rough analysis of the use of a bipolar transistor in the saturation region, the approximate equivalent circuit shown in figure 3-7 can be used. For a more exact analysis of the transistor used in this manner, refer to the values listed in Table 3-1. R_c is the load through which current is to flow. When the transistor is used as a switch, the base bias may not be needed because the input signal itself (v_s) may be sufficient to turn the transistor on. The resistor R_b may not be in the circuit or it may have a very large value.

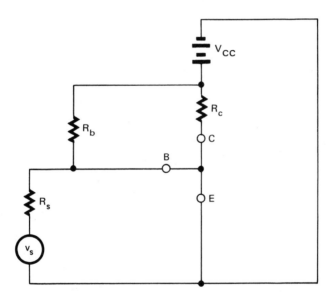

Fig. 3-7   Approximate equivalent circuit of an npn transistor operating in the saturation region

Material	V_{CE} at saturation	V_{BE} at saturation	V_{BE} in active region	V_{BE} at cut in	V_{BE} at cutoff
Silicon	0.3	0.7	0.6	0.5	0
Germanium	0.1	0.3	0.2	0.1	-0.1

Note: pnp transistors have approximately the same values but are opposite in polarity

Table 3-1 Typical npn transistor voltages at 25°C

When the transistor is operating in its cutoff region, the collector current is very nearly zero. (The collector current cannot equal zero because thermally generated current carriers are moving as a result of the back biasing voltage, V_{CC}.) To insure that the transistor is operating in the cutoff region, the base-to-emitter junction is deliberately back biased. Recall that a semiconductor diode must have a forward bias voltage greater than zero if it is to conduct. Therefore, to back bias the base-to-emitter junction, it is necessary only to reduce the base biasing voltage to zero. The transistor then acts as an open switch. As a result, when the transistor is in the cutoff region, it is said to be *cutoff* or *off*. For a rough analysis, an approximate equivalent circuit of the transistor in the cutoff region is shown in figure 3-8. In actual operation, there is some current flowing between the base, emitter, and collector terminals in any cutoff transistor. However, such current is very small and is due entirely to thermally generated current carriers that are moved by the applied back biasing voltages.

When used in the manner just described, the transistor is a voltage controlled switch. (Compare the transistor used in this way to the electromagnetic relay described in Chapter 1.) However, the transistor is more like a current controlled switch, since it seems to be more sensitive to the base current than it is to the base-to-emitter voltage. Regardless of what it is called, the transistor is frequently used as a switch. The only requirements to turn the transistor on are a very small base current and base to emitter voltage (V_{BE} greater than 0.3 volts for a germanium transistor and 0.7 volts for a silicon transistor). Once the transistor is on, the voltage V_{CC} can drive current through the load resistance (R_c in figures 3-7 and 3-8).

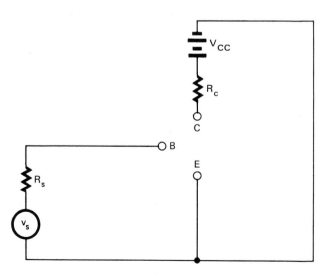

Fig. 3-8 Approximate circuit of an npn transistor operating in the cutoff region

Assume that current is to be passed through a 5 kilohm resistor by means of a transistor switch connected to a 40-volt power source. Is it possible to use the transistor whose characteristics are shown in figure 2-9? What base current is required to close this switch? (R3-4)

The transistor switch has several advantages when compared to the electromagnetic relay. These advantages include the small actuating signal required and the rapid action of the transistor. High speed switching transistors are available which can switch from "off" to "on" in less than a microsecond. The main disadvantage of the transistor switch is that the transistor is never completely off or completely on. This action is unlike the electromagnetic relay which can completely open or close its contacts. Another disadvantage of the transistor switch is that it must be matched carefully to the load. Matching is required because the load affects the biasing of the transistor and thereby alters its switching characteristics.

When a transistor functions as a switch or as a clipper, it is being used as a *nonlinear* amplifier.

In digital logic systems, the transistor is used as a NOT circuit. When a high voltage signal is fed into the base of the transistor circuit shown in figure 2-10, the transistor turns on. As a result, there is a low output voltage from the collector to the emitter. In other words, an input signal causes an output signal to *not* exist. This is also a way of stating that there is a $180°$ phase shift between the input voltage signal and the output voltage signal for a common emitter amplifier.

AMPLIFIER ANALYSIS USING CHARACTERISTIC CURVES

The following quantities frequently must be determined for an amplifier: the input impedance, the output impedance, the current gain, and the voltage gain. These quantities can be calculated if the transistor curves are available and the Q-point and load lines are established.

Input Resistance

As an exercise in determining the quantities listed, assume that the amplifier circuit is the one shown in figure 2-10. The input resistance is the parallel combination of R_b and the resistance expressed by $\Delta v_{BE}/\Delta i_B$. R_b can be neglected because it is so much larger than the transistor base to emitter resistance. Since the base to emitter resistance is nonlinear (refer to figure 2-8), the resistance is different for each value of V_{BE} and I_B. Thus, the resistance depends on the peak-to-peak magnitude of the input signal and on the Q-point location. For this circuit, it is assumed that the base current varies from 0 to 0.20 milliampere for the transistor characteristics shown in figure 2-8. Use figure 2-8 to find the variation in v_{be}. The input resistance for this condition is expressed by the following equation.

$$R_i = \frac{\Delta v_{BE}}{\Delta i_B}$$

Eq. 3.1

$$R_i = \frac{0.22 \text{ V}}{0.20 \text{ mA}} = 1.1 \text{ kilohm}$$

Output Resistance

The output resistance is the parallel combination of R_c and $\Delta V_{CE}/\Delta I_C$. The quantity $\Delta V_{CE}/\Delta I_C$ changes due to the location of the Q-point, the load resistance R_L, and the magnitude of the varying output signal. Using figure 3-1 as an example, assume that $I_{BQ} = 0.04$ milliampere, with the dc load line as shown. R_L is set equal to infinity. For this situation, V_{CE} changes from 7.6 to 0.3 volts as I_C changes from 9.3 to 8 milliamperes.

$$\frac{\Delta V_{CE}}{\Delta I_C} = \frac{7.6 - 0.3}{9.3 - 8.0} = \frac{7.3}{1.3} = 5.61 \text{ kilohms}$$

The amplifier output impedance for the circuit shown in figure 2-11 is expressed by the following equation, assuming the capacitors have a negligible effect:

$$R_o = \frac{R_c \left(\dfrac{\Delta V_{CE}}{\Delta I_C} \right)}{R_c + \dfrac{\Delta V_{CE}}{\Delta I_C}}$$

Eq. 3.2

With R_c equal to 1340 ohms,

$$R_o = \frac{(1.34) \ (5.61)}{1.34 + 5.61} = 1.08 \text{ kilohms}$$

Current Gain

The current gain of the unloaded bipolar transistor itself can be calculated by reading values from the dc load line in figure 3-1. Using Equation 2.1, the current gain is:

$$A_I = \frac{i_{C1} - i_{C2}}{i_{B1} - i_{B2}}$$

Eq. 3.3

$$A_I = \frac{15 - 0}{0.07 - 0} = 214$$

The current gain is modified and redefined when R_L is not equal to infinity. The current through an infinite R_L must be zero. Thus, the current gain from amplifier in to amplifier out is actually zero. Using symbols, this means that the current gain is:

$$A_{IL} = \frac{i_L}{i_b}$$

Eq. 3.4

which is 0 when $i_L = 0$.

For small values of R_L, the overall current gain is quite large. When $R_L = 0$, there is no change in V_{CE} and no ac current through R_c. Thus, the ac load line is vertical. Figure 3-1 does not contain data for base currents larger than 0.05 milliampere when the ac load line is vertical. In addition, base currents larger than 0.05 milliampere may drive the operation of the transistor beyond its maximum power dissipation limit. For this reason, the change in the base current is assumed to be only ± 0.01 milliampere. The overall current gain for this case is determined as follows:

$$A_{IL} = \frac{11.7 - 6.6}{0.05 - 0.03} = \frac{5.1}{0.02} = 255$$

Voltage Gain

The voltage gain for this problem is determined using the ratio of the change in v_{CE} to the change in v_{IN}. Assume that a change in i_B of ± 0.03 milliampere causes V_{BE} to change by 0.10 volt. For an infinite R_L, the voltage gain has a maximum value which is determined by reading along the dc load line. Using figure 3-1, the voltage gain is:

$$A_V = \frac{v_{CE1} - v_{CE2}}{v_{IN1} - v_{IN2}} = \frac{v_{CE1} - v_{CE2}}{v_{BE1} - v_{BE2}}$$

Eq. 3.5

The voltage gain is calculated by substituting in Equation 3.5, using the values read from the curves.

$$A_V = \frac{18 - 0.3}{0.10} = 177$$

An RC coupled amplifier circuit is constructed as shown in figure 2-11. The transistor is described by the curve of figure 2-9. The following values are assumed: $R_c = 4$ kilohms, $R_L = 3$ kilohms, $V_{CC} = 40$ volts, and $I_B = 0.04$ milliampere at the quiescent point. Using the curves given in figure 2-9, *prepare an overlay on which the Q-point is located and the ac load line is drawn. For a peak-to-peak variation in i_b equal to 0.06 milliampere and a peak-to-peak variation in v_{be} equal to 0.20 volt, determine the following quantities:*

    a. output impedance for an infinite R_L
    b. current gain for the unloaded transistor
    c. current gain for a shorted amplifier ($R_L = 0$)
    d. voltage gain for an unloaded amplifier
    e. input impedance
                                             (R3-5)

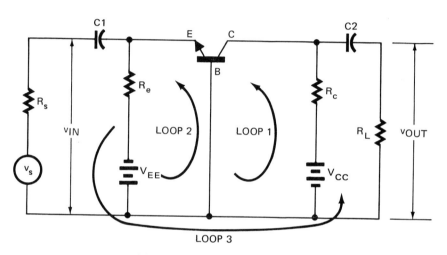

Fig. 3-9 Common base amplifier

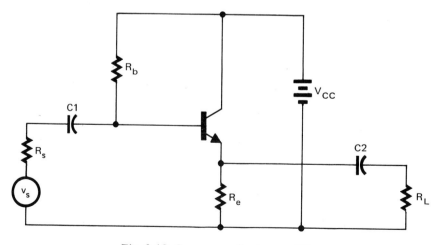

Fig. 3-10 Common collector amplifier

COMMON BASE CIRCUIT CONFIGURATION

The bipolar transistor is most commonly used in the common emitter configuration in a circuit. However, there are two other distinctive configurations that are widely used: the common base and the common collector. These are shown in figures 3-9 and 3-10. Note the similarity of these circuits and the circuit shown in figure 2-6 to the three circuit configurations shown in figure 1-15, using the generalized amplifying device.

The common base (CB) configuration is so called because the base is common to both the input and output circuits. It is helpful to study the performance

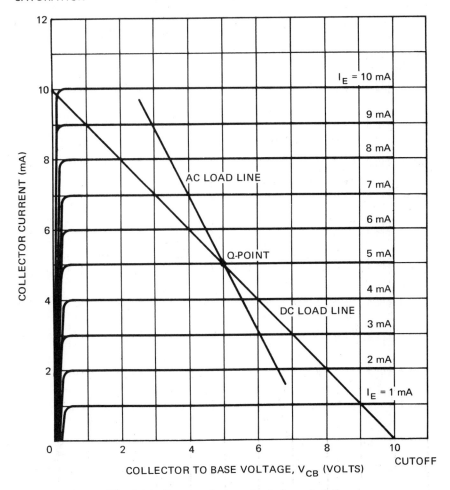

Fig. 3-11 Common base output characteristics

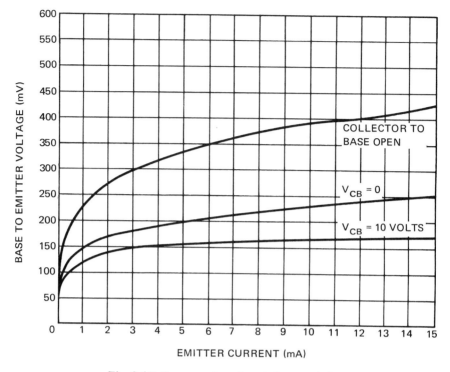

Fig. 3-12 Common base input characteristics

of a transistor connected in its common base mode by using the common base input and output characteristics. Refer to figures 3-11 and 3-12. The saturation, cutoff, and active regions of operation that are identified in figure 3-11 have the same meaning as those defined for the generalized amplifying device in Chapter 1.

In general terms, explain what constitutes saturation, cutoff, and active regions of operation for a transistor. *(R3-6)*

In the active region of the transistor used in the common base configuration, the base-to-emitter diode junction is forward biased and the collector-to-base diode junction is reverse biased. The emitter current is the input current for the common base circuit. Thus, a family of characteristic curves are developed for this transistor with the emitter current as an independent variable.

CURRENT GAIN OF A COMMON BASE CIRCUIT

Considering the common base circuit as an amplifier, the input signal is the emitter current and the output signal is the collector current. The current gain of the common base amplifier can be defined by a formula similar to Equation 1.9.

$$A_I = \frac{i_{C1} - i_{C2}}{i_{E1} - i_{E2}} \qquad\qquad \text{Eq. 3.6}$$

where i_{E1} and i_{E2} are the initial and final values of the emitter current.

If the initial value of the emitter current is zero and the final value is in or near the saturation region, then the current gain A_I is the dc or large signal current gain for the common base connection. If the initial and final values of current are both in the active region for a constant value of the collector-to-base voltage V_{CB}, and there is only a small difference between the initial and final values, then the current gain A_I is the ac or small signal current gain. The large signal current gain is approximately the same as the small signal current gain for a transistor connected in a common base circuit. The common base current gain is known as a (alpha). When the ac signals are small, the common base current gain is known as h_{fb}. This quantity is used to analyze the circuit response to small signals and is one of the h-parameters which are discussed in detail in Chapter 4.

The current gain is measured along the load line drawn on the curves of figure 3-11. An arbitrary Q-point is selected and R_L is assumed to be infinite. What is the approximate value for the current gain of this common base circuit for a peak-to-peak emitter current signal of 1 milliampere? *(R3-7)*

It was pointed out in Chapter 2 that the base current is very small in comparison to the emitter or collector current. By applying Kirchhoff's Current Laws, it can be seen that the emitter and collector currents are very nearly equal. As a result, the current gain of the common base amplifier is approximately equal to unity. The actual values of the current gain range between 0.95 and 0.995.

The value of the common base current gain (a) is related to the common emitter current gain as follows:

$$a = \frac{\beta}{1 + \beta} \qquad\qquad \text{Eq. 3.7}$$

The β for a transistor is given by the specifications as 50. What is the value of a? *(R3-8)*

Determining the Dc Load Line and the Quiescent Point for the Common Base Circuit

The procedure for determining the operating point or quiescent point for the common base amplifier is similar to that for the common emitter amplifier. The circuit is analyzed by using Kirchhoff's Voltage Law, drawing a load line, and locating a point on the load line midway in the active region. A typical load line is shown on the common base curves of figure 3-11. For a common base transistor amplifier circuit such as the one shown in figure 3-9, Kirchhoff's Voltage Law can be applied to Loop 1 to yield:

$$V_{CC} - I_C R_c - V_{CB} = 0 \qquad \text{Eq. 3.8}$$

When $I_C = 0$, $V_{CC} = V_{CB}$. This establishes one end of the load line. When $V_{CB} = 0$,

$$I_C = \frac{V_{CC}}{R_c} \qquad \text{Eq. 3.9}$$

This establishes the other end of the load line.

A load line can be drawn on the curves in figure 3-11. If $V_{CC} = 10$ volts, then one end of the load line is at 10 volts. If $R_c = 1$ kilohm, then at $V_{CB} = 0$,

$$I_C = \frac{V_{CC}}{R_C} = \frac{10}{1} = 10\,mA$$

The other end of the load line then intersects the I_C axis at 10 milliamperes. The slope of the dc load line is 1 mA/V.

The Q-point for the common base circuit can be located readily. At any point in the active region, the emitter current is approximately the same as the collector current, regardless of the base current. To achieve linear amplification, the Q-point is properly located at the approximate center of the active region. For such a Q-point, V_{CB} is about $V_{CC}/2$ which makes $V_{CB} = 5$ volts. Using the curves of figure 3-12, it can be seen that a V_{BE} of about 175 millivolts is expected when $V_{CB} = 5$ volts. Refer to figure 3-9, and apply Kirchhoff's Voltage Law around Loop 2 to obtain:

$$V_{EE} - I_E R_e - V_{BE} = 0 \qquad \text{Eq. 3.10}$$

Equation 3.10 can be solved for I_E. If V_{EE} is 10 volts, and R_e is 2 kilohms, then,

$$I_E = \frac{10 - 0.175}{2} = 5\,mA$$

Operation evidently is taking place in the center of the linear operating region for the transistor.

Ac Load Line for the Common Base Circuit

The ac load line for the common base circuit is determined in a manner similar to methods developed previously. In the circuit in figure 3-9, it is assumed that the output capacitor (C2) is shorted, for the signal frequency used. The resistance used to establish the slope of the load line is the load resistance R_L in parallel with the resistor R_c.

If R_L = 1 kilohm, then the slope is determined as follows.

$$\text{Slope} = \frac{\Delta i_C}{\Delta v_{CB}} = \frac{R_L + R_c}{R_L\,R_c}$$

$$= \frac{1000 + 1000}{(1000)\,(1000)} = 0.002$$

The ac and dc load lines using the above circuit values are drawn on the common base characteristics in figure 3-11.

Amplifier Parameters for the Common Base Circuit

The amplifier parameters (input resistance, output resistance, current gain, and voltage gain) for the circuit of figure 3-9 can be obtained by referring to figure 3-11 and using techniques presented on pages 71 to 74. Assume that there is a peak-to-peak swing of 2 milliamperes in i_E at a constant v_{CB}. The input resistance is R_e in parallel with $\Delta v_{BE}/\Delta i_E$. The input resistance, using values from the curve in figure 3-12, is:

$$R_i = \frac{R_e\left(\dfrac{\Delta V_{BE}}{\Delta I_E}\right)}{R_e + \dfrac{\Delta V_{BE}}{\Delta I_E}}$$

$$R_i = \frac{(9800)\left(\dfrac{180 - 165}{2}\right)}{9800 + \left(\dfrac{180 - 165}{2}\right)} = 7.5 \text{ ohms}$$

The output resistance R_o is computed with no ac load attached. R_o is the resistor R_c in parallel with $\Delta V_{CB}/\Delta I_C$. The curves in figure 3-11 show that v_{CB} changes by 2 volts for a change of 0.002 ampere in i_E. The output resistance is:

$$R_o = \frac{(1000)\,\dfrac{2}{0.002}}{1000 + \dfrac{2}{0.002}} = 500 \text{ ohms}$$

The current gain is given by the ratio $\Delta i_C / \Delta i_E$ and is approximately equal to 1. This value is normal for a common base amplifier.
The voltage gain is determined as follows.

$$A_V = \frac{\Delta v_{CB}}{\Delta v_{EB}} = \frac{2}{0.180 - 0.165} = 133$$

A common base circuit such as the one shown in figure 3-9 is built with V_{CC} = 35 volts and R_c = 3.5 kilohms. Using the transistor whose common base characteristics are shown in the curves of figure 3-13, determine:

a. the dc load line and the Q-point

b. the ac load line if the load resistance R_L = 1500 ohms.   (R3-9)

Calculate the current gain for the amplifier described in problem (R3-9).
(R3-10)

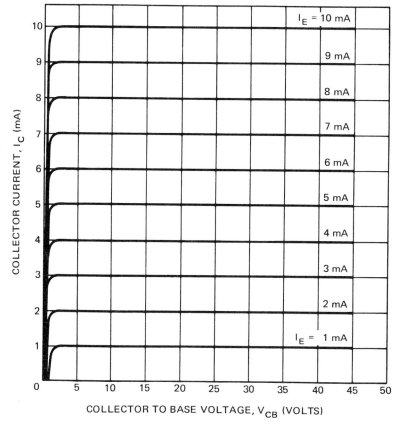

Fig. 3-13 Common base characteristic for use in solving review problems

COMMON COLLECTOR CIRCUIT CONFIGURATION

The third transistor circuit configuration has the collector common to both the input and output terminals. This configuration is known as the common collector (CC) transistor amplifier circuit. Figure 3-10 is an example of a common collector circuit.

The common collector circuit is also known as an *emitter follower* because the emitter-to-common ac voltage is nearly equal to the ac signal voltage. The emitter voltage *follows* the input voltage. This type of amplifier has a voltage gain of almost unity. The common collector amplifier is frequently used as an impedance matching device between a high impedance source and a low-impedance load.

Load Line and Q-point for the Common Collector Circuit

The base current is the input signal for both the common collector circuit and the common emitter circuit. The output current is the emitter current which is the sum of the base and collector currents. The base current is so small in comparison to the collector current, that the output current can be assumed approximately equal to the collector current. This means that the variables needed to analyze the common collector amplifier are the same as the variables needed to analyze the common emitter amplifier; namely, collector current, base current, and collector-to-emitter voltage. Therefore, the common emitter output characteristic curves can also be used to analyze the common collector amplifier circuit. A load line can be determined for this circuit also. Referring to figure 3-10, write Kirchhoff's Voltage Law for the collector circuit, Loop 1.

$$V_{CC} - V_{CE} - (I_B + I_C)R_e = 0 \qquad \text{Eq. 3.11}$$

The Kirchhoff Voltage Law is written for the base circuit, Loop 2, as follows:

$$V_{CC} - I_B R_b - V_{BE} - (I_B + I_C)R_e = 0 \qquad \text{Eq. 3.12}$$

Solve Equation 3.12 for I_B.

$$I_B = \frac{V_{CC} - I_C R_e - V_{BE}}{R_b + R_e} \qquad \text{Eq. 3.13}$$

Substitute Equation 3.13 in Equation 3.11 to obtain:

$$V_{CC} - V_{CE} - \frac{(V_{CC} - I_C R_c - V_{BE})R_e}{(R_b + R_e)} - I_C R_e = 0 \qquad \text{Eq. 3.14}$$

Solve Equation 3.14 for I_C:

$$-I_C = \frac{-V_{CC}R_b - V_{CC}R_e + V_{CE}R_b + V_{CE}R_e + V_{CC}R_e - V_{BE}R_e}{R_e R_b}$$

$$\text{Eq. 3.15}$$

$$I_C = \frac{(V_{CC} - V_{CE})}{R_e} - \frac{(V_{CE} - V_{BE})}{R_b} \qquad \text{Eq. 3.16}$$

When $V_{CE} = 0$, I_C is:

$$I_C = \frac{V_{CC}}{R_e} + \frac{V_{BE}}{R_b} \approx \frac{V_{CC}}{R_e} \qquad \text{Eq. 3.17}$$

This locates one end of the load line. A basis is now provided for the selection of the resistor R_e. R_e is selected to insure that the maximum safe operation levels of collector current and collector power dissipation are not reached while the transistor is in use.

When $I_C = 0$,

$$\frac{V_{CC} - V_{CE}}{R_e} - \frac{V_{CE} - V_{BE}}{R_b} = 0 \qquad \text{Eq. 3.18}$$

If $V_{BE} \approx 0$, then Equation 3.18 can be solved for V_{CE}:

$$V_{CE} = \frac{R_b V_{CC}}{R_b + R_e} \qquad \text{Eq. 3.19}$$

The other end of the load line is therefore located at the point $I_C = 0$ and $V_{CE} = R_b V_{CC}/(R_b + R_e)$.

The slope of the load line is the collector current at $V_{CE} = 0$ divided by V_{CE} at $I_C = 0$. Using Equations 3.17 and 3.19,

$$\text{Slope} = \frac{\dfrac{V_{CC}}{R_e}}{\dfrac{R_b V_{CC}}{R_b + R_e}} = \frac{R_b + R_e}{R_e R_b} \qquad \text{Eq. 3.20}$$

Since R_b is usually very much larger than R_e, the slope reduces very nearly to $1/R_e$ and V_{CE} is approximately V_{CC} at $I_C = 0$.

The base bias is determined by the value of R_b. Equation 3.12 is solved for R_b.

$$R_b = \frac{(V_{CC} - I_C R_e - V_{BE})}{I_B} - R_e \qquad \text{Eq. 3.21}$$

I_C and I_B are selected so that the Q-point is located near the center of the active region. V_{BE} is small and can be neglected for an approximate calculation. V_{CC} is selected at a convenient value that is safely below the maximum allowable value of V_{CE} for the particular transistor used in the circuit. The value of R_e is selected for a satisfactory load line. R_b is calculated directly and set for the desired base bias.

Ac Load Line for the Common Collector Circuit

The ac load line is determined by assuming that the output capacitor is shorted and that the slope of the load line is given approximately by $(R_e + R_L)/R_e R_L$ rather than $1/R_e$. In other words, R_e is replaced by the parallel combination of R_e and R_L for the ac load line. The ac load line is then drawn through the Q-point at the new slope.

Example of the Analysis of a Common Collector Circuit

Assume that a transistor has the common emitter characteristics shown in figure 3-1. It is desired to select a satisfactory load condition for this transistor when it is used in a common collector circuit. The operation of the transistor is placed in the approximate center of its active region for the condition $V_{CC} = 20$ volts and $I_{BQ} = 0.04$ milliampere. A load resistance (R_e) of 1340 ohms is well within the power dissipation capability of the transistor. I_C is determined from Equation 3.17.

$$\dot{I}_C = \frac{20}{1340} = 15 \text{ mA}$$

With this value of I_C, one end of the load line is established. Equation 3.20 reduces to the approximate expression $1/R_e$. Therefore, the slope is:

$$\text{Slope} = \frac{1}{1340} = \frac{I_C \text{ intercept}}{V_{CE} \text{ intercept}} \qquad \text{Eq. 3.22}$$

The other end of the load line is defined at $V_{CE} = V_{CC} = 20$ volts. The Q-point is located at an I_B of 0.04 milliampere. At this point, $I_C = 9.3$ milliamperes. Let $V_{BE} = 0.7$ volt. Equation 3.21 is used to solve for R_b.

$$R_b = \frac{20 - (9.3)(1.34) - 0.7}{0.04} - 1.34 = 169.7 \text{ kilohms}$$

If the load resistance in this example is 1000 ohms, then the ac load line is drawn through the Q-point with a slope of:

$$\frac{1000 + 1340}{(1000)(1340)} = 0.00175$$

Using the same techniques covered in pages 71 to 74, the amplifier parameters R_o, A_I, and A_V can be determined. The common collector (CC) circuit of figure 3-10 and the common emitter curves shown in figure 3-1 can be used to calculate these parameters for any magnitude of the input signal.

The transistor described by the characteristics shown in figure 2-9 below is used in a common collector circuit (such as figure 3-10). For $V_{CC} = 40$ volts, $R_e = 5$ kilohms, and a quiescent base current of 0.03 milliampere, determine the end points and the slope of the dc load line for this circuit. Draw the dc load line on a copy of the characteristic curves. (R3-11)

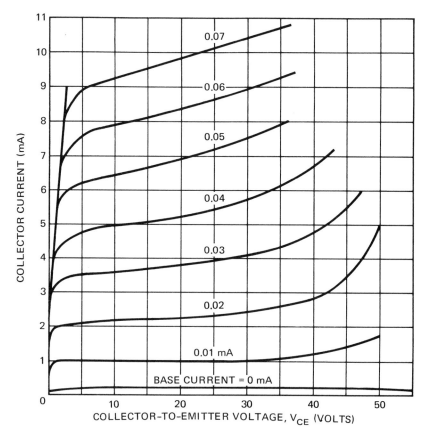

Determine the slope of the ac load line using the results of question (R3-11). The load resistance (R_L) is 3000 ohms. Draw the ac load line on the characteristic curves used for (R3-11). (R3-12)

What is the value of R_b using the results of questions (R3-11) and (R3-12) if V_{BE} = 0.70 volt? (R3-13)

What is the current gain for the circuit in question (R3-11)? Assume Δi_B = 0.04 milliampere peak-to-peak. (R3-14)

MULTISTAGE AMPLIFIERS

An amplifier usually has more than one stage of amplification. When several stages are used, the output of one stage feeds the input of the next stage, figure 3-14. This arrangement is known as *cascading* amplifier stages. The gain A of a multistage amplifier is the product of the gains of the individual stages as shown in the following equation.

$$A = (A_1)(A_2)(A_3)$$ Eq. 3.23

Equation 3.23 expresses the gain A of a multistage amplifier having three stages with gains A_1, A_2, and A_3.

The current gain of stage 1 of a two-stage amplifier is 75. If the second stage of the amplifier has a gain of 50, what is the magnitude of the overall gain of the multistage unit? (R3-15)

The individual stages of a multistage amplifier may be common emitter, common base, or common collector circuits or combinations of these circuits. To match source and load impedances to the amplifier, either common base or common collector stages may be used as the beginning or ending stages of the

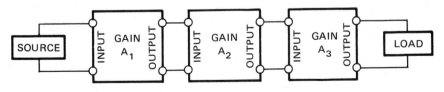

Fig. 3-14 Cascading of amplifiers

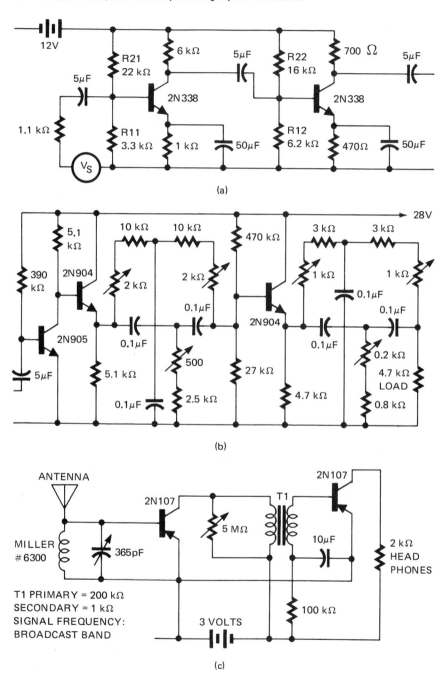

Fig. 3-15 Multistage amplifiers

amplifier. The intermediate amplifier stages generally are common emitter circuits. Some typical multistage amplifiers are shown in figure 3-15.

The method of setting the operating conditions for ac coupled multistage amplifiers is similar to the procedures used for single stage amplifiers. Since the capacitors in the circuit in figure 3-15 (a) effectively block direct current, the quiescent point can be determined separately for each stage of amplification, using the methods previously described. The setting of the ac load lines is more difficult because the equivalent input resistance of all of the stages following the first stage must be known before the load resistance of the first stage can be calculated. The calculation of the input resistances of transistor amplifiers for small signals varying with time is covered in Chapter 4.

How many stages of amplification are contained in the amplifier circuits shown in figure 3-15? Identify these stages by indicating whether they are CC, CE, or CB stages. *(R3-16)*

DIRECT COUPLED AMPLIFIER CONFIGURATIONS

The emitter coupled differential amplifier circuit shown in figure 3-16 is a direct coupled amplifier. This circuit amplifies the difference between two signals and is frequently used to amplify very low frequency or very slowly changing signals. The differential amplifier eliminates an inherent problem in

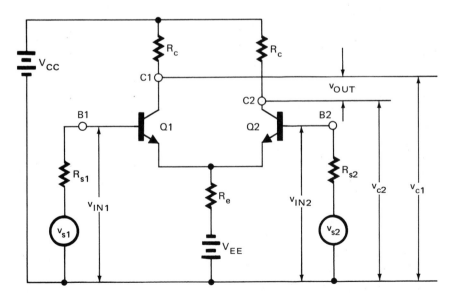

Fig. 3-16 Emitter coupled differential amplifier

amplifiying slowly changing signals: the change in transistor characteristics with temperature. In figure 3-16, the transistors Q1 and Q2 are selected so that they have the same set of characteristic curves and the same sensitivity to temperature. The bias circuits for Q1 and Q2 are the same. As a result, both transistors operate at the same Q-point. The terminal B2 is the input of Q2 and is connected to ground through a simulated source resistance. If the input signal is applied between B1 and common, and the output signal is the difference between the voltages at C1 and C2, then the circuit cancels the effect of the drift of the transistor characteristics due to temperature changes.

As a means of analyzing the bias requirements for the differential amplifier, the circuit shown in figure 3-17 represents the connections to one of the transistors from figure 3-16. Kirchhoff's Voltage Law is applied around Loop 1 (the collector and emitter circuit). I_B is assumed to be negligible.

$$V_{CC} + V_{EE} - V_{CE} - I_C(2R_e + R_c) = 0 \qquad \text{Eq. 3.24}$$

The emitter resistor appears in Equation 3.24 as $2R_e$ because the collector currents of both transistors are flowing through it. Solve for V_{CE} when $I_C = 0$:

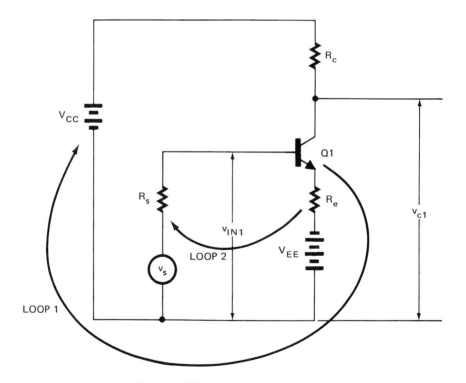

Fig. 3-17 Half of the differential amplifier shown in figure 3-16

$$V_{CE} = V_{CC} + V_{EE} \qquad \text{Eq. 3.25}$$

This value of V_{CE} (at $I_C = 0$) fixes one end of the load line. The other end of the load line is determined when $V_{CE} = 0$

$$I_C = \frac{V_{CC} + V_{EE}}{2R_e + R_c} \qquad \text{Eq. 3.26}$$

The slope of the load line is:

$$\text{Slope} = \frac{1}{2R_e + R_c} \qquad \text{Eq. 3.27}$$

As an example of this analysis, let $R_c = 1.5$ kilohms, $R_e = 1.4$ kilohms, $V_{CC} = 13$ volts, and $V_{EE} = 12$ volts. Thus,

$$I_C = \frac{25}{4.3} = 5.8 \text{ mA}$$

When $I_C = 0$, $V_{CE} = 13 + 12 = 25$ volts. Using a set of curves such as those in figure 3-1, a load line can be drawn having a slope of:

$$\frac{1}{2.8 \text{ k}\Omega + 1.5 \text{ k}\Omega} = 0.000233$$

R_s is assumed to be small enough to be neglected. For the base bias current only, assume that $v_s = 0$ and $I_B R_s$ is very small. Kirchhoff's Voltage Law applied around the base circuit, Loop 2, yields:

$$V_{EE} - V_{BE} - 2(I_B + I_C)R_e = 0 \qquad \text{Eq. 3.28}$$

Chapter 2 defined the h-parameter h_{FE}:

$$h_{FE} = \frac{I_C}{I_B} \qquad \text{Eq. 3.29}$$

A value for h_{FE} can be determined either by measurement or by referring to the manufacturer's quoted value. If $h_{FE} = 214$, then $I_C = 214 \, I_B$.
Substitute Equation 3.29 in Equation 3.28 and use values from the previous example. For $V_{BE} = 0$, solve for I_B:

$$12 - 2(I_B + 214 I_B)(1.4) = 0$$

$$I_B = 0.0199 \text{ mA}$$

This centers the Q-point in the active region. The Q-point can be centered more exactly by changing the resistance R_e slightly. Actually, the larger the value of R_e, the better the differential amplifier is able to reject the influence of drift.

The Q-point just determined was calculated for $R_s = 0$. If R_s is not equal to zero, it must be included in the writing of Kirchhoff's Voltage Law for the base circuit (Loop 2).

Input signals of a number of different types can be applied to the differential amplifier. For example, a signal from a current source or a voltage source can be applied to Q1 or Q2 or both. As an alternative, a current signal can be applied between the Q1 and Q2 inputs.

As an example of a voltage signal applied to Q1, let $v_{s1} = 2$ volts peak-to-peak, with $R_s \approx 0$. R_e is still 1.4 kilohms. The new value of I_B, when $v_{s1} = +1$ volt, is determined by combining Equations 3.28 and 3.29:

$$I_B = \frac{V_{EE} + 1}{2R_c(h_{FE} + 1)} = \frac{13}{(2.8)(215)} = 0.0216 \, \text{mA}$$

The effect of this signal on the output is the same as for a single-stage common emitter amplifier. The change in base current is $0.0217 - 0.0199 = 0.0018$ milliampere. There is no change in the base current for Q2, other than the change that may be caused by temperature or a low value of load resistance. If the load resistance connected between the outputs C1 and C2 is large enough, it does not affect the difference in outputs of the two stages. Thus, the characteristic curves given in figure 3-1 can be used. A change of 0.004 milliampere causes a change of about $4.4 - 3.9 = 0.5$ milliampere. The current gain is determined as follows:

$$A_I = \frac{0.5}{0.0018} = 277$$

Other combinations of input signals and load and source resistances involve more complex circuit analyses. Such exercises are more suitable for an advanced discussion of amplifiers.

Assume that the transistor whose characteristics are shown in figure 3-1 is to have a load line and Q-point similar to those established previously; that is, the collector bias resistance is 1.34 kilohms, $V_{EE} = 6$ volts, and $V_{CC} = 20$ volts. A differential amplifier circuit (figure 3-16) is to be constructed. R_e is selected to be 450 ohms and the input signal is from a voltage source. ($R_s = 0$.) Assume $V_{BE} = 0$. Draw the dc load line and determine the Q-point for the transistors. (R3-17)

Assume $R_s = 1000$ ohms and $V_{BE} = 0$ for the transistor circuit of question (R3-17). Using Kirchhoff's Voltage Law, solve for the quiescent base current I_{BQ} for each of the transistors. (R3-18)

Darlington Circuits

Another useful direct coupled amplifier is the Darlington circuit shown in figures 3-18, 3-19, or 3-20, pages 92 and 93. The Darlington circuit shown in figure 3-18 is available as an integrated circuit having collector, emitter, and base terminals (as in an ordinary transistor). Such a circuit can be used as a common emitter circuit with exceptionally high amplification or as a two-stage common collector amplifier having an extremely high input impedance.

Refer to figure 3-18 to identify the current symbols. An expression for i_{C2} is given as follows:

$$i_{C2} = h_{FE2} i_{B2} \qquad \text{Eq. 3.30}$$

However,

$$i_{B2} = i_{C1} + i_{B1} \qquad \text{Eq. 3.31}$$

$$i_{B2} = h_{FE1} i_{B1} + i_{B1} = (h_{FE1} + 1) i_{B1} \qquad \text{Eq. 3.32}$$

Substituting Equation 3.32 in Equation 3.30 yields a new expression for i_{C2}:

$$i_{C2} = h_{FE2}(h_{FE1} + 1) i_{B1} \approx h_{FE2} h_{FE1} i_{B1} \qquad \text{Eq. 3.33}$$

If h_{FE1} and h_{FE2} are closely matched, then $h_{FE1} \approx h_{FE2}$, and

$$i_{C2} = (h_{FE})^2 i_{B1} \qquad \text{Eq. 3.34}$$

A high quality transistor frequently has an h_{FE} value greater than 100. If the matched transistors in the circuit shown in figure 3-18 have an h_{FE} of 100, then $(h_{FE})^2 = 10,000$. It can be seen that this circuit may have an advantage for a high-gain amplifier application.

Assume that two transistors are connected as in figure 3-18. If Q1 has an h_{FE} of 50 and Q2 has an h_{FE} of 75, what is the overall current gain of the circuit? *(R3-19)*

Input Impedance of the Darlington Circuit

Figure 3-19 is a useful circuit which acts like two common collector circuits connected in cascade. Recall that a common collector circuit has a high input resistance. It can be shown that as a result of this two-transistor connection, the input resistance is improved still further. The voltage drop across R_e is:

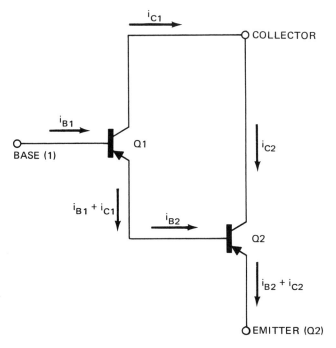

Fig. 3-18 Darlington connection

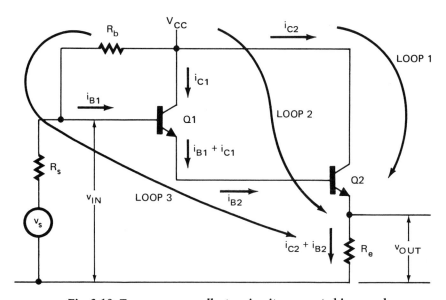

Fig. 3-19 Two common collector circuits connected in cascade

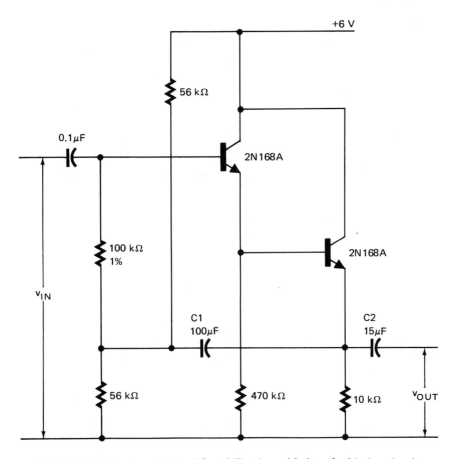

Fig. 3-20 Darlington circuit with stabilization added to the biasing circuit

$$v_{out} = (i_{C2} + i_{B2})R_e \qquad \text{Eq. 3.35}$$

It was determined previously (Equations 3.30 and 3.31) that $i_{B2} = i_{C1} + i_{B1}$ and $i_{C2} = h_{FE2}i_{B2}$. Equation 3.29 ($h_{FE} = I_C/I_B$) can be expressed as $i_{C1} = h_{FE1}i_{B1}$ and combined with Equation 3.31 to form:

$$i_{B2} = i_{B1}(1 + h_{FE1}) \qquad \text{Eq. 3.36}$$

Combining Equation 3.30 with Equations 3.35 and 3.36 yields an expression for v_{OUT}.

$$v_{OUT} = (h_{FE2}i_{B2} + i_{B2})R_e = i_{B2}(h_{FE2} + 1)R_e$$
$$v_{OUT} = i_{B1}(1 + h_{FE1})(1 + h_{FE2})R_e \qquad \text{Eq. 3.37}$$

If the transistors are selected to have the same high value of h_{FE}, v_{OUT} reduces to:

$$v_{OUT} \approx i_{B1}(h_{FE})^2 R_e \qquad \text{Eq. 3.38}$$

The input voltage to the circuit is determined as follows:

$$v_{IN} = v_{BE1} + v_{BE2} + i_{B1}(h_{FE})^2 R_e \qquad \text{Eq. 3.39}$$

Since v_{BE1} and v_{BE2} are small numbers, v_{IN} reduces to:

$$v_{IN} \approx i_{B1}(h_{FE})^2 R_e$$

But, v_{IN} is also:

$$v_{IN} = i_{B1} \text{ times input resistance} \qquad \text{Eq. 3.40}$$

Therefore, the apparent input impedance of the circuit is $(h_{FE})^2 R_e$. This value of input impedance is very large and can reach several megohms. Thus, the circuit shown in figure 3-19 can be used in applications requiring a high input impedance.

If $v_{BE1} = v_{BE2} = 0.2$ volt, $R_e = 1000$ ohms, h_{FE} for transistor Q1 = 100, and h_{FE} for transistor Q2 = 90 in the circuit of figure 3-19, compute the input impedance, assuming $R_b = \infty$. *(R3-20)*

Setting Bias Conditions for the Darlington Circuit

An application requires a circuit such as the one shown in figure 3-19. The transistors Q1 and Q2 are of the same type. Kirchhoff's Voltage Law is applied around Loop 1 for the condition $I_C \approx I_E$.

$$V_{CC} - V_{CE2} - I_{C2}R_e = 0 \qquad \text{Eq. 3.41}$$

By setting $I_{C2} = 0$, $V_{CC} = V_{CE2}$ and one end of the load line is established. By setting $V_{CE2} = 0$, $I_C = V_{CC}/R_e$ for the I_C intercept and the other end of the load line is established. V_{CE1} determines I_{B2} and I_{B1} determines V_{CE1}. Kirchhoff's Voltage Law is then applied around Loop 2 to yield:

$$V_{CC} - V_{CE1} - V_{BE2} - I_{C2}R_e = 0 \qquad \text{Eq. 3.42}$$

Because V_{BE} is so small, it can be neglected and Equation 3.42 becomes the same as Equation 3.41. As a result, Q1 is operating on approximately the same load line as Q2.

Kirchhoff's Voltage Law is then applied around Loop 3.

$$V_{CC} - I_{B1}R_b - V_{BE1} - V_{BE2} - (I_{C2} + I_{B2})R_e = 0$$

Eq. 3.43

However, I_{C2} is:

$$I_{C2} = h_{FE2}I_{B2}$$

Eq. 3.44

I_{B2} locates the quiescent point for Q2. V_{BE1} and V_{BE2} are small. They can be taken from the common emitter input curves or values from Table 3-1 can be used.

$$I_{B2} = I_{C1} + I_{B1}$$

Eq. 3.45

and I_{B1} is small. Therefore,

$$I_{B2} \approx I_{C1}$$

$$I_{B2} \approx h_{FE1}I_{B1}$$

Eq. 3.46

I_{B1} is selected to provide a value of I_{B2} such that Q2 is in the active region. It is evident that unless two vastly different transistors are used, I_{B2} is set near the saturation point for Q2 and I_{B1} is set near the cutoff point for Q1.

An expression for R_b can be obtained using Equation 3.43. Recall from Equations 3.44 and 3.46 that

$$I_{C2} = h_{FE2}I_{B2} = h_{FE2}h_{FE1}I_{B1}$$

Eq. 3.47

Substituting Equations 3.47 and 3.46 into Equation 3.42,

$$V_{CC} - I_{B1}R_b - V_{BE1} - V_{BE2} - (h_{FE1} + h_{FE2}h_{FE1})I_{B1}R_e = 0$$

Eq. 3.48

But $V_{BE1} = V_{BE2} \approx 0$ and $h_{FE1} \ll h_{FE2}h_{FE1}$. Then,

$$V_{CC} - I_{B1}R_b - h_{FE2}h_{FE1}I_{B1}R_e = 0$$

Eq. 3.49

Equation 3.49 can be used to calculate R_b.

As an example, assume that both transistors in the circuit of figure 3-19 have the characteristics shown in figure 3-1. The following values are used: R_e = 1 kilohm and V_{CC} = 20 volts. The problem is to determine the operating specifications for this amplifier.

According to Equation 3.41, $V_{CC} - V_{CE2} - I_{C2}R_e = 0$. If I_{C2} is set equal to 0, $V_{CC} = V_{CE2} = 20$ volts. At $V_{CE2} = 0$, $I_{C2} = V_{CC}/R_e = 20/1 = 20$ milliampere. Both Q1 and Q2 are on the same load line. I_{B2} is approximately the same as I_{C1}. Assume I_{CQ2} is 9.3 milliamperes and $I_{B2} = 0.04$ milliampere at the Q-point in figure 3-1.

Assume that

$$h_{FE1} = \frac{I_{CQ1}}{I_{BQ1}} = \frac{I_{BQ2}}{I_{BQ1}} = 200$$

Therefore,

$$I_{BQ1} = \frac{0.04}{200} = 0.0002 \text{ mA}$$

In addition to I_{B1}, what quiescent currents are set by the value of R_b? (R3-21)

What is the dc current gain in the previous example? (R3-22)

The circuit in figure 3-19 does not provide stabilization against variations in component parameters. Stabilization is provided in the circuit in figure 3-20 by the addition of an emitter self-bias circuit to the basic circuit of figure 3-18. The use of capacitor C2 in this circuit prevents a drop in the input resistance due to the bias stabilization.

Another direct coupled amplifier circuit is the push-pull amplifier circuit shown in figure 3-21. Note that there is no base bias in the circuit. When there is no input signal, the base current in either transistor is zero. When there is an input signal, the polarity of the signal indicates which transistor is conducting. The base-to-emitter junction of one transistor is forward biased and the base to emitter junction of the second transistor is reverse biased. For a symmetrical sinusoidal or square wave signal, the transistors will alternate between off and on. That is, one transistor will be off and the other will be on. Since the entire active operating region of each transistor is available above zero base current, a greater gain can be obtained with this circuit as compared to that available from one transistor alone. The current through the load resistance will be slightly distorted because of the mismatch of the transistors. However, the distortion is small enough so that the signal is a fairly accurate reproduction of the input signal. Push-pull amplifiers are explained in greater detail in Chapter 12.

Explain why no base bias is required for the push-pull amplifier, such as the one shown in the circuit of figure 3-21. (R3-23)

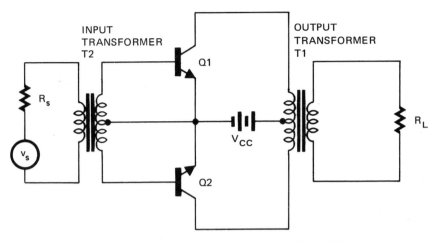

Fig. 3-21 Transformer coupled push-pull amplifier

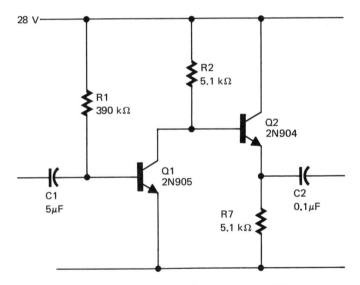

Fig. 3-22 Direct coupled two-stage amplifier

Figure 3-22 shows two common emitter amplifier stages connected to form a direct coupled, two-stage amplifier. Although it appears that the bias level of one stage is influenced by the second stage, a study of the circuit shows that the base-to-emitter junctions of both transistors are forward biased and thus act like virtual shorts across the base to emitter terminals. As a result, the base bias of Q1 does not affect the biasing of Q2 and the base bias of Q2 does not affect the biasing of Q1.

LABORATORY EXPERIMENTS

The following laboratory experiments were tested and found useful for illustrating the theory in this section.

LABORATORY EXPERIMENT 3-1

OBJECTIVE

To build and operate a common base amplifier.

EQUIPMENT AND MATERIALS REQUIRED

1　Volt-ohm-milliammeter
1　Transistor, pnp, 2N1541
2　Resistors, 65 ohms, 10 watts
2　Power supplies, dc, 0 to 20 V dc adjustable, 1 ampere maximum
1　Signal generator, audio frequency, variable amplitude and frequency
1　Oscilloscope, dual channel
3　Capacitors, 20 microfarads, 50 volts

PROCEDURE

1. Connect the listed components according to the common base circuit shown in figure 3-23.

2. Adjust V_{EE} to achieve an undistorted ouput voltage using a 1000-hertz sinusoidal input signal.

3. Measure the voltage gain of the amplifier. Sketch the input and output waveforms and record the V_{EE} value used.

4. Determine the approximate value of ac alpha (α) by measuring the ac voltage developed across the two 65-ohm resistors. Use the VOM.

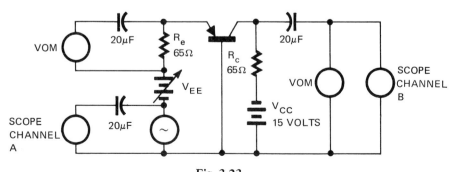

Fig. 3-23

$$a = \frac{I_c}{I_e} = \frac{\text{ac voltage across collector resistor}}{\text{ac voltage across emitter resistor}}$$

$$a = \frac{65\,I_c}{65\,I_e}$$

where $65\,I_c$ = the voltage across the collector resistor
$65\,I_e$ = the voltage across the emitter resistor.

LABORATORY EXPERIMENT 3-2

OBJECTIVE

To build and operate a common collector amplifier

EQUIPMENT AND MATERIALS REQUIRED

1	Transistor, pnp, 2N404
1	Resistance substitution box (Heathkit 1N-37 or equivalent)
1	Resistor, 2 kilohms, 1/2 watt
1	Power supply, 0-30 V dc, regulated
2	Capacitors, 20 microfarads, 50 V dc
1	Oscilloscope, dual channel
1	Decade resistance box (Heathkit 1N-17 or equivalent)

PROCEDURE

1. Build a common collector amplifier. Refer to figure 3-24 for guidance. Note that this circuit has the same biasing as the circuit for Laboratory Experiment 2-3, except that the collector bias resistor is in the emitter circuit.

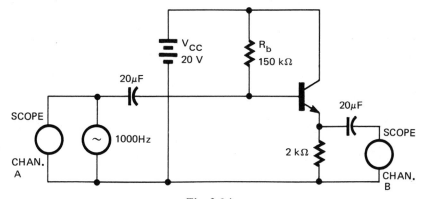

Fig. 3-24

2. The amplifier is to be operated as a common collector amplifier with a sinusoidal signal of 1000 Hz. The only load is the oscilloscope. Readjust the bias resistors and the input signal as necessary to obtain an undistorted sinusoidal output. Note that the voltage gain is low. This is normal. Sketch the input and output signals to scale.

3. Measure and calculate the voltage gain as a ratio and then convert this value to voltage decibels.

4. Measure and calculate the input resistance. The procedure is as follows:

   a. Insert a variable resistance (the decade resistance box) in series with the input signal as shown in figure 3-25.
   b. Adjust the series resistance to zero and measure the input signal voltage.
   c. Increase the series resistance until the input signal is one-half its amplitude for the series resistance value of 0 (step b).
   d. The value of the new series resistance (as read on the dial setting of the decade resistance box) is the input resistance of the amplifier.

5. Measure and calculate the output resistance. The procedure is as follows:

   a. Apply a variable resistance (decade box) load to the amplifier as shown in figure 3-26.
   b. Set the load resistance at infinity by temporarily disconnecting the decade box. Measure the amplifier output voltage.
   c. Adjust the load resistance downward until the output voltage is reduced by one-half. The resulting load resistance (as indicated on the decade box) is the output resistance of the amplifier.

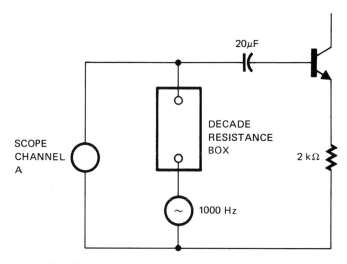

Fig. 3-25 Circuit for measuring input resistance

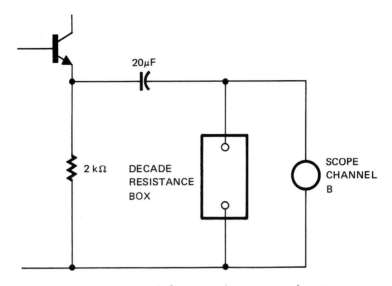

Fig. 3-26 Circuit for measuring output resistance

LABORATORY EXPERIMENT 3-3

OBJECTIVE

To build, analyze, and operate a difference amplifier.

EQUIPMENT AND MATERIALS REQUIRED

2	Transistors, 2N1808
5	Resistance substitution boxes (Heathkit 1N-37 or equivalent)
2	Power supplies, dc, 0-20 V dc
1	Oscilloscope, dual channel
1	Volt-ohm-milliammeter

PROCEDURE

1. Construct a differential amplifier using the circuit of figure 3-27, page 102.

2. Assume that both transistor curves are the same as in figure 2-7. Calculate the approximate quiescent point of each transistor.

3. Calculate h_{FE} at this operating point. Use this value to calculate V_{CE} for each transistor.

4. Measure V_{CE} and verify the calculated value of V_{CE} (step 3).

5. Calculate the approximate value of the dc base-to-common voltage of transistor Q1 (V_1).

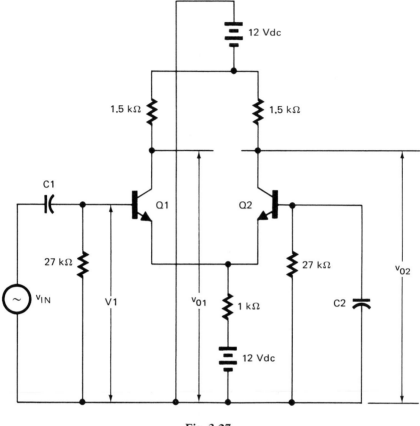

Fig. 3-27

6. Verify the calculated value of V_1 by direct measurement.

7. Apply a small sinusoidal voltage, v_{in}, and measure v_{o1}, v_{o2}, and $v_{o1} - v_{o2}$. Use the VOM. Use a level of v_{in} to provide a sinusoidal voltage at v_{o1} and v_{o2}.

8. Calculate the voltage gains v_{o1}/v_{in} and $(v_{o1} - v_{o2})/v_{in}$.

EXTENDED STUDY TOPICS

1. Define in your own words the meaning of *common emitter, common base,* and *common collector.*

2. What features are characteristic of the Darlington circuit?

3. Can the differential amplifier be used for high-frequency operation even though it is a dc amplifier? Explain.

4. A sinusoidal signal is often used in amplifier testing because it is unchanged in shape by the circuit unless the circuit is nonlinear. Complete problem (R3-3) for a triangular signal with the same peak value and note the distortion.

5. The clipping of a sinusoidal signal when cutoff or saturation occurs, can be shown by the data of problem (R3-2). Apply an input signal which is twice the amplitude for this problem or 0.14 milliampere peak-to-peak. Sketch the clipped waveform. How is this waveform converted to a square wave? Describe a circuit that is used to convert the waveform to a square wave.

6. The load resistance to which an electromagnetic relay supplies current does not affect the signal required to actuate the relay. It is necessary only to insure that the current through the contacts does not exceed the relay rating. Why is it not just as simple to use a transistor as a switch for a particular load resistance?

7. Compare the advantages and disadvantages of the transistor as a switch with the relay as a switch.

8. The current gain of a transistor is calculated by dividing the change in the output current by the change in the input current for any one of three conditions: (a) the ac load resistance at zero, (b) the current changes taken from the characteristic curves along the dc load line, and (c) the current changes taken from the characteristic curves along the ac load line. Explain any differences in the current gain values for the three conditions.

9. Express the common emitter transistor parameter (h_{FE}) in terms of the common emitter base parameter (a).

10. Which circuit configuration is called the emitter follower? Why is it called by this name?

4
Equivalent Circuits and h-Parameters for Bipolar Transistors

OBJECTIVES

After studying this chapter, the student will be able to:

- define and use transistor h-parameters
- develop equivalent circuits and perform an ac analysis of transistor amplifiers
- analyze the performance of a transistor amplifier when a small time varying signal is applied

DEFINITION OF h-PARAMETERS

The characteristic curves of the bipolar transistor show that it is a non-linear device. For small signals, however, the transistor can be treated as a linear device having a predictable performance. This means that it is able to reproduce the waveform of an input signal faithfully at the output terminals. The h-parameters and analysis techniques that are developed in the first part of this chapter are based on the application of small medium-frequency input signals. Later chapters will deal with the application of high- and low-frequency signals.

For this discussion, the transistor is considered to be a black box with two input terminals and two output terminals, figure 4-1. Bias conditions can be ignored since ac signals only are being considered. The transistor is considered to be operating in its active region as a result of the bias voltages applied to it. An equivalent circuit is formed which is subjected to a time varying input signal and produces an output signal which is a complex function of the input. A series of parameters known as *h-parameters* are introduced as the constants to be

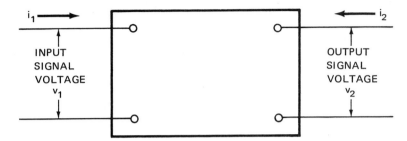

Fig. 4-1   The transistor considered as an amplifying device
with two input terminals and two output terminals

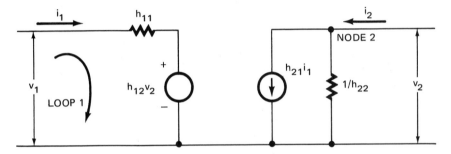

Fig. 4-2 Generalized equivalent amplifier circuit model

used in the equivalent circuit. The input signal may be either a voltage or a
current. However, since the bipolar transistor is basically a current-operated
device, the input signal generally is considered to be a current.

Because many amplifiers contain energy storage devices, such as capacitors, coils, or transformers, the application of a varying (ac) signal to the circuit
causes a phase shift of the signal as well as attenuation or amplification. For
an ac signal, then, *impedance* rather than resistance is used to identify the
parameter restricting the flow of current.

A generalized equivalent circuit of the black box shown in figure 4-1 is
given in figure 4-2. The lower case symbols v and i are used throughout the
analysis to represent instantaneous peak-to-peak variations in voltage and current. Subscript 1 is used for input quantities; subscript 2 is used for output
quantities. Capital letters (V and I) represent dc or average values of the voltages and currents.

The four generalized h-parameters are defined by the following expressions.

1.   Input impedance with the output terminals short circuited is defined by

$$h_{11} = \frac{v_1}{i_1} \text{ with } v_2 = 0$$

Eq. 4.1

2. Reciprocal of the voltage gain with the input terminals open circuited is defined by

$$h_{12} = \frac{v_1}{v_2} \text{ with } i_1 = 0 \qquad \text{Eq. 4.2}$$

3. Current gain with the output terminals shorted is defined by

$$h_{21} = \frac{i_2}{i_1} \text{ with } v_2 = 0 \qquad \text{Eq. 4.3}$$

4. Reciprocal of the output impedance with the input terminals open circuited is defined by

$$h_{22} = \frac{i_2}{v_2} \text{ with } i_1 = 0 \qquad \text{Eq. 4.4}$$

For the present, the h-parameters are considered to be independent of frequency. However, for some amplifiers under certain conditions, the h-parameters have both reactive and real parts. As a result, the parameters are complex numbers. (See the Appendix for a discussion of complex numbers.) The amplifier circuit of figure 4-2 can be defined by two equations that satisfy Kirchhoff's Current and Voltage Laws.

$$v_1 = h_{11}i_1 + h_{12}v_2 \qquad \text{Eq. 4.5}$$

$$i_2 = h_{21}i_1 + h_{22}v_2 \qquad \text{Eq. 4.6}$$

h-Parameters for a Common Emitter Amplifier

A common emitter transistor model is determined based on the above equations. This model is shown in figure 4-3. It can be pictured as a circuit consisting of the impedances, voltage sources, and current sources developed from the model circuit in figure 4-2.

The complete amplifier, of which the circuit of figure 4-3 is a part, could be the ac coupled common emitter amplifier circuit shown in figure 2-11. For this discussion, the signal frequency is assumed to be in a range that insures that the reactance of the coupling capacitors is negligibly small and the reactance of the shunt capacitance in any of the circuit components is negligibly large. Recall that the bias conditions were set previously and only the effects of time varying input signals are of interest. Thus, the amplifier circuit can be drawn like the one shown in figure 4-4, for ac signals only. This ac equivalent circuit ignores the biasing circuits and the coupling capacitors. Z_L is the equivalent ac load impedance. Z_s is the Thévenin equivalent impedance of any preceding stages of amplification or other circuit that makes up the input signal source. The

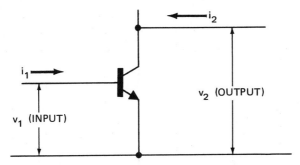

Fig. 4-3 Common emitter transistor model

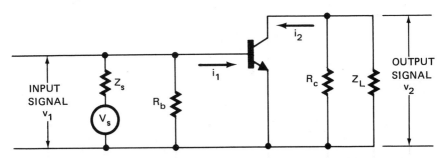

Fig. 4-4 Exact ac circuit for the rc coupled amplifier of figure 2-11, for operation at intermediate frequency signals

Thévenin equivalent voltage (v_s) represents the amplifier input signal from whatever kind of source it comes.

The values of v_1 and i_1 are determined by the input signal source characteristics and the transistor input characteristics. The values of v_2 and i_2 are determined by the output signal load characteristics and the transistor output characteristics. The input and output characteristics of a transistor were defined separately in Chapter 2.

Write Kirchhoff's Voltage Law for Loop 1 of the circuit in figure 4-2. Compare this expression with Equation 4.5. (R4-1)

Write Kirchhoff's Current Law for Node 2 of the circuit shown in figure 4-2. Compare this expression with Equation 4.6. (R4-2)

For the circuit in figure 4-4, the transistor is operating at its Q-point and the parameters are determined for operation around that Q-point. The following values are true at the Q-point: V_{CE} is at V_{CEQ}; I_C is at I_{CQ}; and I_B is at I_{BQ}.

The output signal varies about the Q-point. The h-parameters are calculated for a small value of the output signal. The parameters can be measured approximately on an actual transistor. (Refer to Laboratory Experiment 4-1.)

Since h_{21} is defined when the output terminals are shorted, the ac load line must be vertical. In figure 4-5, the ac load line must be drawn so that it passes through the Q-point. The resulting value of h_{21} can be determined using the following procedure and values taken from the characteristic curves in figure 4-5.

$$h_{21} = h_{fe} = \frac{i_c}{i_b} = \frac{i_{C2} - i_{C1}}{i_{B2} - i_{B1}} \qquad \text{Eq. 4.7}$$

$$h_{21} = \frac{10 - 8.75}{0.0633 - 0.0565} = 184$$

The current values are positive because an npn transistor (2N4074) is used in this example. If a pnp transistor is used, the resulting current values are negative.

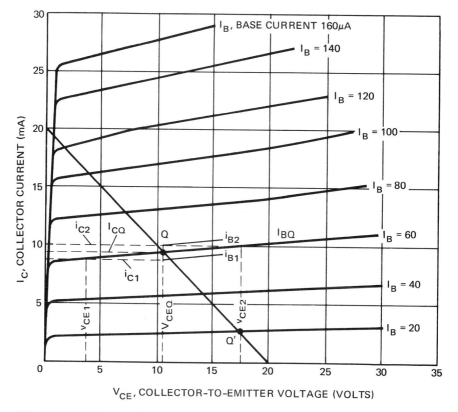

Fig. 4-5 Calculation of h_{fe} and h_{oe} using common emitter characteristic curves

The parameter h_{22} is defined when the input terminals are open. As a result, there is no input signal current and the base current is constant at the quiescent value, I_{BQ}. The value of h_{22} is determined by the use of the following equation.

$$h_{22} = h_{oe} = \frac{i_c}{v_{ce}} = \frac{i_{C2} - i_{C1}}{v_{CE2} - v_{CE1}} \qquad \text{Eq. 4.8}$$

$$h_{22} = \frac{10 - 8.75}{17.5 - 3.75} = \frac{1.25}{13.75} = 0.0909 \ \frac{mA}{V}$$

The values inserted in Equation 4.8 are read from the characteristic curves. Note that the voltages are positive with respect to common. If a pnp transistor is used in the example, then all of the values are negative.

The parameter h_{11} can be determined from the input characteristic curves. Use the input characteristic curves in figure 4-6. h_{11} is the input impedance

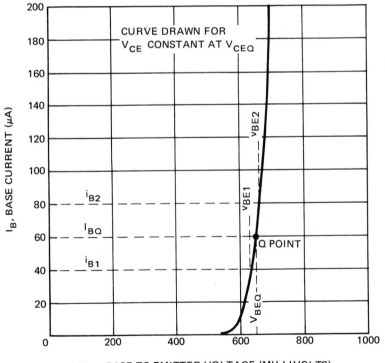

Fig. 4-6 Calculation of h_{ie} using common emitter input characteristic curves

with the output terminals shorted. At the Q-point, V_{BE} is at V_{BEQ}, I_B is at I_{BQ}, and V_{CE} is at V_{CEQ}. h_{11} is determined as follows:

$$h_{11} = h_{ie} = \frac{v_{be}}{i_b} = \frac{v_{BE2} - v_{BE1}}{i_{B2} - i_{B1}} \qquad \text{Eq. 4.9}$$

$$h_{11} = \frac{0.660 - 0.630}{0.000080 - 0.000040} = 750 \text{ ohms}$$

The values substituted in Equation 4.9 are read from the input characteristic curves.

The parameter h_{12} is defined with the input open. As a result, I_B must be the value at the Q-point (I_{BQ}). The expression for h_{12} is,

$$h_{12} = h_{re} = \frac{v_{BE2} - v_{BE1}}{v_{CE2} - v_{CE1}} \qquad \text{Eq. 4.10}$$

Since V_{BE} changes so little with V_{CE}, it cannot be read easily from the curves. However, the manufacturers' specifications state that a typical value of h_{12} for this transistor is 0.000125.

A similar set of h-parameters can be calculated graphically from common base characteristic curves. These parameters are called h_{fb}, h_{ob}, h_{ib}, and h_{rb}.

Refer to figure 2-9. Draw a dc load line with a slope of 1/5000. Establish a Q-point of I_{BQ} = 0.03 milliampere with V_{CC} = 40 volts. Measure graphically the values of h_{fe} and h_{oe} for this transistor.   *(R4-3)*

What is the difference between h_{fe} and h_{FE}?   *(R4-4)*

Why are the directions of the currents through npn and pnp transistors different?   *(R4-5)*

Locate the point on figure 4-5 marked Q'. Determine graphically the values of h_{fe} and h_{oe} for this transistor at the new operating point. Discuss why the values are different at this point from those at the original operating point Q.   *(R4-6)*

EQUIVALENT CIRCUITS USING h-PARAMETERS

The h-parameters were defined earlier in this chapter using the common emitter circuit and the common emitter input and output characteristics as examples. However, as implied in the previous discussion, the h-parameters are defined for the transistor as a black box amplifier, which can take many forms.

The h-parameters are defined in technical literature for all three transistor circuit connections: common emitter, common base, and common collector. Different subscripts are used for each circuit configuration. The first subscript represents the general definition of the h-parameter. The second subscript defines the transistor circuit configuration for which the h-parameter is used. Table 4-1 lists the h-parameters and their subscript designations for the three circuit connections. Table 4-2, page 112, lists the common circuit connections (common emitter, common base, and common collector), their equivalent circuits, and the network equations describing the equivalent circuits. The same current directions are assumed for all three circuits. Table 4-3, page 113, lists typical values of the various h-parameters for selected transistors.

Defining h-Parameters for one Circuit in Terms of h-Parameters for Another Circuit

An examination of the CE, CC, and CB circuits given in Table 4-2 will show that, with respect to a discussion of h-parameters, one circuit can be converted to another circuit. For example, compare the output side for the CE and CC circuits shown in figure 4-7, page 113. It is evident that h_{fe} must be the same parameter as h_{fc} and h_{oc} must be the same parameter as h_{oe}. Now examine the input side of the CE and CC circuits, figure 4-8, page 113. Table 4-3 indicates that the value of h_{rc} is usually about unity and that h_{re} is a very small number. If it is assumed that $h_{rc} \approx 1$ and $h_{re} \approx 0$, the input circuits shown in

h-parameter	Common Base	Common Emitter	Common Collector
h_{11} Input Impedance	h_{ib}	h_{ie}	h_{ic}
h_{12} Reversed Voltage Gain	h_{rb}	h_{re}	h_{rc}
h_{21} Forward Current Gain	h_{fb}	h_{fe}	h_{fc}
h_{22} Output Admittance	h_{ob}	h_{oe}	h_{oc}

Table 4-1 h-Parameter nomenclature

Table 4-2

(a) COMMON CONNECTION	(b) EQUIVALENT CIRCUIT	(c) NETWORK EQUATIONS
CE	CE	CE $v_{be} = h_{ie}i_b + h_{re}v_{ce}$ $i_c = h_{fe}i_b + h_{oe}v_{ce}$
CC	CC	CC $v_{bc} = h_{ic}i_b + h_{rc}v_{ec}$ $i_e = h_{fc}i_b + h_{oc}v_{ec}$
CB	CB	CB $v_{eb} = h_{ib}i_e + h_{rb}v_{cb}$ $i_c = h_{fb}i_e + h_{ob}v_{cb}$

Parameter		Transistor Type			
		2N1808	2N1711	2N338	2N4074
h_{11} (ohms)	h_{ib} h_{ie} h_{ic}	5.2 450 450	29 3600 3600	600	750
h_{12}	h_{rb} h_{re} h_{rc}	0.00123 0.00084 0.9992	0.0005 0.00041 0.9996	0.0002	0.000125
h_{21}	h_{fb} h_{fe} h_{fc}	-0.99 86 -87	-0.992 124 -125	170	184
h_{22} (mhos)	h_{ob} h_{oe} h_{oc}	$4.6\,(10)^{-6}$ 0.0004 0.0004	$0.25\,(10)^{-6}$ 0.000031 0.000031	0.000055	0.0000909
$r_{bb}{}'$ (ohms)		100			

Table 4-3 Typical values of h-parameters

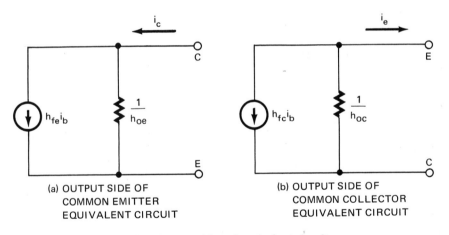

(a) OUTPUT SIDE OF
COMMON EMITTER
EQUIVALENT CIRCUIT

(b) OUTPUT SIDE OF
COMMON COLLECTOR
EQUIVALENT CIRCUIT

Fig. 4-7 Output sides of equivalent circuits

figure 4-8 can be changed to the circuits of figure 4-9, page 114. These circuits, in turn, can be considered as one approximate equivalent circuit, figure 4–10, page 144, which can be used in the analysis of either the CE or CC circuit. Using the approximation, it can be seen that $h_{ie} = h_{ic}$.

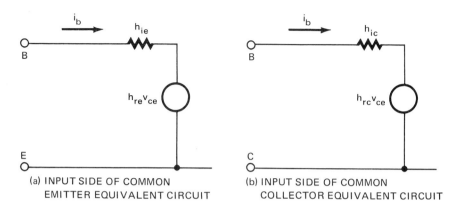

(a) INPUT SIDE OF COMMON
    EMITTER EQUIVALENT CIRCUIT

(b) INPUT SIDE OF COMMON
    COLLECTOR EQUIVALENT CIRCUIT

Fig. 4-8 Input sides of equivalent circuits

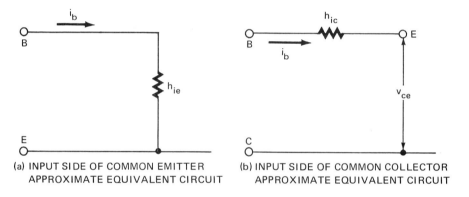

(a) INPUT SIDE OF COMMON EMITTER
    APPROXIMATE EQUIVALENT CIRCUIT

(b) INPUT SIDE OF COMMON COLLECTOR
    APPROXIMATE EQUIVALENT CIRCUIT

Fig. 4-9 Approximation of circuits in figure 4-8

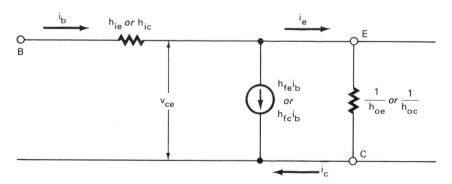

Fig. 4-10 Approximate equivalent circuit to be used to analyze
either the CC or CE circuit

Symbol	Common Emitter	Common Collector	Common Base
h_{11}, h_{ie}	h_{ie}	h_{ic}	$\dfrac{h_{ib}}{1 + h_{fb}}$
h_{12}, h_{re}	h_{re}	$1 - h_{rc}$	$\dfrac{h_{ib} h_{ob}}{1 + h_{fb}} - h_{rb}$
h_{21}, h_{fe}	h_{fe}	$-1 - h_{fc}$	$\dfrac{-h_{fb}}{1 + h_{fb}}$
h_{22}, h_{oe}	h_{oe}	h_{oc}	$\dfrac{h_{ob}}{1 + h_{fb}}$
h_{11}, h_{ib}	$\dfrac{h_{ie}}{1 + h_{fe}}$	$\dfrac{-h_{ic}}{h_{fc}}$	h_{ib}
h_{12}, h_{rb}	$\dfrac{h_{ie} h_{oe}}{1 + h_{fe}} - h_{re}$	$h_{rc} - \dfrac{h_{ic} h_{oc}}{h_{fc}} - 1$	h_{rb}
h_{21}, h_{fb}	$\dfrac{-h_{fe}}{1 + h_{fe}}$	$\dfrac{-1 + h_{fc}}{h_{fc}}$	h_{fb}
h_{22}, h_{ob}	$\dfrac{h_{oe}}{1 + h_{fe}}$	$\dfrac{-h_{oc}}{h_{fc}}$	h_{ob}
h_{11}, h_{ic}	h_{ie}	h_{ic}	$\dfrac{h_{ib}}{1 + h_{fb}}$
h_{12}, h_{rc}	$1 - h_{re}$	h_{rc}	1
h_{21}, h_{fc}	$-1 - h_{fe}$	h_{fc}	$\dfrac{-1}{1 + h_{fb}}$
h_{22}, h_{oc}	h_{oe}	h_{oc}	$\dfrac{h_{ob}}{1 + h_{fb}}$

Table 4-4 h-Parameter conversion equations

Further examination of the various equivalent circuits will reveal addition-al possibilities for identifying similarities between the circuits. For example, it can be shown that the h-parameters for the CE circuit can be identified and de-fined in terms of the parameters for the CB circuit. It should not be surprising that the parameters can be manipulated in this manner since they all define conditions for the same device. The only difference between the CB, CC, and CE circuits is in the way in which the ouput and input terminals are specified. However, many parameter conversions are more complex than those just dis-cussed and shown in figures 4-7, 4-8, 4-9, and 4-10. Some more complex con-versions require the use of Thévenin's Theorem and Kirchhoff's Voltage and Current Laws. Table 4-4 lists the h-parameters for each circuit defined in terms of h-parameters for each of the other circuits.

A manufacturer lists the following constants for a transistor which is being operated in the center of its active region.

h_{ie} = *450 ohms*

h_{fe} = *100*

Using Table 4-4, compute h_{ib}, h_{fc}, h_{ic}, and h_{fb}.   (R4-7)

Use Table 4-2 and the text to define, in terms of the statements on pages 105 and 106, the following h-parameters: h_{ib}, h_{rc}, h_{fb}, and h_{oe}. Use appropri-ate transistor terminal names rather than input or output.   (R4-8)

Verify the value of h_{fc} given in Table 4-3 for a 2N1808 transistor using the value of h_{fe} given in the same table.   (R4-9)

A manufacturer lists the following values of the common base h-param-eters for a transistor.

h_{ib} = *21.6 ohms*

h_{fb} = *-0.98*

h_{rb} = *0.00029*

h_{ob} = *0.49 micromho*

Using the listed values and Table 4-4, compute the values of the four common emitter h-parameters for this transistor.   (R4-10)

EQUIVALENT CIRCUIT APPROXIMATIONS FOR THE CE CIRCUIT

The equivalent circuit models in Table 4-2 are exact equivalent circuits. Therefore, an exact analysis can be made. However, an exact analysis is unnecessary, in view of the fact that the linear parameters used are based on a linear representation of a device that is not linear. Many approximations can be made which are more accurate than the assumptions upon which these linear parameters are based.

Consider the common emitter amplifier circuit shown in figure 2-11. The exact equivalent circuit of this amplifier, using h-parameters, is shown in figure 4-11. This model can be reduced to the circuit of figure 4-12. The internal impedance of the signal source can affect the amplifier performance. Therefore, this impedance and the signal input voltage (v_s) are included in the circuit. The bias resistors can also affect amplifier performance. For example, the base bias resistance for the common emitter amplifier causes the input current of the amplifier to be different from the input or base current of the transistor. The shunting effect of the collector bias resistor causes the amplifier output current to be different from the output current of the transistor. Actually, the output current of the transistor is the transistor collector current. A study of Table 4-3 shows that because h_{re} is so small, $h_{re}v_{ce}$ can be neglected. In comparison with typical values of load impedance and collector bias resistance, h_{oe} also can be neglected. Using the simplified circuit of figure 4-12, amplifier parameters such as gain and impedance can be determined to a sufficient degree of accuracy. The current gain for the circuit in figure 4-12 is:

$$A_I = \frac{i_L}{i_{in}}$$

Eq. 4.11

In general, the base bias resistor (R_b) is so large in comparison with h_{ie} that $i_{in} \approx i_b$. The load impedance is shunted by the collector bias resistor R_c. Applying the current divider theorem to this circuit yields the following expression.

$$i_L = (-i_c)\frac{R_c}{Z_L + R_c}$$

Eq. 4.12

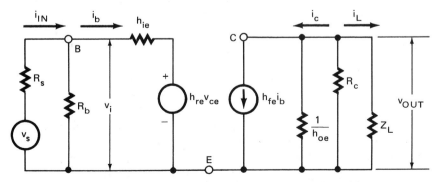

Fig. 4-11 Ac equivalent circuit for RC coupled amplifier of figure 2-11

The direction of the load current is selected so that it is opposite to the direction of the collector current. This means that a minus sign is placed in front of i_c. Figure 4-12 is a model for the circuit of figure 2-11 which uses an npn transistor. The collector current is:

$$i_c = h_{fe}i_b \qquad \text{Eq. 4.13}$$

When Equations 4.13 and 4.12 are substituted into Equation 4.11, the resulting expression is the current gain for the common emitter amplifier.

$$A_I = (-h_{fe}) \frac{R_c}{Z_L + R_c} \qquad \text{Eq. 4.14}$$

The current gain is negative because of the assumed direction of load current. If Z_L is very small with respect to R_c, then the magnitude of the current gain reduces to the h-parameter, h_{fe}. This is because the ouput terminals are shorted and, by definition, h_{fe} is the short-circuit current gain.

The voltage gain is:

$$A_V = \frac{v_{out}}{v_{in}} \qquad \text{Eq. 4.15}$$

where

$$v_{out} = i_L Z_L \qquad \text{Eq. 4.16}$$

and

$$v_{in} = i_b h_{ie} \qquad \text{Eq. 4.17}$$

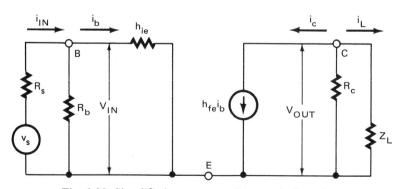

Fig. 4-12 Simplified common emitter equivalent circuit

Substitute Equations 4.12 and 4.13 into Equation 4.16 to obtain the following expression for v_{out}.

$$v_{out} = \frac{-h_{fe}i_b Z_L R_c}{Z_L + R_c}$$

Eq. 4.18

Equations 4.18 and 4.17 can now be substituted into Equation 4.15 to yield an expression for the voltage gain.

$$A_V = \frac{-h_{fe}R_c Z_L}{(Z_L + R_c)h_{ie}}$$

Eq. 4.19

If R_c is large with respect to Z_L, Equation 4.19 is reduced to:

$$A_V = \frac{-h_{fe}Z_L}{h_{ie}}$$

Eq. 4.20

On the other hand, if the amplifier is operating essentially at no load, then $Z_L \approx \infty$ and Equation 4.19 is reduced to:

$$A_V = \frac{-h_{fe}R_c}{h_{ie}}$$

Eq. 4.21

For the input impedance, the base bias resistance R_b is usually much larger than h_{ie}. Therefore,

$$Z_i = \frac{v_{in}}{i_{in}} \approx \frac{v_{in}}{i_b}$$

Eq. 4.22

Substitute Equation 4.17 in Equation 4.22 to obtain:

$$Z_i = h_{ie}$$

Eq. 4.23

The impedance normally connected between the collector and emitter is very much smaller than $1/h_{oe}$. The smaller value is due to the fact that R_c must be small enough to provide sufficient collector bias current. The actual output resistance of the amplifier is very nearly equal to the collector bias resistance R_c.

The expressions just determined for A_I, A_V, Z_i, and the output impedance Z_o are obtained using the approximate equivalent circuit shown in figure 4-12. The values resulting from these expressions are usually in error by less than 10% from the values obtained using the exact equivalent circuit of figure 4-11. The exact equations can be obtained from an analysis of the circuit in figure 4-11. This procedure is generally true for the CB and CC approximate equivalent circuits as well as for the CE circuits.

Example of the Use of an Equivalent Circuit

As an example of the use of an equivalent circuit in the analysis of a problem, assume that the amplifier of figure 2-11 has the following component values: R_c = 1000 ohms, C2 = 10 microfarads, V_{CC} = 20 volts, R_b = 200 kilohms, Z_s = 1000 ohms resistive, and C1 = 10 microfarads. The load impedance (Z_L) is 1000 ohms resistive. The input signal is sinusoidal at a frequency of 1000 hertz. The peak-to-peak value of the input signal is small enough to permit the use of the h-parameter approximate equivalent circuit. The transistor used in the circuit is assumed to have the following values: h_{re} = 0.000125, h_{ie} = 750 ohms, h_{oe} = 90.9 micromhos, and h_{fe} = 184. The reactance of capacitors C1 and C2 is determined as follows:

$$X_c = \frac{1}{2 \pi fC} \qquad \text{Eq. 4.24}$$

$$X_c = \frac{1}{6.28(1000)(10)(10)^{-6}} = 15.9 \text{ ohms}$$

The impedances of the amplifier circuit are so much removed from this level of reactance that the input signal can be considered to be passed from the signal source directly into the amplifier.

The resistance R_b for this example is so much larger than h_{ie} that it may be neglected. Thus, the approximate input impedance is $Z_i \approx h_{ie}$ = 750 ohms.

The output impedance with no load on the amplifier is R_c in parallel with $1/h_{oe}$.

$$\frac{1}{h_{oe}} = \frac{1}{0.0000909} = 11,000 \text{ ohms}$$

Since $1/h_{oe}$ is large compared to R_c, it can be neglected in the approximate equivalent circuit. The output impedance is $Z_o \approx R_c$ = 1000 ohms.

The current gain is:

$$A_I = \frac{(-h_{fe})R_c}{Z_L + R_c}$$

$$= \frac{(-184)(1000)}{1000 + 1000} = -92$$

The voltage gain is:

$$A_V = \frac{(-h_{fe})R_c Z_L}{(Z_L + R_c)h_{ie}}$$

$$= \frac{(-184)(1000)(1000)}{(1000 + 1000)(750)} = -123$$

The amplifier in figure 2-11 has the following component values: R_c = *5000 ohms,* Z_L = *4000 ohms resistive, C1* = *C2* = *20 microfarads,* V_{CC} = *40 volts,* R_b = *1.5 megohms,* Z_s = *1000 ohms resistive, and frequency f* = *1000 hertz. The transistor used in the circuit has the following h-parameters:* h_{re} = *0.00041,* h_{ie} = *3600 ohms,* h_{oe} = *0.000031 mho, and* h_{fe} = *124.*

    a. Is it possible to neglect the reactances of C1 and C2?
    b. Compute the current gain.
    c. Compute the voltage gain.
    d. Compute the input and output impedances.   *(R4-11)*

    An input signal (*v_s*) *of 5 sin 10t volts is applied to the amplifier described in review problem (R4-11). Can the h-parameter technique presented in the chapter to this point be used to analyze the circuit? Can the method presented in Chapter 3 (pages 61 to 74) be used for the analysis? Explain.*   *(R4-12)*

THE APPROXIMATE EQUIVALENT CC AMPLIFIER CIRCUIT

    An approximation can be made for the common collector circuit. Consider the common collector amplifier circuit of figure 3-10. This circuit can be represented in h-parameter notation as shown in figure 4-13. The direction of the load current is the same as in the preceding analysis: from the transistor terminal to common. In this case, i_c has the same direction as i_L. This common collector circuit can be simplified because h_{oc} is usually very large when compared to the load impedance Z_L and the bias resistor R_e. The parameter h_{rc} is very close to unity, making $h_{rc}v_{ec} \approx v_{ec}$. This means that the circuit can be simplified to the one shown in figure 4-14, page 122. The amplifier parameters

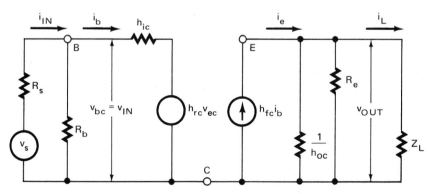

Fig. 4-13 Ac equivalent circuit for RC coupled common collector amplifier

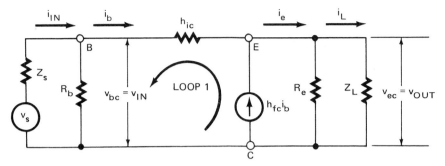

Fig. 4-14 Simplified common collector equivalent circuit

can be determined readily from figure 4-14. The current gain is given by Equation 4.11.

$$A_I = \frac{i_L}{i_{in}}$$

Since $R_b \gg h_{ic}$, $i_{in} \approx i_b$. The load current is:

$$i_L = \frac{i_e R_e}{R_e + Z_L} \qquad \text{Eq. 4.25}$$

But,

$$i_e = h_{fc} i_b \qquad \text{Eq. 4.26}$$

Substituting Equations 4.25 and 4.26 into Equation 4.11 yields:

$$A_I = \frac{h_{fc} R_e}{R_e + Z_L} \qquad \text{Eq. 4.27}$$

When the amplifier operates at no load, $Z_L = \infty$, and $A_I = 0$. When the amplifier output terminals are shorted, $Z_L = 0$ and $A_I = h_{fc}$, as it should since h_{fc} is defined as the short circuit current gain.

The voltage gain is given by Equation 4.15:

$$A_V = \frac{v_{out}}{v_{in}}$$

where

$$v_{out} = i_L Z_L \qquad \text{Eq. 4.28}$$

Writing Kirchhoff's Voltage Law for the circuit of figure 4-14 yields:

$$v_{in} = i_b h_{ic} + \frac{i_e R_e Z_L}{R_e + Z_L}$$

Eq. 4.29

However, Equation 4.26 defined i_e:

$$i_e = h_{fc} i_b$$

Substituting this expression for i_e in Equation 4.29 yields the following:

$$v_{in} = i_b h_{ic} + \frac{h_{fc} i_b R_e Z_L}{R_e + Z_L}$$

$$v_{in} = i_b \left(h_{ic} + \frac{h_{fc} R_e Z_L}{R_e + Z_L} \right)$$

Eq. 4.30

Applying the current divider theorem results in the following expression.

$$i_L = \frac{i_e R_e}{R_e + Z_L}$$

Eq. 4.31

Combine Equations 4.26, 4.31, 4.28, 4.30, and 4.15 to obtain an expression for the voltage gain:

$$A_V = \frac{(h_{fc})(i_b)\dfrac{R_e Z_L}{R_e + Z_L}}{i_b \left(h_{ic} + \dfrac{h_{fc} R_e Z_L}{R_e + Z_L} \right)}$$

$$A_V = \frac{h_{fc} R_e Z_L}{h_{ic}(R_e + Z_L) + h_{fc} R_e Z_L}$$

Eq. 4.32

When there is no load on the amplifier, $Z_L \approx \infty$ and A_V reduces to:

$$A_V = \frac{h_{fc} R_e}{h_{ic} + h_{fc} R_e}$$

Eq. 4.33

However, since $h_{fc} R_e \gg h_{ic}$, the voltage gain (A_V) is approximately equal to 1. When the output terminals are shorted, $Z_L \approx 0$. For this condition, the voltage gain (A_V) must be zero.

The input impedance is:

$$Z_i = \frac{v_{in}}{i_{in}} \approx \frac{v_{in}}{i_b}$$

Eq. 4.34

Substituting Equation 4.30 into Equation 4.34 yields:

$$Z_i = h_{ic} + \frac{h_{fc}R_e Z_L}{R_e + Z_L}$$

Eq. 4.35

If the amplifier is operated at no load, then:

$$Z_i = h_{ic} + h_{fc}R_e$$

Eq. 4.36

This input impedance is a very high value for most transistors. If the amplifier output terminals are shorted, then the input impedance becomes:

$$Z_i = h_{ic}$$

Eq. 4.37

When no ac load is connected, the amplifier output impedance depends on the signal source impedance as well as the transistor parameters. Using figure 4-14, look back into the circuit from the terminals E and C. The impedance out to R_e is:

$$\frac{v_{ce}}{i_e} = \frac{v_{ec}}{i_b + h_{fc}i_b} = \frac{v_{ec}}{(1 + h_{fc})i_b}$$

Eq. 4.38

Apply Kirchhoff's Voltage Law to Loop 1 to obtain the following expression:

$$v_{ec} - i_b\left(h_{ic} + \frac{Z_s R_b}{Z_s + R_b}\right) = 0$$

$$v_{ec} = i_b\left(h_{ic} + \frac{Z_s R_b}{Z_s + R_b}\right)$$

Eq. 4.39

Substitute Equation 4.39 into the right-hand side of Equation 4.38.

$$\frac{v_{ec}}{i_e} = \frac{h_{ic} + \dfrac{Z_s R_b}{Z_s + R_b}}{1 + h_{fc}}$$

Eq. 4.40

The amplifier output impedance is v_{ec}/i_e in parallel with R_e.

$$Z_o = \frac{\left(\dfrac{V_{ec}}{i_e}\right) R_e}{R_e + \left(\dfrac{V_{ec}}{i_e}\right)}$$

$$Z_o = \frac{\left(\dfrac{h_{ic} + \dfrac{Z_s R_b}{Z_s + R_b}}{1 + h_{fc}}\right) R_e}{R_e + \dfrac{h_{ic} + \dfrac{Z_s R_b}{Z_s + R_b}}{1 + h_{fc}}}$$

Eq. 4.41

When $Z_s = \infty$,

$$Z_o = \frac{\left(\dfrac{h_{ic} + R_b}{1 + h_{fc}}\right) R_e}{R_e + \dfrac{h_{ic} + R_b}{1 + h_{fc}}}$$

Eq. 4.42

When $Z_s = 0$, R_e is almost shorted out, and

$$Z_o \approx \frac{h_{ic}}{1 + h_{fc}}$$

Eq. 4.43

Example of the Use of the CC Equivalent Circuit

To illustrate this analysis, assume that the 2N4074 transistor is being used in the CC circuit of figure 3-10. The following parameter values are true for the circuit: $R_e = 1000$ ohms, $Z_L = 1000$ ohms resistive, $R_b = 200$ kilohms, $Z_s = 1000$ ohms resistive, $V_{CC} = 20$ volts, and $C1 = C2 = 10$ microfarads. The input signal applied to this amplifier is the same as the signal applied previously to the CE amplifier.

No new information is needed for the common collector amplifier since $h_{ie} = h_{ic} = 750$ ohms, and $h_{fe} \approx h_{fc} = 184$. The approximate equivalent circuit is the same as the circuit in figure 4-14. The amplifier specifications are determined according to the various equations just developed. The current gain is:

$$A_I = \frac{h_{fc} R_e}{R_e + Z_L} = \frac{(184)(1000)}{1000 + 1000} = 92$$

The voltage gain (A_V) is determined from Equation 4.32.

$$A_V = \frac{h_{fc} R_e Z_L}{h_{ic}(R_e + Z_L) + h_{fc} R_e Z_L}$$

$$= \frac{(184)(1000)(1000)}{(750)(1000 + 1000) + (184)(1000)(1000)}$$

$$\approx 1$$

The input impedance is given by Equation 4.35.

$$Z_i = h_{ic} + \frac{h_{fc} R_e Z_L}{R_e + Z_L}$$

$$= 750 + \frac{(185)(1000)(1000)}{1000 + 1000} = 93.3 \text{ kilohms}$$

From Equation 4.41, the output impedance of the amplifier is:

$$Z_o = \frac{\left(\dfrac{h_{ic} + \dfrac{R_b Z_s}{R_b + Z_s}}{1 + h_{fc}} \right) R_e}{R_e + \dfrac{h_{ic} + \dfrac{Z_s R_b}{Z_s + R_b}}{1 + h_{fc}}}$$

$$= \frac{\left(\dfrac{750 + \dfrac{(200{,}000)(1000)}{200{,}000 + 1000}}{1 + 185} \right) 1000}{1000 + \left(\dfrac{750 + \dfrac{(200{,}000)(1000)}{200{,}000 + 1000}}{1 + 185} \right)}$$

$$= 9.29 \text{ ohms}$$

If the input signal is an ideal voltage source, then $Z_s = 0$. As a result, the output impedance becomes:

$$Z_o = \frac{h_{ic}}{1 + h_{fc}} = \frac{750}{1 + 185} = 4.03 \text{ ohms}$$

If the input signal is an ideal current source, $Z_s = \infty$. The output impedance for this condition is:

$$Z_o = \frac{\left(\dfrac{h_{ic} + R_b}{1 + h_{fc}}\right) R_e}{R_e + \dfrac{h_{ic} + R_b}{1 + h_{fc}}}$$

$$= \frac{\left(\dfrac{750 + 200{,}000}{1 + 185}\right) 1000}{1000 + \dfrac{750 + 200{,}000}{1 + 185}} = 519 \text{ ohms}$$

A 2N1711 transistor is used in a common collector amplifier such as the one shown in figure 3-10. The component values for this circuit are as follows: $R_e = 5000\ ohms$, $Z_L = 3000\ ohms\ resistive$, $R_b = 1.5\ megohms$, $Z_s = 1000\ ohms$ resistive, $V_{CC} = 40\ volts$, and $C2 = C1 = 20\ microfarads$. A small sinusoidal input signal is applied at v_s.

- *a. Using the h-parameters listed in Table 4-3, compute the voltage gain, current gain, and input impedance for this circuit.*
- *b. Assuming that Z_L is disconnected, compute the approximate output· impedance.* *(R4-13)*

THE APPROXIMATE EQUIVALENT CB AMPLIFIER CIRCUIT

To analyze the common base amplifier, consider the circuit shown in figure 3-9. Using the appropriate h-parameters, the ac equivalent circuit is shown in figure 4-15, page 128. An approximate equivalent circuit for the CB configuration can be developed. This can be seen by noticing, in Table 4-3, that both h_{rb} and h_{ob} are very small. The term $h_{rb}v_c$ is small enough to be neglected and $1/h_{ob}$ is very large. This means that the equivalent circuit can be simplified to that of figure 4-16, page 128. This circuit can be used to determine the common base amplifier parameters. The general form of the current gain was expressed by Equation 4.11:

$$A_I = \frac{i_L}{i_{in}}$$

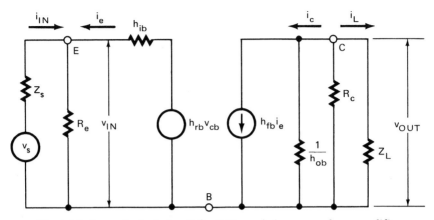

Fig. 4-15 Ac equivalent circuit for RC coupled common base amplifier

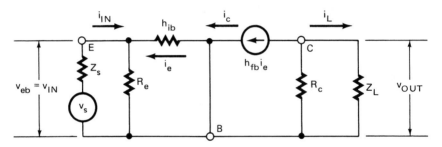

Fig. 4-16 Simplified common base circuit

According to the current divider theorem,

$$i_L = - \frac{i_c R_c}{R_c + Z_L}$$ Eq. 4.44

and

$$i_e = \frac{i_{in} R_e}{R_e + h_{ib}}$$

Therefore,

$$i_{in} = \frac{-i_e(R_e + h_{ib})}{R_e}$$ Eq. 4.45

Table 4-2 defines i_c as:

$$i_c = h_{fb} i_e \qquad \text{Eq. 4.46}$$

Substitute Equations 4.46, 4.45, and 4.44 into Equation 4.11 to obtain the current gain:

$$A_I = \frac{h_{fb} R_c R_e}{(R_c + Z_L)(R_e + h_{ib})} \qquad \text{Eq. 4.47}$$

If the output terminals are shorted, the current gain becomes:

$$A_I = \frac{h_{fb} R_e}{R_e + h_{ib}} \qquad \text{Eq. 4.48}$$

If the amplifier is operated without a load, the current gain is zero because $Z_L \approx \infty$.

The voltage gain was expressed by Equation 4.15:

$$A_V = \frac{v_{out}}{v_{in}}$$

An expression for v_{out} is given by

$$v_{out} = i_L Z_L \qquad \text{Eq. 4.49}$$

Substitute Equations 4.44 and 4.46 into Equation 4.49 to obtain v_{out}.

$$v_{out} = \frac{-h_{fb} i_e R_c Z_L}{R_c + Z_L} \qquad \text{Eq. 4.50}$$

An expression for v_{in} is obtained by applying Ohm's Law to the circuit of figure 4-16.

$$v_{in} = -i_e h_{ib} \qquad \text{Eq. 4.51}$$

Substitute Equations 4.50 and 4.51 into Equation 4.15.

$$A_v = \frac{h_{fb} R_c Z_L}{h_{ib}(R_c + Z_L)} \qquad \text{Eq. 4.52}$$

A further analysis of figure 4-16 shows that the input impedance is the parallel combination of h_{ib} and R_e, resulting in

$$Z_i = \frac{h_{ib} R_e}{R_e + h_{ib}} \qquad \text{Eq. 4.53}$$

In addition, the output impedance is:

$$Z_o = R_c \qquad \text{Eq. 4.54}$$

Example of the Analysis of a CB Amplifier

To illustrate the analysis of a CB amplifier, consider the circuit shown in figure 3-9. A 2N4074 transistor is used in this circuit. The following parameter values apply to the circuit: $R_e = 1000$ ohms, $Z_s = 1000$ ohms resistive, C1 = C2 = 10 microfarads, $R_c = 1000$ ohms, $V_{CC} = V_{EE} = 10$ volts, and $Z_L = 1000$ ohms resistive. A small sinusoidal input signal (v_s) of 1000 hertz is applied. No new information is needed to determine the h-parameters for the 2N4074 transistor, even though only the common emitter h-parameters are available. Table 4-4 gives the following information.

$$h_{fb} = \frac{-h_{fe}}{1 + h_{fe}} \qquad \text{Eq. 4.55}$$

$$h_{ib} = \frac{h_{ie}}{1 + h_{fe}} \qquad \text{Eq. 4.56}$$

$$h_{ob} = \frac{h_{oe}}{1 + h_{fe}} \qquad \text{Eq. 4.57}$$

The parameter h_{fb} is also known as the transistor alpha (a). A value for a may be given by the manufacturer as a separate transistor parameter between 0.95 and 0.995. For the 2N4074 transistor,

$$h_{fb} = \frac{-184}{1 + 184} = -0.995$$

$$h_{ib} = \frac{750}{1 + 184} = 4.05 \text{ ohms}$$

$$h_{ob} = \frac{0.0909}{1 + 184} = 0.000491 \text{ milliampere per volt}$$

The parameter h_{ob} is even smaller than h_{oe}; $1/h_{ob} = 1/0.000491$. This is 3.1 megohms! Therefore, the simplified equivalent circuit of figure 4-16 can be used. The amplifier parameters are determined as follows:

$$A_I = \frac{h_{fb} R_c R_e}{(R_c + Z_L)(R_e + h_{ib})} = \frac{-0.995(1000)(1000)}{(1000 + 1000)(1000 + 4.05)} = -0.496$$

$$A_V = \frac{h_{fb} R_c Z_L}{h_{ib}(R_c + Z_L)} = \frac{-0.995(1000)(1000)}{4.05(1000 + 1000)} = -122$$

If the amplifier input is connected to a negligibly small load, then the voltage gain A_V is $(-0.995)(1000)/4.05 = -281$. The input impedance is

$$Z_i = \frac{h_{ib} R_e}{R_e + h_{ib}} \approx h_{ib} = 4.05 \text{ ohms}$$

The output impedance is $Z_o = R_c = 1000$ ohms.

A 2N1711 transistor is used in the CB circuit shown in figure 3-9. The following parameter values are used in the circuit: $Z_s = 1000$ ohms resistive, $V_{CC} = 50$ volts, $V_{EE} = 10$ volts, $R_c = 5000$ ohms, $R_e = 20,000$ ohms, $Z_L = 3000$ ohms resistive, $C2 = C1 = 10$ microfarads, and $f = 1000$ hertz. Compute the current gain, voltage gain, and input impedance for this circuit. (R4-14)

For the amplifier described in question (R4-14), compute the voltage gain at no load. (R4-15)

For the amplifier described in question (R4-14), compute the current gain with the amplifier output terminals shorted. (R4-16)

COMPARISON OF TRANSISTOR AMPLIFIER CONFIGURATIONS

It is interesting to compare these parameters for the three amplifier circuit configurations, Table 4-5, page 132. The values of A_I and A_V are calculated using the equations developed on pages 116 to 132. The values of Z_o and Z_i are calculated for $Z_L = \infty$ and 0 and $Z_s = 0$ and ∞. The values for these extreme circuit conditions must be determined by the use of the exact equivalent circuits of figures 4-11, 4-13, and 4-15. The equations from which they were calculated are as follows:

Quantity	CE	CC	CB
A_I	-92	92	-0.496
A_I with $Z_L = 0$ (maximum current gain)	-184	185	-0.991
A_V	-123	0.996	-122
A_V with $Z_L = \infty$ (maximum voltage gain)	-245	0.996	-245
Z_o	906 ohms	635 ohms	1,000 ohms
Z_o with $Z_s = 0$ and $Z_L = \infty$	823 ohms	3.8 ohms	938 ohms
Z_o with $Z_s = \infty$ and $Z_L = \infty$	977 ohms	519 ohms	1,000 ohms
Z_i	739 ohms	92,700 ohms	4.1 ohms
Z_i with $Z_L = 0$	750 ohms	750 ohms	4.1 ohms
Z_i with $Z_L = \infty$	497 ohms	184,000 ohms	4.3 ohms

Values used in the calculations:

$$h_{fe} = 184 \qquad R_c = 1,000 \text{ ohms}$$
$$h_{re} = 0.000125 \qquad R_b = 200,000 \text{ ohms}$$
$$h_{ie} = 750 \text{ ohms} \qquad R_e = 1,000 \text{ ohms}$$
$$h_{oe} = 90.9 \text{ micromhos}$$

When not 0 or infinity:

$$Z_L = 1000 \text{ ohms resistive}$$
$$Z_s = 1000 \text{ ohms resistive}$$

Table 4-5 Transistor Amplifier Specifications

$$A_I = \cfrac{h_{f_-}}{1 + \cfrac{h_{o_-} R_c Z_L}{R_c + Z_L}} \qquad \text{Eq. 4.58}$$

$$Z_i = h_{i_-} - \cfrac{h_{f_-} h_{r_-}}{h_{o_-} + \cfrac{R_c + Z_L}{R_c Z_L}} \qquad \text{Eq. 4.59}$$

$$Z_o = \cfrac{1}{h_{o\_} - \cfrac{h_{f\_} h_{r\_}}{h_{i\_} + \cfrac{R_b Z_i}{R_b + Z_i}}}$$

Eq. 4.60

In addition, Equations 4.19, 4.20, 4.21, 4.32, 4.33, and 4.52 are also used.

Note in Equations 4.58, 4.59, and 4.60 that only the first subscript of the h-parameters is used. Each equation can be used for the CC, CB, or CE circuit by inserting the proper subscripts. For example, $h_{f\_}$ may be used as h_{fe}, h_{fc}, or h_{fb}, depending on the circuit to be analyzed. The student is invited to derive some of these equations from the exact ac equivalent circuit, using the appropriate network theorems.

According to the data in Table 4-5, the common emitter amplifier has a high voltage gain, a high current gain, a moderately low value of input impedance, and a moderately high value of output impedance. Note that the actual levels of these quantities depend not only on the transistor, but also on the components that make up the rest of the amplifier. The common emitter amplifier has an ouput impedance that depends to a great extent on the collector bias resistor. The common emitter amplifier is a popular circuit for amplifying voltage, current, and power.

The common collector amplifier has a high current gain, a voltage gain of unity, and input and output impedances that vary widely according to the values of the load and source impedances. The output impedance of this amplifier is affected by the source impedance and the input impedance is affected by the load impedance. When using large values of biasing and load impedance, it is possible to have an extremely high input impedance for the common collector circuit. It is also possible to have an extremely low output impedance when the input signal is driven by an ideal voltage source. The common collector amplifier (also known as an emitter follower) is widely used to match a high-impedance source to a low-impedance load. The CC circuit is useful when used alone and in combination with other semiconductor circuits.

The common base amplifier has a low current gain and a moderately high voltage gain. Table 4-5 indicates that this circuit has a high output impedance and a low input impedance. The common base amplifier has the lowest input impedance of the three circuit configurations. As a result, this circuit tends to load a signal source to a greater degree than the other two configurations. This factor and the low current gain of the circuit mean that it is the least used of the three types. It may be used to match a low-impedance source to a high-impedance load, but so may either of the other two connections. The common base amplifier is sometimes used as the last stage of the constant current signal generator or in a tuned amplifier (see Chapter 12).

Complete the following table by inserting CB, CC, and/or CE in the appropriate spaces.   (R4-17)

	High	Low	Moderately High or Low
A_V			
A_I			
Z_I			
Z_L			

USE OF APPROXIMATE TRANSISTOR PARAMETERS

Manufacturers frequently provide values of h_{fe} for their transistors and none of the other parameters. It should be realized that the manufacturer expects that, for the first look at a proposed circuit, an approximate feel for the transistor's performance is all that is needed. For example, in figure 4-12, h_{ie} is the only other parameter needed. This value is the input impedance for the approximate equivalent circuit.

According to semiconductor diode theory, the incremental diffusion resistance of a forward biased diode is given by:

$$r_e = \frac{kT}{qIn}$$

Eq. 4.61

where

k = Boltzmann's constant, $1.38(10)^{-23}$ watt-second per degree Celsius
q = the charge on an electron, $1.6(10)^{-19}$ coulomb
n = constant of 1 for germanium and about 1.5 for silicon
T = temperature, in degrees Kelvin
I = current, in amperes

When typical values at room temperature are substituted, Equation 4.61 is reduced to:

$$r_e = \frac{0.026}{I} \text{ ohms}$$

Eq. 4.62

where I is measured in amperes and n is assumed to be unity.

Verify that Equation 4.62 is obtained from Equation 4.61.   (R4-18)

Calculate the value of r_e for the forward biased diode represented by figure 2-3 at a diode current of 1 and 3 milliamperes. Use Equation 4.62 and compare the calculated value with the incremental resistance value read directly from the graph and calculated by Ohm's Law. (R4-19)

Since the base-to-emitter junction of a transistor operating in the active region behaves like a forward biased diode, Equation 4.62 is also applicable to transistor performance. As in the forward biased diode, the emitter current flows through the base-to-emitter junction shown in figure 4-17. This diode

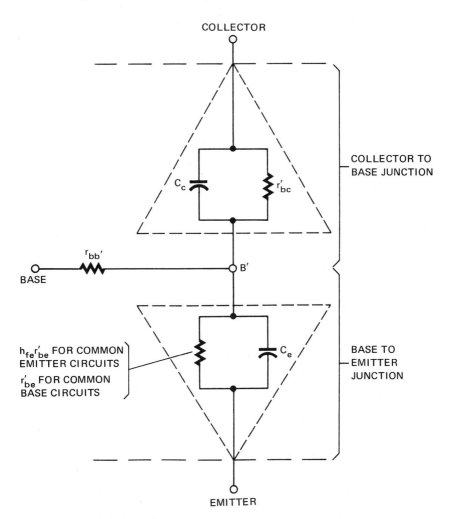

Fig. 4-17 Diode representation of high-frequency transistor

representation of the transistor includes shunt capacitance across the junctions. Such capacitance is effective only at high signal frequencies and will be covered further in Chapter 5. For the present discussion, only r_{bb}' and $r_b'_e$ are considered.

The transistor parameter r_{bb}' is known as the base spreading resistance. This parameter is the ohmic resistance of the base section of the transistor between the base terminal and the diode junctions in the transistor. This resistance is analogous to the grid resistance in a vacuum tube.

The parameter $r_b'_e$ is the incremental diffusion resistance of the forward biased base-to-emitter junction. This resistance is nonlinear. The nonlinearity can be seen by examining the semiconductor diode volt-ampere curve shown in figure 2-3 and the transistor base-to-emitter characteristic shown in figure 2-8. An approximate representation of $r_b'_e$ is:

$$r_b'_e = \frac{0.026}{I_E} \qquad \text{Eq. 4.63}$$

where I_E is the quiescent emitter current in amperes. As long as I_E is not excessively high, the value of $r_b'_e$ is usually large enough to permit r_{bb}' to be neglected. Thus,

$$r_b'_e \approx \frac{v_{be}}{i_e} \qquad \text{Eq. 4.64}$$

Recall from Equation 4.9 that the h-parameter

$$h_{ie} = \frac{v_{be}}{i_b}$$

by definition. Therefore, if r_{bb}' can be neglected, $h_{ie} = r_b'_e$. Equation 4.9 can be rewritten as:

$$h_{ie} = \frac{v_{be}}{i_e} \cdot \frac{i_e}{i_b} \qquad \text{Eq. 4.65}$$

Assuming that r_{bb}' can be neglected and $i_e \approx i_c$, substitute Equation 4.64 in Equation 4.65 to obtain the following expression.

$$h_{ie} = (r_b'_e)\frac{i_c}{i_b} \approx r_b'_e h_{fe} \qquad \text{Eq. 4.66}$$

An approximate value of h_{ie} is obtained by substituting Equation 4.63 into Equation 4.66, yielding

$$h_{ie} = \frac{0.026h_{fe}}{I_E} \approx \frac{0.026h_{fe}}{I_C} \qquad \text{Eq. 4.67}$$

where I_C is the quiescent value of the collector current and is almost equal to the quiescent value of emitter current I_E. As a result, the use of the simplified common emitter equivalent circuit of figure 4-12 means that the only parameter needed to make initial calculations is h_{fe}.

Calculate h_{ie} for the transistor shown in figure 4-5 for a quiescent collector current of 10 milliamperes. Use Equation 4.67. Compare this result with the value of h_{ie} calculated for the transistor described on page 110.

(R4-20)

Since h_{ic} and h_{ie} are the same and $h_{fe} \approx h_{fc}$, the CC circuit can be analyzed using figure 4-14. Again, the analysis can be made knowing only the single h-parameter h_{fe}.

The CB circuit can be analyzed when only h_{fe} is available as a transistor constant and using the simplified circuit of figure 4-16. For example, using the equations in Table 4-4,

$$h_{ib} = \frac{h_{ie}}{1 + h_{fe}} \qquad \text{Eq. 4.68}$$

and

$$h_{fb} = \frac{-h_{fe}}{1 + h_{fe}} \qquad \text{Eq. 4.69}$$

Using the information given on pages 134 to 137, compute h_{fb} and h_{ib}, using an h_{fe} of 100 for a quiescent base current of 0.1 milliampere.   *(R4-21)*

SMALL SIGNAL ANALYSIS OF MULTISTAGE AMPLIFIERS

The h-parameters are useful in the analysis of multistage transistor amplifiers. This is due to the fact that the inherently low internal impedance of transistors results in the output impedance of one stage being affected by both the input impedance and the output impedance of the next stage. The analysis is greatly simplified by reducing the transistor circuit to an equivalent circuit containing only impedances, voltage sources, and current sources.

As an example of the analysis of a multistage amplifier, consider the circuit shown in figure 4-18, page 138. It is desired to calculate the current gain,

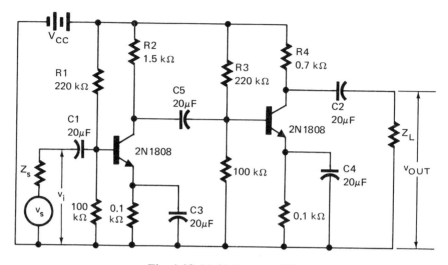

Fig. 4-18 Multistage amplifier

voltage gain, input resistance, and ouput resistance at intermediate frequencies for this circuit. There are many conditions under which the analysis can be made. The input signal source impedance Z_s may be high or low, depending upon the signal generator characteristics. The load Z_L may be high or low, depending upon the characteristics of the load. Each of these options will produce different values for the amplifier parameters. By selecting a voltage source to supply a signal to an unloaded amplifier, all of the necessary transistor parameters will be examined for a likely situation involving the use of such an amplifier.

To make a small signal analysis, an ac equivalent circuit is used for the multistage amplifier. This is shown in figure 4-19. Note that the emitter resistors are not included in the equivalent circuit. Since the capacitors C3 and C4 have a negligibly small impedance at the standard signal frequencies, the emitter resistors are shorted out for the ac signal. It is assumed that the capacitors C1, C2, and C5 also have a negligibly small impedance at this frequency and do not have to be included in the analysis. The bias voltage source, being a dc power supply, is replaced by a wire.

The circuit in figure 4-19 can be analyzed without simplifying it further. The appropriate network theorems can be used to make the analysis. These theorems are reviewed in the Appendix. It is sometimes necessary to perform a complete analysis of this circuit. However, that amount of accuracy usually is not required. To illustrate the method of analysis, a simplified form of the circuit may be used.

A study of the values of the equivalent circuit show that the circuit can be simplified to the form shown in figure 4-20. R_{b2} and R_{b1} are omitted because they are so large when compared to other resistances in the circuit.

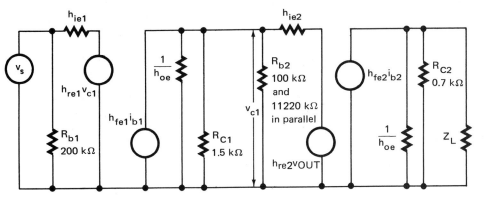

Fig. 4-19 Ac equivalent circuit of the multistage amplifier shown in figure 4-18

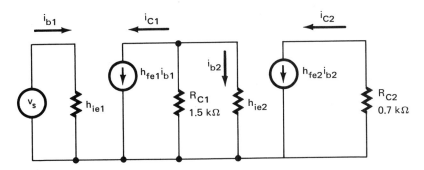

Fig. 4-20 Approximate equivalent circuit of the multistage amplifier of figure 4-18

A complete and accurate analysis requires that all of the h-parameters be known. As discussed on pages 134 to 137, an adequate analysis in many cases can be made knowing only the approximate quiescent current and the value of h_{fe}. Whether the method is simple or complex, it reduces to a solution based on the analysis of an electrical network consisting of a number of impedances and current sources. Standard network theorems are used. Thus, after the device characteristics are defined (Chapter 2), methods of applying the characteristics are presented (Chapters 2 and 3), and the circuit parameters are defined (pages 104 to 137), the problem reduces to the analysis of a simple circuit such as the one shown in figure 4-20.

Example of the Analysis of a Multistage Amplifier

As an example of a typical analysis, assume that the only information available on the transistors is that $h_{fe} = 86$ and the quiescent collector current is

3.5 milliamperes for both transistors. Although each transistor is loaded differently and is operated at a different Q-point (with each probably having a different h_{fe}), they *are* the same type of transistor. Therefore, a single approximate value of h_{fe} and h_{ie} can be used for both transistors. The results will not be significantly changed by this assumption. The analyst must determine if a more accurate analysis is necessary or if an approximate feel for the circuit is sufficient. The technician often may find that if a more accurate analysis is desired, it is more convenient to buy inexpensive sample components and breadboard the circuit in the laboratory. The test results are irrefutable and may take less time to obtain than is required to complete a detailed analysis.

With I_{CQ} = 3.5 milliamperes, Equation 4.64 can be used to find h_{ie}.

$$h_{ie} = \frac{0.025h_{fe}}{I_{CQ}} = \frac{(0.025)(86)}{0.0035} = 614 \text{ ohms}$$

The voltage gain is

$$A_V = \frac{v_{out}}{v_{in}} = \frac{v_{out}}{v_s}$$

where

$$v_{out} = i_{c2}R_{c2} = h_{fe2}i_{b2}R_{c2}$$

$$= (86)(700)i_{b2}$$

$$= 60,200i_{b2}$$

According to the current divider theorem:

$$i_{b2} = \frac{i_{c1}R_{c1}}{h_{ie2} + R_{c1}} = \frac{1500i_{c1}}{1500 + 614} = 0.709i_{c1}$$

But,

$$i_{c1} = h_{fe1}i_{b1} = 86i_{b1}$$

Therefore,

$$v_{out} = 60,200i_{b2} = (60,200)(0.709)i_{c1}$$

$$= 42,700i_{c1} = (42,700)(86)i_{b1} = (3.67)(10)^6i_{b1}$$

$$v_{in} = v_s = i_{b1}h_{ie1} = 614i_{b1}$$

$$A_V = \frac{(3.67)(10)^6i_{b1}}{614i_{b1}} = 5980$$

The voltage gain can also be found by multiplying the gains of the individual stages, A_{V1} and A_{V2}. A_{V2} is determined first:

$$A_{V2} = \frac{700i_{c2}}{614i_{b2}} = \frac{(700)(86)i_{b2}}{614i_{b2}} = 98$$

A_{V1} can be found by forming an equation in terms of the base currents:

$$A_{V1} = \frac{614i_{b2}}{614i_{b1}} = \frac{i_{b2}}{i_{b1}}$$

The current divider theorem is used to find the base current of the second transistor (i_{b2}). The base current is

$$i_{b2} = 0.709i_{c1}$$

Thus,

$$A_{V1} = \frac{0.709i_{c1}}{i_{b1}}$$

$$= \frac{(0.709)(86)i_{b1}}{i_{b1}}$$

$$= 60.9$$

The overall voltage gain is determined as follows:

$$A_V = (A_{V1})(A_{V2}) = (60.9)(98) = 5968$$

The difference in the two values of A_V is due to the rounding off of all numbers to three significant figures. This value is the maximum voltage gain possible for the condition of no load on the amplifier. If an ac load is connected ($Z_L \neq \infty$), then R_{c2} is replaced in the calculations by the equivalent parallel resistance. As a result, the voltage gain is reduced.

The current gain is zero for an amplifier with no load. The maximum current gain occurs when the output terminals are shorted. The current gain of the second stage is:

$$\frac{i_{c2}}{i_{b2}} = h_{fe2} = 86$$

The current through R_{c1} causes the equivalent of another stage of current gain.

$$\frac{i_{b2}}{i_{c1}} = 0.709$$

The gain of the first stage is:

$$\frac{i_{c1}}{i_{b1}} = h_{fe1} = 86.$$

The overall current gain is the same as that of the individual stages multiplied together.

$$\frac{i_{c2}}{i_{b2}} \cdot \frac{i_{b2}}{i_{c1}} \cdot \frac{i_{c1}}{i_{b1}} = \frac{i_{c2}}{i_{b1}} = A_I$$

$$A_I = (86)(0.709)(86) = 5243$$

This gain is reduced if the load impedance Z_L is increased from zero. For this situation, there is another stage of amplification, determined by i_L/i_{c2}. If $Z_L = 500$ ohms, then

$$\frac{i_L}{i_{c2}} = \frac{700}{500 + 700}$$

$$= \frac{700}{1200}$$

$$= 0.583$$

The overall current gain is determined as follows:

$$A_I = 5243 \cdot 0.583 = 3057$$

The impedance is:

$$Z_i = \frac{v_{in}}{i_{b1}} = \frac{v_s}{i_{b1}}$$

$$= \frac{h_{ie1} i_{b1}}{i_{b1}} = h_{ie}$$

$$= 614 \text{ ohms}$$

If Z_L is disconnected, the output impedance is:

$$Z_o = \frac{v_{out}}{i_{c2}}$$

$$= \frac{i_{c2}R_{c2}}{i_{c2}}$$

$$= 700 \text{ ohms}$$

One parameter of occasional interest is the power gain, A_p. For a resistive load, the power gain is:

$$A_p = \frac{\text{output}}{\text{input}} = \frac{v_{out}i_L}{v_{in}i_{b1}} \qquad \text{Eq. 4.70}$$

The power gain can be written in terms of the voltage and current gains as follows:

$$A_p = \frac{v_{out}}{v_{in}} \cdot \frac{i_L}{i_{b1}} = A_V A_I \qquad \text{Eq. 4.71}$$

Using values as in the previous example, and for a load resistance of 1000 ohms, recalculate A_V and A_I, and determine the power gain, A_p. (R4-22)

Refer to the schematic in figure 3-15(a) of the text. Assume the following values: $h_{fe} = 170$, $Z_s = 1100$ ohms resistive, and $Z_L = 1000$ ohms resistive.
a. Draw the ac equivalent circuit and the approximate equivalent circuit. It may not be possible to ignore the base biasing resistors in this circuit (R_{12}, R_{11}, R_{22}, and R_{21}) to obtain an accurate evaluation of the parameters. A general rule is that a resistor cannot be neglected unless it is different by more than a factor of ten from the other resistances in the circuit.
b. Calculate the value of A_I, A_V, Z_i, Z_o, and A_p for this two-stage circuit. (R4-23)

LABORATORY EXPERIMENTS

The following experiments were tested in an electronics laboratory:

LABORATORY EXPERIMENT 4-1

OBJECTIVE

To make an approximate measurement of the h-parameters for a common emitter circuit.

EQUIPMENT AND MATERIALS REQUIRED

1   Transistor, 2N404
2   Capacitors, 10 microfarads, 50 volts
2   Resistance substitution boxes (Heathkit Model EU-28A or equivalent)
1   Decade resistance box (Heathkit Model 1N-37 or equivalent)
1   Volt-ohm-milliammeter, high-impedance FET type, with most sensitive full-scale setting at least as low as 10 millivolts
1   Signal generator, capable of low level, adjustable, sinusoidal inputs between 1000 and 10 kilohertz
1   Oscilloscope, dual channel
1   Resistor, 10 ohms, 1/2 watt, 1%
1   Resistor, 10 kilohms, 1/4 watt, 1%
1   Regulated power supply, 0-30 V dc, 1/2 ampere

PROCEDURE

1. Construct a fixed bias common emitter transistor circuit, figure 4-21. Use a sinusoidal input of about 2 kilohertz. The amplitude of the input is to be small enough to insure that the transistor produces linear amplification. Adjust the bias resistors and voltages if necessary to achieve satisfactory operation.

2. The first h-parameter to be measured is h_{ie}. This is the small signal input resistance. This parameter is defined as follows:

$$h_{ie} = \frac{v_{be}}{i_b}, \text{ with } v_{ce} = 0 \qquad \text{Eq. 4.72}$$

To measure h_{ie}:
   a. Change the circuit to the one shown in figure 4-22.
   b. Adjust the decade resistance box to R = 0.
   c. Adjust the signal generator output voltage to a low level. The voltage should be as low as possible and still make it possible to measure v_s accurately with the equipment available.
   d. Adjust the decade resistance box to a level that reduces v_s to one-half of its value when R = 0.
   e. Record the decade box resistance. This value is an approximate value of h_{ie}.

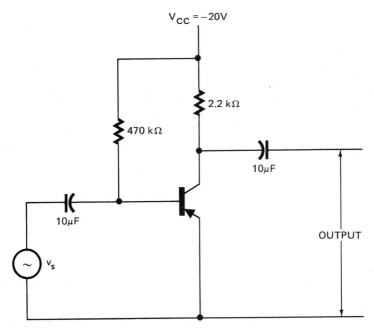

Fig. 4-21 Common emitter amplifier

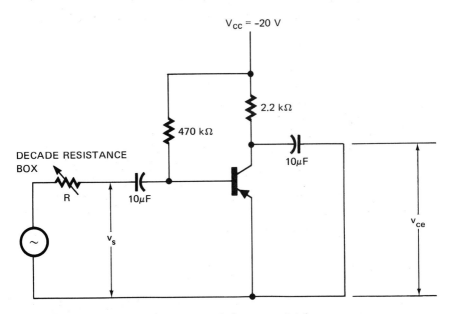

Fig. 4-22 Circuit for measuring h_{ie}

3. An approximate value of the small signal short circuit current gain, h_{fe}, is measured.   This parameter is described by the following equation:

$$h_{fe} = \frac{i_c}{i_b}, \text{with } v_{ce} = 0$$

Eq. 4.73

To measure h_{fe}:

a. Change the circuit to the one shown in figure 4-23. The load resistance of 10 ohms represents a virtual short circuit for the ac output signal and $v_{ce} \approx 0$.

b. Adjust the signal generator output voltage to as low a level as possible without sacrificing the accuracy of the reading for v_1 and v_{ce} using a sensitive volt-ohm-milliammeter. By Ohm's Law, v_1 and v_{ce} can be used to determine i_c and i_b. The values of i_c and i_b can be obtained by:

$$i_c = \frac{v_{ce}}{10}$$

Eq. 4.74

and

$$i_b = \frac{v_1}{10^4}$$

Eq. 4.75

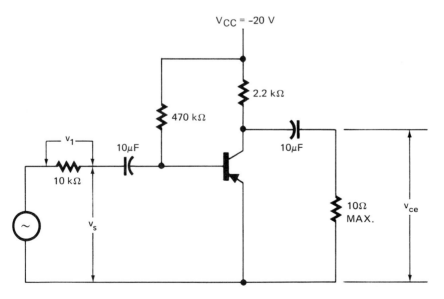

Fig. 4-23 Circuit for measuring h_{fe}

The parameter h_{fe} can be calculated using the following expression.

$$h_{fe} = 1000 \frac{v_{ce}}{v_1}$$

Eq. 4.76

Record the values and calculate h_{fe}.

4. The parameter h_{oe} is in parallel with R_c. The value of h_{oe} can be deduced by measuring the ouput resistance of the amplifier (figure 4-21) with the input circuit open. The parameter h_{oe} is defined as follows:

$$h_{oe} = \frac{i_c}{v_{ce}}, \text{ with } i_b = 0$$

Eq. 4.77

To measure h_{oe}:
   a. Change the circuit to the one shown in figure 4-24.
   b. Do not make an ac connection to the base terminals. The decade resistance box is to be set at zero ohms. Apply a small signal input voltage as shown in figure 4-24. The signal amplitude should be large enough to be measurable with a sensitive volt-ohm-milliammeter.
   c. Adjust the decade box resistance so that the signal input voltage is reduced to one-half the value it was at R = 0. Record the new value of R. This value must also be the resistance of the circuit to which the

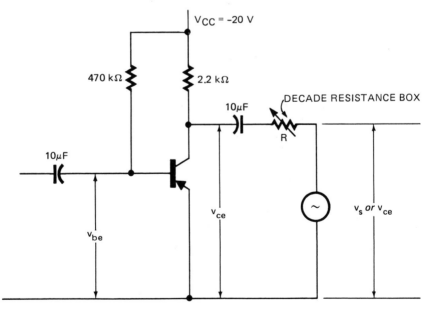

Fig. 4-24 Circuit for measuring h_{oe}

voltage is applied. The parameter h_{oe} is the small signal conductance of the transistor between its collector and emitter terminals. The value of h_{oe} can be computed, using the parallel circuit theorem, as follows:

$$\frac{\dfrac{1}{h_{oe}}\, R_c}{\dfrac{1}{h_{oe}} + R_c} = R \qquad \text{Eq. 4.78}$$

Solving for h_{oe}:

$$h_{oe} = \frac{R_c - R}{RR_c} \qquad \text{Eq. 4.79}$$

The fourth h-parameter, h_{re}, is expressed as follows:

$$h_{re} = \frac{v_{be}}{v_{ce}}, \text{ with } i_b = 0$$

This quantity is too difficult for the student to measure because of the large change in v_{ce} required for a measurable change of v_{be}.

LABORATORY EXPERIMENT 4-2

OBJECTIVE

To construct a two-stage transistor amplifier and evaluate its performance.

EQUIPMENT AND MATERIALS REQUIRED

2	Transistors, 2N404
5	Capacitors, 10 microfarads, nonpolarized, 50 volts
4	Resistance substitution boxes, (Heathkit EU-28A or equivalent)
1	Decade resistance box (Heathkit Model DR-1 or equivalent)
1	Volt-ohm-milliammeter, high-impedance FET type, with most sensitive full-scale setting at least as low as 10 millivolts.
1	Signal generator, capable of low level, adjustable inputs between 10 hertz and 10 kilohertz
1	Oscilloscope, dual channel, with sensitivity to 5 millivolts/centimeter on the display graticule.
2	Resistors, 100 ohms, 1/2 watt, 5%
2	Resistors, 100 kilohms, 1/2 watt, 5%
1	Power supply, regulated, 0-30 volts dc, 1/2 ampere

PROCEDURE

1. Construct a two-stage transistor amplifier, using the circuit shown in figure 4-18. This circuit uses emitter self-bias for stabilization. Due to differences between transistors, a better amplifier may be obtained by adjusting resistors R1, R2, R3, and/or R4. Adjust these resistors as desired for best performance. Observe the following precautions.

   a. The capacitor values are arbitrary. Smaller values may be used, but it may be necessary to increase the signal frequency beyond 1 or 2 kilohertz to prevent a loss of gain due to high capacitive impedances.

   b. The input signal to a two-stage amplifier is very much smaller than the input to a single-stage amplifier. Thus, a large resistance may be required in series with the signal source to *swamp* out a part of the signal. A small signal must be used so that the second stage is not overdriven.

   c. The construction of a two-stage amplifier requires that the following factors be considered: the loading effects between stages, the differences between the transistors, and differences in the operating points of the individual transistors. Basically, however, a two-stage amplifier occurs when the input of one amplifier is connected to the output of another amplifier. To save laboratory time, an efficient way of approaching the experiment is for one laboratory partner to build and test one stage while another partner builds and tests the second stage. A third partner, if there is one, can act as the "quality control technician" to insure that the circuit is built properly.

2. The total gain of a multistage amplifier is obtained by multiplying the gains of the individual stages. The total phase shift of a multistage amplifier is the sum of the phase shifts of the individual stages. Use a dual-channel oscilloscope to verify these statements. Measure and record the stage voltage gains and the overall voltage gain. Describe the phase shifts through the stages that can be detected with the oscilloscope. Note that there is phase inversion in each stage. This means that the phase of the second stage output is nearly the same as the input to the first stage.

3. Determine the current gain of the amplifier. As discussed previously, the current gain is the output current divided by the input current. The output current is the ac output voltage divided by the load resistance.

The input current can be obtained by the following method. Insert a decade resistance box in series with the amplifier input line, as shown in figure 4-25, page 150. Adjust the decade resistance until it is at a convenient level, such as 10 kilohms. The input current is the voltage drop across this resistance. (Use the low level ac voltage scale on the volt-ohm-milliammeter.) Once the output and input currents are known, the current gain can be calculated.

4. Measure the input impedance, using the method outlined in Laboratory Experiment 3-2, step 4.

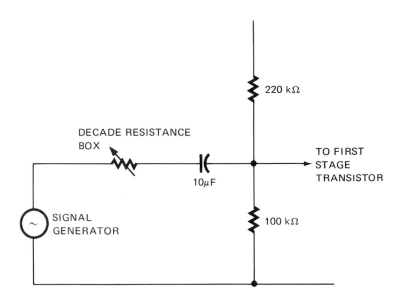

Fig. 4-25 Circuit for measuring input current

5. Measure the output impedance, using the method outlined in Laboratory Experiment 3-2, step 5.

6. Calculate the power gain of the amplifier, using the values of voltage and current gain that have been determined.

EXTENDED STUDY TOPICS

1. Explain the relationship between β, h_{FE}, and h_{fe}.

2. Why is it necessary to use a small voltage or current signal when a transistor circuit is analyzed by means of h-parameters?

3. Are the h-parameters the same regardless of the Q-point selected? Why?

4. The base-to-emitter voltage (V_{BE}) is frequently assumed to be a constant value or zero in transistor circuit calculations. Why?

5. Select an h-parameter for a common base circuit. Define this parameter in terms of the h-parameters for a common emitter circuit. It will be necessary to draw an equivalent circuit in a common base configuration based on the use of common emitter h-parameters. Check the results obtained with the information given in Table 4-4.

6. Common base and common collector circuits are usually restricted to either the first or last stage of a multistage amplifier. These circuits are

never used as an intermediate stage. What defects do the CB and CC circuits have that make them unsuitable for use as an intermediate stage?

7. How is it determined that an approximate equivalent circuit can be formed from an exact equivalent circuit? What criteria are followed in neglecting or modifying resistances or transistor parameters?

8. What criteria are followed to decide if coupling capacitors can be neglected in the small signal analysis of RC coupled amplifiers?

9. Why is the equation for the incremental diffusion resistance of a forward biased semiconductor diode (Equation 4.61) applicable also to the analysis of a bipolar transistor?

10. What is the effect of temperature on the incremental diffusion resistance?

11. What two h-parameters must be available for the analysis of a transistor and its use in a circuit? (These parameters may be obtained either by calculation or by experiment.)

12. Which of the h-parameters are negative numbers? What do these negative values mean when computing the current or voltage gain?

13. It is possible for a sinusoidally varying signal to be shifted in phase between the input and ouput of a transistor amplifier. The amplifier current and voltage gains have both a phase and a magnitude. If it is assumed that the coupling capacitors do not contribute to the phase shift, determine the phase shift for the following:
   a. A_V, CB circuit
   b. A_I, CB circuit
   c. A_V, CC circuit
   d. A_I, CE circuit
   e. A_V, CE circuit
   f. A_I, CC circuit

5
Using Bipolar Transistors at High and Low Frequencies

OBJECTIVES

After studying this chapter, the student will be able to:

- specify low-frequency equivalent circuits and circuit analysis techniques
- define high-frequency transistor parameters
- specify high-frequency equivalent circuits and circuit analysis techniques

SENSITIVITY OF TRANSISTOR AMPLIFIERS TO FREQUENCY

The analysis of transistor amplifiers in previous chapters was based on amplifier operation at a particular range of frequencies. At these frequencies, there is no attenuation or alteration of the signal because of circuit capacitances or inductances.

In actual practice, an amplifier may be required to perform at very low and very high frequencies. Because of this possibility, amplifier performance for the extreme frequency ranges must be analyzed. An analysis of this type makes use of a frequency response graph such as the one shown in figure 5-1. To obtain this graph, an input signal is applied to the amplifier. This signal varies with time according to the expressions:

$$v_{in} = V_m \sin 2 \pi f t$$

or

$$v_{in} = V_m \sin \omega t \qquad \text{Eq. 5.1}$$

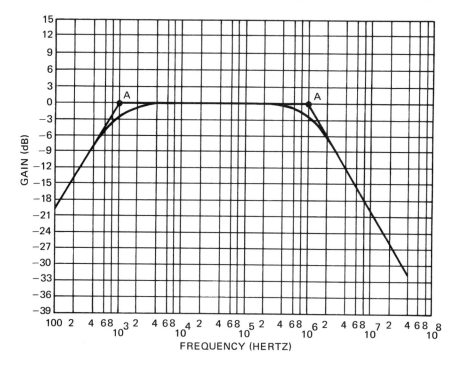

Fig. 5-1 Amplifier frequency response

where

V_m = maximum or peak value of the signal
ω = frequency in radians per second
f = frequency in hertz
t = time in seconds

The input signal is held at a constant peak amplitude and its frequency is varied. The output signal amplitude is measured at each frequency. The gain or amplification is calculated and plotted as a function of frequency. The gain is usually plotted on semilogarithmic graph paper in units of decibels as in figure 5-1 (see the Appendix for a discussion of decibels). If the gain is to be plotted as a direct ratio of the output to the input, then log-log gragh paper is used.

At high frequencies, shunt capacitances are present within the transistor and other circuit components. Such capacitances cause a drop in the gain. The reason for this can be seen by referring to the equation for capacitive reactance:

$$X_c = \frac{1}{\omega C} = \frac{1}{2 \pi f C}$$

Eq. 5.2

At high frequencies, any capacitance connected directly or indirectly across the output terminals has a low reactance. As a result, the capacitance acts as a partial short circuit. There is a resulting drop in the voltage output and current is bypassed around (rather than through) the load resistance.

For ac coupled amplifiers, there is also a drop in the voltage output and the apparent gain at low frequencies. For RC coupled amplifiers (figure 2-11), the reactance formed by coupling capacitors C1 and C2 restricts the flow of signal and load currents at low frequencies. According to Equation 5.2, the reactance is high at low frequencies.

Transformer coupled amplifiers also display poor performance at low frequencies. A transformer coupled amplifier is shown in the schematic of figure 2-12. Electrical energy is transferred from the primary winding to the secondary winding in this circuit. Transformers depend on a rapidly varying signal for efficient operation. The voltage across the secondary winding terminals (E_2) is the result of magnetic induction which is governed by:

$$E_2 = 4.44 \, \phi_m f N_2 \qquad \text{Eq. 5.3}$$

where

ϕ_m = the maximum value of the magnetic flux for the primary and secondary windings
f = the signal frequency
N_2 = the number of turns in the transformer secondary winding.

It can be seen from this equation that the lower the frequency, the lower is the voltage induced in the secondary. Thus, for a transformer coupled amplifier, such as the one shown in figure 2-12, the voltage output and overall gain are lower at low frequencies.

The amplifier considered in review problem (R4-12) had an input coupling capacitance of 20 microfarads and was operated at 1000 hertz. Calculate the capacitive reactance at 1000 hertz.

If the signal frequency is lowered to 10 hertz, what is the resulting capacitive reactance? At which of the two frequencies should the amplifier be operated? Why? (R5-1)

Examine figure 5-1. Why does a lower gain result at very low frequencies? Why does a lower gain result at very high frequencies? (R5-2)

Three regions of amplifier operation can be defined with regard to the frequency spectrum: low, intermediate, and high. The first analysis that will be made is concerned with the performance of a transistor amplifier at low frequencies.

AMPLIFIER PERFORMANCE AT LOW FREQUENCIES

The RC coupled amplifier shown in figure 2-11 is to be considered. The approximate equivalent circuit for this amplifier when operated at low frequency is given in figure 5-2. The input impedance is given as;

$$Z_i = \frac{1}{j\,\omega\,C1} + \frac{h_{ie}R_b}{R_b + h_{ie}}$$

Eq. 5.4

where j is the operator indicating the imaginary part of a complex number. (Refer to the Appendix for the definition and use of the j-operator.)

The output impedance for the circuit is determined by the bias resistor R_c and the coupling capacitor. The bias resistance is a factor because R_c is "inside" the amplifier output terminals. As a result, the output impedance is given approximately by

$$Z_o = R_c + \frac{1}{j\,\omega\,C2}$$

Eq. 5.5

For a resistive load (R_L), the current gain is given by Equation 4.11:

$$A_I = \frac{i_L}{i_{in}}$$

According to the current divider theorem, i_L is:

$$i_L = \frac{-h_{fe}i_b R_c}{R_c + \dfrac{1}{j\,\omega\,C2} + R_L}$$

Eq. 5.6

For the case where R_b is very large,

$$i_{in} = i_b$$

Eq. 5.7

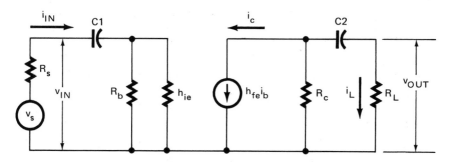

Fig. 5-2 Simplified common emitter equivalent circuit for low-frequency operation

When Equations 5.7 and 5.6 are substituted into Equation 4.11 and the resulting expression is simplified, A_I becomes:

$$A_I = \frac{-h_{fe}R_c j \omega C2}{(R_c + R_L)j \omega C2 + 1} \qquad \text{Eq. 5.8}$$

The voltage gain is given by Equation 4.15:

$$A_V = \frac{v_{out}}{v_{in}}$$

If a resistive load is assumed and i_L is expressed by Equation 5.6, then:

$$v_{out} = i_L R_L = \frac{-h_{fe}i_b R_c R_L}{R_c + R_L + \dfrac{1}{j \omega C2}} \qquad \text{Eq. 5.9}$$

Writing Kirchhoff's Voltage Law for the input circuit,

$$v_{in} = \frac{i_{in}h_{ie}R_b}{R_b + h_{ie}} + \frac{i_{in}}{j \omega C1}$$

$$\approx i_b\left(h_{ie} + \frac{1}{j \omega C1}\right) \qquad \text{Eq. 5.10}$$

Equations 5.10 and 5.9 can be substituted into the equation for the voltage gain (Equation 4.15) and the resulting equation simplified to yield the following:

$$A_V = \frac{-h_{fe}R_c R_L (j \omega)^2 C1 C2}{(h_{ie}j \omega C1 + 1)[(R_c + R_L)j \omega C2 + 1]} \qquad \text{Eq. 5.11}$$

An examination of Equations 5.8 and 5.11 shows how A_I and A_V vary with frequency.

At a frequency of $\omega = 2\pi f = 0$ (or very nearly zero), the current gain and the voltage gain are both zero or very close to it. At low frequencies,

$$A_I \approx -h_{fe}R_c j \omega C2 \qquad \text{Eq. 5.12}$$

and

$$A_V \approx -h_{fe}R_c R_L (j \omega)^2 C1 C2 \qquad \text{Eq. 5.13}$$

As the frequency increases, the denominators of Equations 5.8 and 5.11 increase. As the frequency increases still further, A_I and A_V become less sensitive to frequency. At high frequencies, A_V and A_I reduce to the forms of Equations 4.19 and 4.14. Note that in this case A_V and A_I are not sensitive to frequency. The amplifier is considered to have a flat response at the higher frequencies. At extremely high frequencies, the shunt capacitances become effective and the gain is reduced. This effect is discussed more completely when high-frequency parameters are discussed.

In equation 5.8, at the point where $(R_c + R_L)j\,\omega C2 = 1$, the denominator is $1 + j1$ and the current gain is down by 3 decibels. (Refer to figure 5-1.) Note that in the intermediate range of frequencies, the gain curve is flat but begins to drop off at point A. When the dropoff equals 3 decibels, the point is reached at which the denominator of the equation for A_I is $1 + j1$. This point is known as the *half-power point* or *cutoff frequency*. At frequencies below this value, the amplifier response is considered to be too sensitive to the signal frequency. As a result, the amplifier should not be used at such frequency levels.

The amplifier shown in figure 2-11 is to be considered. It is subjected to a sinusoidal input signal having a peak magnitude within the linear operating region of the transistor. What is the lowest frequency at which this amplifier can be operated and still have a flat frequency response curve for the current gain?

The first step in the analysis is to determine where the "flat" gain ends and where the gain begins deteriorating because of the frequency level of the incoming signal. A standard criterion for locating this transition is the point at which the gain drops by 3 decibels.

$$\text{Decibels} = 20 \log \frac{(\text{output signal})}{(\text{input signal})}$$

$$-3 = 20 \log A_I$$

$$\log A_I = \frac{-3}{20} = -0.15$$

$$A_I = 0.707$$

This means that the point where A_I is down by 3 decibels is equivalent to 70.7% of its value for the flat portion of the frequency response curve. This point occurs when the denominator of Equation 5.8 is $1 + j1$. Thus,

$$\omega(R_c + R_L)C2 = 1 \qquad\qquad \text{Eq. 5.14}$$

Example of a Low Frequency Analysis

By assigning values to the components of the amplifier shown in figure 2-11, Equation 5.14 can be solved for f. For $R_L = 1000$ ohms, $R_c = 1340$ ohms, $C2 = 10$ microfarads, and $\omega = 2\pi f$,

$$f = \frac{1}{C2(R_c + R_L)2\pi}$$

$$= \frac{1}{(10)(10)^{-6}(2340)(6.28)}$$

$$= 6.8 \text{ hertz}$$

At this level of frequency, the current gain is 3 decibels down from its level in the intermediate frequency range.

Suppose it is desired to limit the input signal frequency according to the point at which the *voltage gain* drops off by 3 decibels. What will be the lowest frequency that the input signal can have?

The solution to this problem is complicated by the fact that there are two terms in the denominator of Equation 5.11. It must be determined which term has the deciding effect. Assume h_{ie} is 450 ohms. The remaining circuit values are the same as those used in the current gain calculation.

For the first term:

$$h_{ie}C1 = (450)(10)(10)^{-6} = 0.0045$$

For the second term:

$$(R_c + R_L)C2 = (2340)(10)(10)^{-6} = 0.0234$$

It is evident that $j\omega(R_c + R_L)C2$ will reach 1 at a lower frequency than $j\omega h_{ie}C1$. Therefore, the limiting value of frequency for voltage gain is produced by the "$h_{ie}C1$" term, or $(0.0045)(2\pi f) = 1$ and $f = 35.4$ Hz.

The equations derived for the current gain and voltage gain were based on a resistive load. The resistive load was assumed to simplify the explanation. For some applications, however, the load contains reactance. Such a load must be included in any calculations as an impedance Z_L. Z_L will be a complex number having both an imaginary part and a real part. For this case, the equations are more complicated and the calculations are more difficult.

The analysis of a circuit containing a reactive load follows the same basic pattern as the analysis of a circuit with an assumed resistive load. Standard circuit analysis techniques are used. These statements also are true for the other circuits analyzed in this section. The equations can always be extended and redefined to accommodate a reactive load. Z_L is used in the following discussions as a generalized term.

Using the amplifier specifications listed in review problem (R4-11), calculate the frequency at which the current gain drops by 3 decibels from the level in the intermediate frequency range.   *(R5-3)*

Using the amplifier specifications given for review problem (R4-11), calculate the frequency at which the voltage gain drops by 3 decibels from the level in the intermediate frequency range.   *(R5-4)*

Assume that ω is very large (approaching infinity) and show that Equations 5.8 and 5.11 reduce to Equations 4.19 and 4.14, respectively.   *(R5-5)*

A frequency response curve is to be prepared for the amplifier specified in review problem (R4-11). Using the results of (R4-11) and (R5-3), prepare the graph for this amplifier at low and intermediate range frequencies, using a decibel scale. Substitute the amplifier parameter values into Equation 5.8 for the very low frequency region. (Hint: the denominator of Equation 5.8 is equal to one at very low frequencies.)   *(R5-6)*

Using Equation 5.8, verify that the current gain of the amplifier of review problem (R4-11) actually does drop by 3 decibels at the frequency calculated in review problem (R5-3).   *(R5-7)*

Repeat review problem (R5-6) for the voltage gain. In Equation 5.11, note that the denominator becomes unity for very low frequency values. (R5-8)

Using Equation 5.11, verify that the voltage gain of the amplifier of review problem (R4-11) actually does drop by 3 decibels at the frequency calculated in review problem (R5-3). Although there may be an effect of the $(R_L + R_c)C2$ term in the denominator at this frequency, this effect is very slight.   *(R5-9)*

HIGH-FREQUENCY TRANSISTOR PARAMETERS

At high frequencies, there is an internal transistor capacitance which takes effect across the base to emitter junction. This capacitance does not have any effect at low or intermediate frequencies. A second capacitance also takes effect at high frequencies; it is the capacitiance from the base to the collector. The high-frequency transistor can be shown using the two-diode representation given in figure 4-17.

The following discussion further defines the symbols indicated in figure 4-17.

The capacitor C_e is the incremental capacitance across the base-to-emitter junction. This capacitance represents the rate of change of charge for the current carriers injected into the base from the emitter. This action occurs across the junction of any forward biased diode. When the transistor specifications give a value for the diffusion capacitance C_{De}, the value may be used as the approximate value for C_e. Occasionally, the transistor specifications will list a value for the emitter transmission capacitance, C_{Te}. This value, however, cannot be assumed to be an approximate value for C_e. The relationship between the capacitances is defined as:

$$C_e = C_{De} + C_{Te} \qquad\qquad \text{Eq. 5.15}$$

C_{Te} is a much smaller number than C_{De}. As a result, C_{Te} is frequently neglected in an analysis. Transistor manufacturers may list a capacitance value as C_{ib}, or "input capacitance." This term also is generally assumed to be equivalent to C_e.

The capacitance C_c is a capacitor formed by an apparent dielectric in the region of the reverse biased collector-to-base junction. This apparent dielectric or insulating material is caused by the lack of mobile charges in the junction area of a reverse biased diode. Recall that a capacitor is formed by placing an insulating material between two layers of conducting material. This condition exists between the collector and base of the transistor. Manufacturers refer to the following terms when reporting transistor collector-to-base capacitances:

• Collector transition capacitance, C_{Tc}

• Common base output capacitance, C_{ob}

• Output capacitance

• Collector capacitance

All of these terms can be assumed approximately equal to C_c.

The resistance $r_b{'}_c$ occurs between the collector and base of a transistor when the junction is reverse biased. This resistance is a finite value because of thermally generated current carriers.

A circuit now exists between the collector and the actual junction between the base and the collector. This junction is node B′ in figure 4-17. It is the main difference between the circuits for the high-frequency transistor model and the low-frequency model. The presence of this node requires a knowledge of the base spreading resistance ($r_{bb}{'}$). However, $r_{bb}{'}$ cannot be measured by normal classroom experimental techniques. Node B′ is an internal connection not available to the experimenter. A circuit analysis in the intermediate-frequency region does not require a knowledge of r_{bb}' because the experimental value of h_{ie} includes this value. If h_{ie} is used, its value must be determined at the most probable operating signal frequency.

The following additional terms and values are used by manufacturers to specify the high-frequency capabilities of transistors.

- f_{hfb} (or f_α): the alpha cutoff frequency.
  This is the frequency at which the common base current gain drops by 3 decibels (or, in other words, when the gain drops by a factor of 0.707).

- f_β: the beta cutoff frequency.
  This is the frequency at which the common emitter current gain drops by a factor of 0.707 (or, drops by 3 decibels). The following equation relates f_β and f_α.

$$f_\beta = f_\alpha(1 + h_{fb}) \qquad \text{Eq. 5.16}$$

- f_T: the frequency at which the short circuit common emitter gain (h_{fe}) falls to unity. The relationship between f_T and f_β is:

$$f_T = h_{fe}f_\beta \qquad \text{Eq. 5.17}$$

Often, f_T is called the *current gain bandwidth product* because of the relationship expressed in Equation 5.17.

- Q_{SB} or Q_B: the stored base charge or the average charge stored in the base of a forward biased base-to-emitter junction.

Q_{SB} is given in coulombs (C) or microcoulombs (μC). It is used to determine the emitter diffusion capacitance (C_{De}).

$$C_{De} = \frac{Q_{SB}}{V_{BE}} \qquad \text{Eq. 5.18}$$

A manufacturer lists the stored base charge of a transistor as 1500 micro-microcoulombs ($\mu\mu C$). Figure 2-8 represents the common emitter input characteristics for this transistor. The value of V_{CE} is assumed at normal rated values and the base current = 0.10 milliampere. What is the value of the emitter diffusion capacitance?   *(R5-10)*

Using the results of review problem (R5-10), what is the approximate value of the total incremental capacitance across the base-to-emitter junction?

(R5-11)

The value of the alpha cutoff frequency for a transistor is 3.5 megahertz. The short circuit common emitter current gain is 50. What is the approximate frequency at which the short circuit common emitter current gain falls to unity? (Hint: use the approximation that h_{fb} is unity, according to Table 4-3.) (R5-12)

A transistor specification lists an input capacitance of 20 picofarads (pF). What is the approximate value of the incremental capacitance across the base-to-emitter junction?   (R5-13)

What is the beta cutoff frequency of a transistor when the alpha cutoff frequency is 3 megahertz and the short circuit common base current gain is 0.99?   (R5-14)

The common base output capacitance of a transistor is 2 picofarads. What is the collector-to-base capacitance of the transistor?   (R5-15)

The following values are given for a transistor: the reverse biased collector-to-base resistance of a transistor = 4 megohms, the collector capacitance is 2 picofarads, and the base spreading resistance = 100 ohms. What is the imped-ance between the collector and base of this transistor when the signal frequency is 10,000 hertz? What is the impedance when the signal frequency is 10 megahertz?   (R5-16)

HIGH-FREQUENCY EQUIVALENT CIRCUITS
FOR COMMON EMITTER AMPLIFIERS

To analyze an equivalent cicuit for a common emitter amplifier operating at high frequencies, consider the amplifier circuit shown in figure 2-11. The h-parameter equivalent circuit for this amplifier at intermediate frequency levels is shown in figure 4-11. The equivalent circuit is modified in figure 5-3 to in-clude the high-frequency effects. The quantities and parameters Z_s, v_s, Z_L, R_c, R_b, h_{oe}, and h_{fe} have the same definitions for figure 5-3 as for figure 4-11.

To design a common emitter amplifier, the impedance formed by $r_b{}'_c$ and C_c is changed to obtain an approximate equivalent circuit that can be analyzed easily. This relationship can be defined by the use of *Miller's Theorem*. This theorem was developed as a means of converting vacuum triode interelectrode capacitances to a form that could be used to make an equivalent circuit

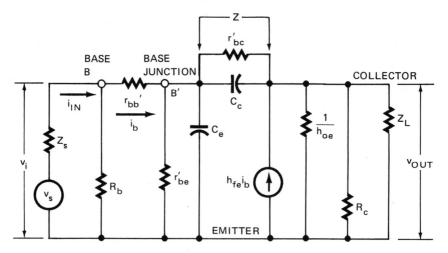

Fig. 5-3 High-frequency equivalent circuit for common emitter amplifier

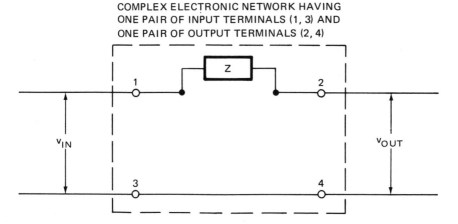

Fig. 5-4 Complex network having an impedance Z between the input and output terminals

for analyzing vacuum tube amplifiers at high frequencies. The use of Miller's Theorem can be extended to the analysis of transistor amplifiers. Consider a circuit such as the one shown in figure 5-4 where an impedance Z is placed between the input and output terminals. According to Miller's Theorem, such a circuit can be converted to the circuit shown in figure 5-5, page 164.

Miller's Theorem says that the impedance Z between the input and output terminals can be used to express an impedance across the input terminals (Z_1) and an impedance across the output terminals (Z_2):

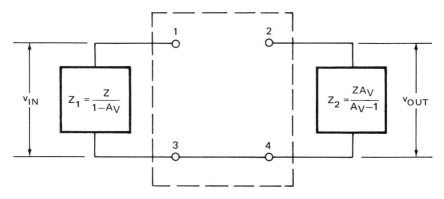

Fig. 5-5 Complex network of figure 5-4 transformed to equivalent circuit by the use of Miller's Theorem

$$Z_1 = \frac{Z}{1 - A_V} \qquad \text{Eq. 5.19}$$

$$Z_2 = \frac{ZA_V}{A_V - 1} \qquad \text{Eq. 5.20}$$

A_V is the voltage gain or transfer function of the network as expressed by Equation 4.15:

$$A_V = \frac{v_{out}}{v_{in}}$$

Measurements of v_{out} and v_{in} are made in the intermediate frequency range of the amplifier.

The value of A_V for a single-stage transistor amplifier is -100 in the intermediate-frequency range. The impedance between the collector and base of the transistor is $1 + j.3$ megohms at a signal frequency of 1 megahertz. For this signal frequency, what is the value of the Miller input and output impedances? (Note that this impedance is a complex number having both magnitude and phase.) Comment on the relative magnitudes of the input and output impedances. (R5-17)

In the circuit of figure 5-3, Z is the parallel circuit formed by $r_{b'c}$ and the capacitive reactance, $1/j\omega C_c$. Z is defined by:

$$Z = \frac{r_{b'c}}{j\omega C_c r_{b'c} + 1} \qquad \text{Eq. 5.21}$$

The value of $r_b'_c$ is a very large number when compared to the reactance of C_c in the high frequency region. Thus, Equation 5.21 is changed to:

$$Z \approx \frac{1}{j\omega C_c} \qquad \text{Eq. 5.22}$$

A capacitance C_{eqi} is defined so that the following expression is obtained for Z_1.

$$Z_1 = \frac{1}{j\omega C_{eqi}} \qquad \text{Eq. 5.23}$$

Substitute Equations 5.23 and 5.22 into Equation 5.19 and solve for C_{eqi}.

$$C_{eqi} = C_c(1 - A_V) \qquad \text{Eq. 5.24}$$

Equation 5.24 defines C_{eqi} as the Miller capacitance for the input side of the equivalent circuit.

A quantity C_{eqo} can be defined so that the following expression for Z_2 is obtained:

$$Z_2 = \frac{1}{j\omega C_{eqo}} \qquad \text{Eq. 5.25}$$

Now substitute Equations 5.25 and 5.22 into Equation 5.20 and solve for C_{eqo}.

$$C_{eqo} = \frac{C_c(A_V - 1)}{A_V} \qquad \text{Eq. 5.26}$$

Recall that A_V is expressed in terms of the transistor amplifier parameters by Equations 4.19, 4.20, or 4.21.

$$A_V = \frac{-h_{fe}R_c Z_L}{h_{ie}(R_c + Z_L)}$$

For heavy loads, $A_V = \dfrac{-h_{fe}Z_L}{h_{ie}}$

For light loads, $A_V = \dfrac{-h_{fe}R_c}{h_{ie}}$

The high-frequency equivalent circuit (figure 5-3) is transformed by Miller's Theorem into the circuit shown in figure 5-6, page 166. The quantity $r_b'_c$

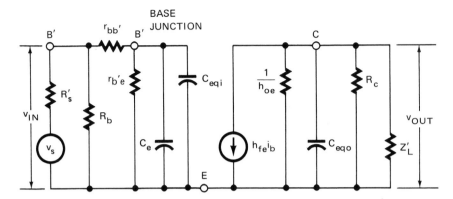

Fig. 5-6 High-frequency equivalent circuit of figure 5-3, after transformation by Miller's Theorem

is neglected because when it is translated through Miller's Theorem at high frequencies, its value has a negligible effect on the results.

By substituting Equations 4.19, 4.20, or 4.21 into Equations 5.24 and 5.26 and including C_e, expressions are obtained for the equivalent capacitances in terms of the intermediate frequency amplifier parameters.

$$C_o = C_{eqo} = \frac{C_c(A_V - 1)}{A_V} \qquad \text{Eq. 5.27}$$

$$C_i = C_e + C_c(1 - A_V) \qquad \text{Eq. 5.28}$$

Example of the Use of the Equations

To determine these equivalent capacitances, consider the transistor of figure 5-3. The following values are to be used in an analysis: C_e = 100 picofarads, C_c = 12 picofarads, $r_b'c$ = 4 megohms, h_{fe} = 86, and h_{ie} = 450 ohms. The transistor is to be used in a common emitter amplifier such as the one shown in figure 2-11. The values of R_c and Z_L are 1340 ohms and 1000 + j0 ohms, respectively. (The reactive part of Z_L frequently is zero or nearly so. Thus, a j0 term for the imaginary part is not unusual.) The Miller's Theorem representation is to be used to compute the input and output capacitances. The input capacitance is determined from Equation 5.28, using

$$A_V = \frac{-h_{fe}R_cZ_L}{h_{ie}(R_c + Z_L)} = \frac{-(86)(1000)(1340)}{450(1000 + 1340)} = -109.5$$

The assumed values are substituted in Equation 5.28 to obtain C_i.

$$C_i = (100)(10)^{-12} + (12)(10)^{-12}(1 + 109.5)$$

$$= 1425(10)^{-12} \text{ farads}$$

$$= 0.001425 \text{ microfarads}$$

The output capacitance is determined as follows:

$$C_o = \frac{C_c(A_V - 1)}{A_V} = \frac{(12)(10)^{-12}(-109.5 - 1)}{-109.5}$$

$$= 12.1(10)^{-12} \text{ farads}$$

$$= 12.1 \text{ picofarads}$$

To determine if it is possible to neglect $r_b{'}_c$ and express Z by Equation 5.22, the following analysis is used, assuming a signal frequency of 1000 hertz.

$$Z = \frac{r_b{'}_c}{j\omega C_c r_b{'}_c + 1} = \frac{4(10)^6}{j(6,280)(12)(10)^{-12}(4)(10)^6 + 1}$$

$$= \frac{4(10)^6}{j0.302 + 1}$$

$$= 3.82(10)^6 \; \underline{/-16.8°} \quad \text{ohms}$$

This expression shows that $r_b{'}_c$ cannot be neglected, but Z can! The signal frequency is low enough to insure that the transistor is operating well within the intermediate frequency region of its frequency response curve.

If the signal frequency is 1 megahertz (rather than 1000 hertz), then Z is equal to:

$$Z = \frac{4(10)^6}{j302 + 1} \approx \frac{4(10)^6}{j302}$$

$$= 13,250 \; \underline{/-90°} \quad \text{ohms}$$

In this case, Z is almost entirely reactive. Thus, at a signal frequency of 1 megahertz, Equation 5.22 can be used. The high-frequency representation requires only that C_o and C_i be added to the equivalent circuit.

A common emitter amplifier is connected in a collector-to-base bias circuit as shown in figure 2-16. For this circuit, $h_{ie} = 450$ ohms, $h_{fe} = 100$, $R_c = 1000$ ohms, $Z_L = \infty$, $R_b = 200$ kilohms, and $r_{bb}' = 0$. Use Miller's theorem to obtain an ac equivalent circuit and then compute the voltage gain. The signal frequencies are assumed to be within the intermediate frequency range.   (R5-18)

The supplier of a 2N1711 transistor lists an input capacitance (C_{ib}) of 20 picofarads and an output capacitance (C_{ob}) of 4 picofarads. (The characteristics of a 2N1711 are given in Table 4-3.) This device is to be used in a common emitter amplifier application. For this circuit, R_c is 2500 ohms and the ac load impedance is 2000 ohms resistive. Find the value of the equivalent common emitter input and output capacitances as expressed by Equations 5.28 and 5.27. Comment on the relative effect of these two capacitances on the high-frequency response of the amplifier.   (R5-19)

A transistor specification lists values for h_{fe} and f_{hfb}. Knowing the voltage and current values at the quiescent point, list the high-frequency parameters that can be determined approximately from a knowledge of these values.(R5-20)

Approximations for High-Frequency Analysis

The combination of resistance and capacitances in the input side of the circuit (to the left of node E in figure 5-6), usually has a predominant effect at high frequencies. In general, the high-frequency limit of operation of the transistor is due to the capacitance on the input side of the equivalent circuit and not to C_o. In practice, C_i is about 40 times C_o so it may not be necessary to consider C_o.

Equation 4.66 gives h_{ie} for $r_{bb}' \approx 0$:

$$h_{ie} \approx r_{b'e}h_{fe}$$

If it is not possible to neglect r_{bb}', then it must be included as a resistance in series with the resistance expressed by Equation 4.66. At intermediate frequencies, the exact value of the input resistance for a common emitter circuit, including the base spreading resistance, is:

$$h_{ie} = r_{bb}' + h_{fe}r_{b'e} \qquad \text{Eq. 5.29}$$

At high frequencies, the input resistance is an impedance. C_i represents a capacitive reactance and serves as a shunt for the base current. As a result, the ability of the transistor to amplify is reduced.

The analysis of this circuit is simplified by the fact that the transistor is a *current* operated device. As a result, a series resistance, such as $r_{bb}{}'$, usually has a negligible effect. In many cases, $r_{bb}{}'$ can be assumed to be zero because it is small when compared to the incremental diffusion resistance. It is fortunate that this resistance is small because it cannot be measured easily or calculated accurately. Equation 4.66 can still be used for the analysis. The parameter h_{oe} can be omitted from the equivalent circuit in the approximation of high-frequency operation. Recall that this parameter was also omitted in the approximation of low-frequency operation.

The approximate equivalent circuit for high-frequency operation reduces to the form shown in figure 5-7. The amplifier parameters for high-frequency operation can now be written.

The current gain for the common emitter model of figure 5-7 is given by Equation 4.11.

$$A_I = \frac{i_L}{i_{in}}$$

The effect of frequency on the current gain is changed very little by the output capacitance and so the output capacitance can be omitted from the analysis. The relationship between the collector current (i_c) and the load current (i_L) is:

$$i_L = \frac{-i_c R_c}{R_c + Z_L} \qquad \text{Eq. 5.30}$$

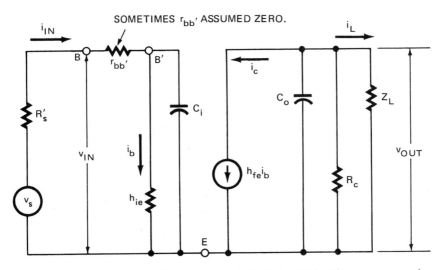

Fig. 5-7 **Approximate high-frequency equivalent circuit for the common emitter amplifier**

The effect of the input capacitance can be included by considering it as a capacitive reactance expressed by $1/j\omega C_i$ at the signal input frequency. According to the current divider theorem,

$$i_b = \frac{(i_{in})\dfrac{1}{j\omega C_i}}{h_{ie} + \dfrac{1}{j\omega C_i}}$$ Eq. 5.31

From the definition of h_{fe} in Equation 4.7,

$$i_c = h_{fe}i_b$$ Eq. 5.32

Substitute Equations 5.32 and 5.31 into Equation 5.30 and simplify the result to obtain an expression for the load current i_L:

$$i_L = \frac{-i_{in}h_{fe}R_c}{(j\omega h_{ie}C_i + 1)(R_c + Z_L)}$$ Eq. 5.33

Equation 5.33 is then substituted into Equation 4.11 to obtain the current gain:

$$A_I = \frac{-h_{fe}R_c}{(j\omega h_{ie}C_i + 1)(R_c + Z_L)}$$ Eq. 5.34

Demonstration of the Use of the Equations

The manufacturer's specifications for a transistor state that the stored base charge is 1000 picocoulombs and the output capacitance is 20 picofarads. The remaining transistor parameters are listed in Table 4-3 for the 2N1808 transistor. The transistor is to be used in the circuit shown in figure 2-11. The problem is to find the maximum frequency at which this amplifier can be used, if the current gain is not to be affected by the signal frequency.

The first step in determining the frequency is to judge how far the current gain (A_I) must drop before it is certain that the maximum frequency limit is reached. The usual limit is 3 decibels down from the gain level in the intermediate frequency region. This statement is equivalent to saying that the gain is allowed to decrease by a factor of 0.707 as the result of an increasing frequency. In other words, the current gain A_I, as expressed by Equation 5.34, is reduced by

$$\frac{1}{j2\pi f h_{ie}C_i + 1} = 0.707$$

In this case, $j2\pi f h_{ie} C_i = 1$. (The resulting current gain then is reduced by $1/(1 + j1) = 0.707 \, \underline{/-45°}$.) Solve for the frequency as follows:

$$f = f_\beta = \frac{1}{2\pi h_{ie} C_i}$$

Eq. 5.35

C_i is calculated from Equation 5.28. The value of C_e must be calculated from the relationship between the capacitance and the stored charge (see the Appendix). That is, by combining Equations 5.18 and 5.15 and assuming the emitter transition capacitance (C_{Te}) is negligibly small, C_e is expressed as:

$$C_e = \frac{Q_{SB}}{V_{BE}}$$

Eq. 5.36

The symbols in Equation 5.36 were previously defined. Assume that V_{BE} is about 0.200 volt for a germanium transistor operating in its active region. C_e is:

$$C_e = \frac{1000(10)^{-12} \text{ coulombs}}{0.200 \text{ volt}}$$

$$= 5(10)^{-9} \text{ farad} = 0.005 \text{ microfarad}$$

According to Equation 5.28,

$$C_i = C_e + C_c(1 - A_V)$$

The output capacitance (C_c) = 20 picofarads. The value of $C_c(1 - A_V)$ must be greater than 0.1 times the value of C_e if it is to have any effect. The statement is expressed mathematically as:

$$20(10)^{-12}(1 - A_V) > 0.0005$$

Eq. 5.37

What value of A_V will cause Equation 5.37 to be true? Since $1 - A_V \approx A_V$, the inequality reduces to the following:

$$20(10)^{-12} A_V > 0.0005$$

Thus, the value of A_V is expressed as:

$$A_V > \frac{0.0005}{20(10)^{-12}} > 2.5(10)^7$$

A value of A_V of $2.5(10)^7$ is beyond the capabilities of any single electronic amplifying device! Therefore, C_c is assumed negligible and $C_i = 0.005$ microfarad. Table 4-3 gives a value for h_{ie} of 450 ohms for the 2N1808 transistor. The signal for this amplifier can have a maximum frequency of:

$$f = \frac{1}{(6.28)(450)(0.005)(10)^{-6}} = 70{,}800 \text{ hertz}$$

Simplified Forms of Equations for Gain and Impedance

As indicated in Chapter 5, it is possible to define a frequency (f_T) and relate it to the parameters given in Equation 5.34. This frequency is formed from Equation 5.35 by substituting f_T for f:

$$f_T = \frac{h_{fe}}{2\pi h_{ie} C_i} \qquad \text{Eq. 5.38}$$

This equation is solved for h_{ie}.

$$h_{ie} = \frac{h_{fe}}{2\pi f_T C_i} \qquad \text{Eq. 5.39}$$

An expression for the current gain is obtained by substituting Equation 5.39 into Equation 5.34 and recalling that $\omega = 2\pi f$:

$$A_I = \frac{h_{fe} R_c}{R_c + Z_L}\left(\frac{1}{j\dfrac{f h_{fe}}{f_T} + 1}\right) \qquad \text{Eq. 5.40}$$

Thus, the current gain is expressed in terms of the frequency and the parameter f_T. The parameter f_T is frequently used by transistor manufacturers to indicate the high-frequency capabilities of their transistors.

Examine Equation 5.40 to determine the current gain at various load conditions. If the amplifier operates at no load, then the current gain is zero. If the amplifier output terminals are shorted, the gain is given by:

$$A_I = \frac{h_{fe}}{j\dfrac{f h_{fe}}{f_T} + 1} \qquad \text{Eq. 5.41}$$

If the transistor is operated at a frequency well within the intermediate frequency range, A_I, as expressed by Equation 5.41, reduces to h_{fe} which was shown earlier (Chapter 4) to be the short circuit current gain of a transistor.

The high-frequency characteristics of a transistor are measured under the following conditions: normal room temperature, I_C = 7 milliamperes, and V_{CE} = 15 volts. The characteristics are listed as follows: h_{fe} = 120, h_{ie} = 450 ohms, and C_c = 3 picofarads. At a signal frequency of 8 megahertz, the current gain for a common emitter connection is 15. Calculate the beta cutoff frequency (f_β), the alpha cutoff frequency (f_α), the common emitter short circuit bandwidth (f_T), the input capacitance, the incremental diffusion resistance ($r_b'_e$), and the base spreading resistance ($r_{bb'}$). Assume V_{CC} = 30 V. (R5-21)

A 2N1711 transistor is to be used. The characteristics of this transistor are listed in Table 4-3. The output capacitance is specified as C_{ob} = 25 picofarads and the input capacitance is given as C_{ib} = 80 picofarads. The transistor is to be used in the circuit of figure 5-11. What is the maximum frequency at which this amplifier can be used, if the 3-dB down level of the current gain is to serve as the criteria? (R5-22)

The input impedance, as seen when looking into the base-to-emitter terminals, is the parallel equivalent of h_{ie} and $1/j\omega C_i$. It can be written as

$$Z_i = \frac{h_{ie}}{j2\pi f h_{ie} C_i + 1} \qquad \text{Eq. 5.42}$$

Substitute Equation 5.38 into Equation 5.42 to obtain Z_i in the following form:

$$Z_i = \frac{h_{ie}}{\dfrac{jfh_{fe}}{f_T} + 1} \qquad \text{Eq. 5.43}$$

When the amplifier is operated within its intermediate frequency range, Z_i reduces to h_{ie}. This is to be expected, according to the discussion in Chapter 4.

Sample Calculation Using Simplified Equations

As an example, consider the 2N1808 transistor whose characteristics are listed in Table 4-3. All that is known about its high-frequency capabilities is that the value of f_T is 10 megahertz. The transistor is to be used in an amplifier such as the one shown in figure 2-11 where R_c = 1340 ohms and Z_L = 1000 + j0 ohms. R_b is assumed to be too large for consideration. The problem is to calculate the values of A_I and Z_i for an input signal of 5 megahertz from a voltage source.

Using Equation 5.39:

$$A_I = \frac{h_{fe} R_c}{(j \dfrac{fh_{fe}}{f_T} + 1)(R_c + Z_L)}$$

$$= \frac{(86)(1340)}{\left[j \dfrac{(86)(5)}{(10)} + 1 \right] \left[1340 + 1000 + j0 \right]}$$

$$= 1.14 \text{ at a phase angle of about } -90°.$$

Using Equation 5.43:

$$Z_i = \frac{h_{ie}}{\dfrac{jfh_{fe}}{f_T} + 1} = \frac{450}{\dfrac{j(86)(5)}{10} + 1}$$

$$= 10.5 \text{ ohms (almost completely reactive)}$$

What is the input impedance for the transistor of review problem (R5-22) when the transistor is operated at the maximum operating frequency determined in (R5-22)?   *(R5-23)*

Analysis for the Case of Small r_{bb}'

The voltage gain at high frequencies is not affected by the input capacitance when the base spreading resistance (r_{bb}') is small in comparison to the incremental diffusion resistance ($r_b'_e$). For this situation, the input impedances drop out of the voltage gain equation. A high-frequency dropoff occurs, but at a higher frequency than the current gain. The dropoff is due to the equivalent load impedance which is expressed as:

$$Z_{eqL} = \frac{\dfrac{R_c Z_L}{Z_L + R_c}}{\dfrac{j\omega C_o R_c Z_L}{R_c + Z_L} + 1}$$ Eq. 5.44

The voltage gain is given by Equation 4.19.

$$A_V = \frac{-h_{fe} R_c Z_L}{(R_c + Z_L) h_{ie}}$$

The term $R_c Z_L / (R_c + Z_L)$ was the equivalent load impedance before C_0 was added. This term is now equivalent to Z_{eqL}. The resulting voltage gain for the high-frequency condition is defined as:

$$A_V = \frac{-h_{fe} Z_{eqL}}{h_{ie}} \qquad \text{Eq. 5.45}$$

Substituting Equation 5.44 into Equation 5.45 yields:

$$A_V = \left[\frac{-h_{fe} R_c Z_L}{h_{ie}(R_c + Z_L)} \right] \left(\frac{1}{\dfrac{j\omega C_0 R_c Z_L}{R_c + Z_L} + 1} \right) \qquad \text{Eq. 5.46}$$

The voltage gain at intermediate frequencies usually is large compared to one. When this is true, $C_0 \approx C_c$. C_c is the value normally given in transistor specifications.

Sample Calculation Using the Above Equations

The following analysis considers the amplifier design discussed in the previous paragraphs. This amplifier has the circuit shown in figure 2-11. The transistor used in the circuit has an output capacitance of 20 picofarads. The values given are: $h_{ie} = 450$ ohms and $h_{fe} = 86$. The amplifier is fed by a current source. R_c is 1300 ohms and the load impedance is 1000 ohms resistive. The problem is to determine the maximum frequency at which the amplifier is to be used, if a voltage gain reduction of 3 decibels is the criteria. In other words, the value of A_V at the maximum frequency is reduced by a facter of 0.707. A negligibly small value of r_{bb}' is assumed. The frequency dependent portion of A_V (Equation 5.46) is:

$$\frac{1}{\dfrac{j\omega C_0 R_c Z_L}{R_c + Z_L} + 1} = 0.707$$

This condition occurs when

$$\frac{2\pi f C_0 R_c Z_L}{R_c + Z_L} = 1$$

(Note: $\omega = 2\pi f$.) Solve this expression for frequency,

$$f = \frac{R_c + Z_L}{2\pi C_o R_c Z_L}$$

$$f = \frac{1300 + 1000}{(6.28)(20)(10)^{-12}(1340)(1000)} = 13.6 \text{ megahertz}$$

The transistor considered in the previous analysis is to be used in the circuit shown in figure 2-17. (This circuit was discussed in Chapter 2, beginning on page 44.) A capacitor is normally used to bypass R_e so that $R_e \approx 0$ for high-frequency signals. Let $R_c = 890$ ohms and $Z_L = R_L = 2000$ ohms. The base spreading resistance is a negligibly small number. What is the maximum frequency at which this circuit can be used, if a voltage gain reduction of 3 decibels is to be used as a criteria? (R5-24)

Explain why the input impedance for an amplifier, such as the one shown in figure 2-11, does not appear in the voltage gain equation (Equation 5.46) when the base spreading resistance (r_{bb}') is small compared to the incremental diffusion resistance $(r_b'e)$. (R5-25)

Analysis for the Case of Large r_{bb}'

Now consider the case where r_{bb}' is large when compared to $r_b'e$ and cannot be neglected. The transistor parameter $r_b'e$ is sensitive to the location of its operating point. (See Equation 4.61.) The value of $r_b'e$ is much smaller for a transistor operating near the saturation region than it is for a transistor operating point near the cutoff region. The input impedance is r_{bb}' in series with the quantity defined by Equation 5.43. The parameter h_{ie} is replaced with the value in parallel with C_i, so that the input impedance becomes:

$$Z_i = r_{bb}' + \frac{h_{fe}r_b'e}{\dfrac{jfh_{fe}}{f_T} + 1}$$ Eq. 5.47

This expression for Z_i is used in the equation for the voltage input,

$$v_{in} = i_{in}Z_i$$ Eq. 5.48

The output voltage is given by

$$v_{out} = i_L Z_L \qquad \text{Eq. 5.49}$$

Substitute Equations 5.48 and 5.49 in the expression for the voltage gain.

$$A_V = \frac{v_{out}}{v_{in}} = \frac{i_L}{i_{in}} \cdot \frac{Z_L}{Z_i} = \frac{A_I Z_L}{Z_i} \qquad \text{Eq. 5.50}$$

Since the value of $r_{bb'}$ has no effect on the current gain, the equation for A_I is the same as Equation 5.39. Equations 5.39 and 5.47 are substituted in Equation 5.50. When the terms are rearranged, the resulting expression is:

$$A_V = \frac{h_{fe} Z_L R_c}{(r_{bb'} + h_{fe} r_{b'e})(R_c + Z_L)\left[\dfrac{jh_{fe} fr_{bb'}}{(f_T)(r_{bb'} + h_{fe} r_{b'e})} + 1\right]}$$

Eq. 5.51

This analysis assumes that the output capacitance is so far removed from the input capacitance that only C_i must be considered when $r_{bb'}$ is included. (Although C_i does not appear directly in the equation, it is present by virtue of the defining equation for f_T, (Equation 5.38).

$$f_T = \frac{h_{fe}}{2\pi h_{ie} C_i}$$

An equation can be written for the voltage gain that includes the effect of both C_i and C_o. Such an expression, however, is unnecessarily complicated for most applications. Both C_o and C_i must be included if a complete graph of the frequency response is to be made in the high-frequency region. The only frequency level of importance is the lowest frequency at which dropoff occurs. Above this frequency, the amplifier is no longer useful.

In equation 5.51, the term "$r_{bb'} + h_{fe} r_{b'e}$" is merely h_{ie}, as expressed by Equation 5.29. A value for h_{ie} may have been determined previously at low frequencies. If this value is reasonably accurate, then it may be used in Equation 5.51 as follows:

$$A_V = \frac{h_{fe} R_c Z_L}{h_{ie}(R_c + Z_L)\left(1 + \dfrac{jh_{fe} r_{bb'} f}{h_{ie} f_T}\right)} \qquad \text{Eq. 5.52}$$

This equation reduces to Equation 4.19 for low frequencies.

Example of the Use of Equations

Consider an amplifier circuit such as that shown in figure 2-11 containing a 2N404 transistor. The characteristics of this transistor are listed in Table 4-3. The following values are used: R_c = 1340 ohms, Z_L = 1000 + j0 ohms, and I_{EQ} = 9.6 milliamperes. The value of f_T for this transistor is 10 megahertz. The voltage gain is to be determined for a signal frequency of 5 megahertz.

To calculate A_V, it is necessary to know the value of the base spreading resistance ($r_{bb'}$). This value cannot be measured easily. It is recommended that the manufacturer's specifications be consulted. The value of $r_{bb'}$ from Table 4-3 is 100 ohms. The value of $r_{b'e}$ is computed from Equation 4.63.

$$r_{b'e} = \frac{0.026}{I_E}$$

Using the value of I_E at the Q-point,

$$r_{b'e} = \frac{0.026}{0.0096} = 2.7 \text{ ohms}$$

The values are substituted in Equation 5.51. The results are:

$$A_V = \frac{(86)(1340)(1000)}{[100 + (86)(2.7)](1340 + 1000)\left\{\dfrac{j(100)(86)(5)}{[100 + (86)(2.7)](10)} + 1\right\}}$$

$$= \frac{148}{j17.75 + 1}$$

$$= 8.3 \ \underline{/-90°}$$

The output capacitance of a 2N1711 transistor is 5 picofarads and the input capacitance is 20 picofarads. Table 4-3 lists the appropriate h-parameters for this transistor. The value of $r_{bb'}$ is assumed to be 85 ohms. This transistor is to be used in an amplifier circuit such as the one shown in figure 2-17 except that R_e is bypassed by a large capacitance.

a. Draw the equivalent circuit of this amplifier and make approximations appropriate to the circuit. Assume that R1 and R2 are large enough to be neglected in the equivalent circuit.

b. Assume an input signal of 1 megahertz from a voltage source and an equivalent ac load impedance (RC in parallel with ZL) of 1000 + j0 ohms. Using the equivalent circuit, compute the voltage gain, input impedance, and current gain.   *(R5-26)*

Rework problem (R5-26) for a signal frequency of 1 kilohertz.   *(R5-27)*

Example of a High-Frequency Analysis

Transistor manufacturers frequently list the high-frequency capabilities of their devices by specifying the output capacitance (C_{ob}), f_{hfb}, and h_{fe}.

The "collector-to-base feedback capacitance" or "output capacitance," identified by "$C_b'_c$" or "C_{ob}," respectively, is called by that name because it is the capacitance obtained by operating the transistor in a common base circuit with an open collector circuit. A value for the input capacitance may not be given and may have to be determined.

The following problem shows how the input capacitance is determined. A transistor manufacturer specifies a transistor as follows:

1. Collector-to-base feedback capacitance $(C_b'_c)$ = 9 picofarads.
2. Small signal forward current transfer ratio cutoff frequency (f_{hfb}) = 10 megahertz.
3. h_{fe} = 160 at a quiescent collector current of 1 milliampere.

These values are all the information that is available to apply this transistor as a high-frequency current amplifier up to 1 megahertz. The current gain of this transistor is to be determined at 1 megahertz when used in the common emitter configuration.

The first step is to determine the short circuit current gain bandwidth, f_T. Using Equations 5.17 and 5.16,

$$f_T = h_{fe}f_\beta$$

$$= h_{fe}f_a(1 + h_{fb})   \qquad \text{Eq. 5.53}$$

According to Table 4-3, h_{fb} is very close to -1. Use Equation 4.51 to obtain the value of h_{fb}.

$$h_{fb} = \frac{-h_{fe}}{1 + h_{fe}} = \frac{-160}{161} = -0.9938$$

Substitute the values given in the problem to obtain the bandwidth.

$$f_T = (160)(10)(10)^6(1 - 0.9938) = 9.92 \text{ megahertz}$$

Since the circuit is to be used as a current amplifier, Z_L is assumed to be small when compared to R_c. The current gain can then be determined using Equation 5.41 (the simplified version of Equation 5.40).

$$A_I = \frac{160}{\dfrac{j(1)(160)}{9.92} + 1}$$

$$= 9.94 \text{ at a phase angle of approximately } -90°$$

There is considerable attenuation of the current gain at a frequency of 1 megahertz.

Several additional questions can be asked about the performance of this transistor. For example, what is the maximum frequency at which the amplifier can be operated and still have a flat frequency response? The current gain drops by 3 decibels when $fh_{fe}/f_T = 1$. This expression can be solved for the frequency, f.

$$f = \frac{f_T}{h_{fe}} = \frac{9.92}{160} \text{ megahertz}$$

$$= 62{,}000 \text{ hertz}$$

What are the relative values of the input and output capacitances? The output capacitance is given as 9 picofarads. The input capacitance is determined by using Equation 5.38,

$$f_T = \frac{1}{2\pi h_{ie} C_i}$$

The following approximate relationship (Equation 4.64) is used to determine h_{ie}.

$$h_{ie} \approx h_{fe} r_{b'e} \approx \frac{0.026 h_{fe}}{I_c}$$

A quiescent current of 1 mA is assumed.

$$h_{ie} = \frac{(160)(0.026)}{0.010} = 416 \text{ ohms}$$

Therefore, C_i is determined as follows:

$$C_i = \frac{1}{2\pi h_{ie} f_T}$$

$$= \frac{1}{(6.28)(416)(9.92)(10)^6}$$

$$= 3.56(10)^{-11} = 38.5 \text{ picofarads}$$

This calculation verifies that $C_o \ll C_i$ for this example. Therefore, C_o can be ignored in the calculations to determine the maximum frequency at which the amplifier can be operated.

The following values are listed for a transistor operating at room temperature: $h_{fe} = 100$, $f_T = 200$ *megahertz*, $I_{CQ} = 2$ *milliamperes, and* $r_{bb}' \approx 0$. *Compute* f_β *and* C_i. (R5-28)

HIGH-FREQUENCY ANALYSIS OF COMMON COLLECTOR AND COMMON BASE CIRCUITS

Common Collector Circuit

The previous high-frequency analysis of a transistor amplifier was based entirely on common emitter circuits. However, the analysis can be extended to the common collector (CC) circuit since the only change in the equivalent circuit is that there is a different output terminal designation. That is, the voltage output, instead of being from the collector to the emitter is now from the emitter to the collector. To verify that the equivalent circuit is almost the same, refer to figure 4-10 and the discussion on pages 110 to 116. A high-frequency analysis of the common collector circuit can therefore be made using the same high-frequency parameters that are given for the common emitter circuit.

Common Base Circuit

A high-frequency analysis of the common base circuit is complicated by the fact that the base spreading resistance (r_{bb}') cannot be neglected. An equivalent circuit for the high-frequency CB circuit is shown in figure 5-8, page 182. This circuit is the basic circuit shown in figure 4-15, modified to include the effect of r_{bb}' and the capacitances C_c and C_e.

The use of h-parameters exclusively to analyze this circuit is not possible because of the presence of r_{bb}'. The value of $r_b'_e$ as given by equation 4.63 is frequently too low to assume that all of the voltage drop occurs across $r_b'_e$. In addition, the presence of C_e further complicates the situation. Experience shows that h_{rb} is a very small number and can be neglected. This allows the approximate equivalent circuit of figure 5-9, page 182, to be used.

Expressions for the high-frequency parameters of the common base circuit can be written, based on an analysis of the circuit shown in figure 5-9. For example, the current gain as given in Equation 4.11, $A_I = i_L/i_{in}$, can be determined as a function of transistor and circuit parameters. To determine i_L, a Norton equivalent circuit can be developed with the output terminals being C and B. The quantity i_L is expressed in terms of i_{in} and the circuit impedances and then is inserted into Equation 4.11.

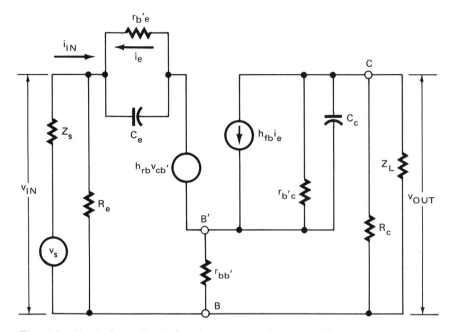

Fig. 5-8 Equivalent circuit for the common base amplifier operated at high frequency

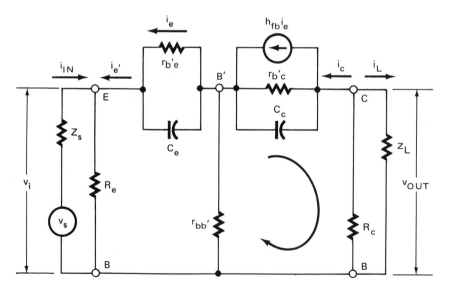

Fig. 5-9 Approximate equivalent circuit for common base amplifier operated at high frequency

There is a simpler method of determining the current gain if all that is desired is the frequency at which the amplifier gain begins to drop a significant amount. Consider the circuit in figure 5-9. The quantities $r_b{}'_c$ and the reactance of C_c can be neglected by assuming that they are very large at the signal frequency as compared to the input impedances ($r_b{}'_e$ and C_e). The load current (i_L) from the current divider theorem is:

$$i_L = \frac{-i_c R_c}{R_c + Z_L}$$

Eq. 5.54

This equation can be expressed in terms of the current source $h_{fb}i_e$ as follows:

$$i_L = \frac{-h_{fb}i_e R_c}{R_c + Z_L}$$

Eq. 5.55

The current (i_e) is shunted by the low reactance of C_e, rather than going entirely through $r_b{}'_e$. The actual current through $r_b{}'_e$ is:

$$i_e = i_e{}'\left(\frac{\frac{1}{j\omega C_e}}{r_b{}'_e + \frac{1}{j\omega C_e}}\right)$$

$$= i_e{}'\left(\frac{1}{j\omega C_e r_b{}'_e + 1}\right)$$

Eq. 5.56

The current i_e alone is effective in producing amplification through the transistor. According to the current divider theorem, the following expression is obtained for i_{in}.

$$i_{in} = {}^-i_e{}'\left[\frac{R_e + \dfrac{r_b{}'_e\left(\dfrac{1}{j\omega C_e}\right)}{r_b{}'_e + \dfrac{1}{j\omega C_e}}}{R_e}\right]$$

Eq. 5.57

By substituting Equation 5.56 into Equation 5.57 and simplifying, i_{in} can be expressed in terms of i_e.

$$i_{in} = {}^-i_e\left[j\omega C_e r_b{}'_e + \left(1 + \frac{r_b{}'_e}{R_e}\right)\right]$$

$$= {}^-i_e\left[j2\pi f C_e r_b{}'_e + \left(1 + \frac{r_b{}'_e}{R_e}\right)\right]$$

Eq. 5.58

Now substitute Equations 5.55 and 5.58 into Equation 4.11 and rearrange the terms to arrive at an expression for A_I:

$$A_I = \frac{h_{fb}R_c}{\left(R_c + Z_L\right)\left(1 + \dfrac{r_{b'e}}{R_e}\right)\left[j2\pi fC_e\left(\dfrac{r_{b'e}}{1 + \dfrac{r_{b'e}}{R_e}}\right) + 1\right]}$$

Eq. 5.59

The alpha cutoff frequency is defined as follows:

$$f_a = \frac{1}{2\pi C_e}\left(\frac{1 + \dfrac{r_{b'e}}{R_e}}{r_{b'e}}\right)$$

Eq. 5.60

Solve for the term in parentheses to obtain:

$$\left(\frac{1 + \dfrac{r_{b'e}}{R_e}}{r_{b'e}}\right) = 2\pi f_a C_e$$

Eq. 5.61

Substitute Equation 5.61 into Equation 5.59 and assume that $r_{b'e} \ll R_e$ to obtain the following:

$$A_I = \frac{h_{fb}R_c}{(R_c + Z_L)(j\dfrac{f}{f_a} + 1)}$$

Eq. 5.62

If the circuit being considered is taken to be an ideal current amplifier, with $Z_L \approx 0$, then A_I reduces to:

$$A_I = \frac{h_{fb}}{\dfrac{jf}{f_a} + 1}$$

Eq. 5.63

Note the similarity of Equation 5.63 to Equation 5.41 for the common emitter circuit. Actually, the exact definition of f_a in the literature includes h_{fb}. If the exact definition is used, Equation 5.63 becomes:

$$A_I = \frac{h_{fb}}{\dfrac{jfh_{fb}}{f_a} + 1}$$

Eq. 5.64

This expression is even more like Equation 5.41 for the CE circuit. However, since $h_{fb} \approx 1$ in this case, it is neglected in the present analysis.

Example of a Common Base Circuit Analysis

Consider the following problem involving a common base amplifier stage, such as the circuit shown in figure 3-9. The ac equivalent circuit of this amplifier is shown in figure 5-8. The collector bias resistor is 1.9 kilohms and Z_L = 2 kilohms. Figure 5-9 represents the approximate circuit for this problem. The transistor is operated at a quiescent current of 10 milliamperes and has an h_{fe} of 100. The input capacitance for the transistor (C_{ib}) is 20 picofarads. What is the current gain of the circuit at 10 megahertz?

The input capacitance C_{ib} can be taken to mean the same as C_e. Equation 4.63 is used to calculate $r_{b'e}$:

$$r_{b'e} = \frac{0.026}{I_E}$$

I_E is approximately equal to I_C. Thus, at the quiescent point, $r_{b'e}$ becomes:

$$r_{b'e} = \frac{0.026}{0.010} = 2.6 \text{ ohms}$$

The first step in this amplifier problem is to solve for f_a. Equation 5.60 can be used. However, since $r_{b'e} \ll R_e$, the term in parentheses reduces to $1/r_{b'e}$. Therefore,

$$f_a \approx \frac{1}{2 \pi C_e r_{b'e}} \qquad \text{Eq. 5.65}$$

Substituting the known values into this equation yields a value for f_a:

$$f_a = \frac{1}{(6.28)(20)(10)^{-12}(2.6)} = 3062 \text{ megahertz}$$

The equations of Table 4-4 can be used to calculate h_{fb}. The parameter h_{fb} is related to h_{fe} by Equation 4.56:

$$h_{fb} = \frac{-h_{fe}}{1 + h_{fe}}$$

Using the assumed value of h_{fe},

$$h_{fb} = \frac{-100}{1 + 100} = -0.99$$

The current gain at 10 megahertz is calculated using Equation 5.63.

$$A_I = \frac{h_{fb}}{\dfrac{jf}{f_a} + 1} = \frac{-0.99}{\dfrac{j10}{3062} + 1}$$

$$= -0.99 \text{ at a phase angle of approximately 180 degrees}$$

The following conditions have been shown to be true for the common base amplifier: (1) f_a has a relatively high magnitude and (2) the frequency has a very minor effect on the current gain at this level of frequency. Because of these factors, the common base amplifier can be used at a higher frequency than either the common emitter circuit or the common collector circuit. However, since the current gain of a common base amplifier is approximately unity, there is little advantage in using this circuit at lower frequencies. The circuit is of greater interest at radio frequencies.

Recalculate the current gain of the previous example for a signal frequency of 100 megahertz.   *(R5-29)*

For the previous example, determine the frequency at which the response as a current amplifier is no longer flat.   *(R5-30)*

A single stage amplifier is to be operated at a frequency of 5 megahertz. The transistor in this amplifier has an output capacitance (C_{ob}) of 12 picofarads and an alpha cutoff frequency of 13 megahertz. Can it be operated at this frequency as a common base amplifier?   *(R5-31)*

A transistor has a base spreading resistance of 100 ohms. It is to be operated as a common base amplifier at a quiescent collector current of 10 milliamperes. Can the base spreading resistance be neglected in the equivalent circuit of this amplifier?   *(R5-32)*

An Approximate Relationship for Multistage Amplifiers

If the various stages of a multistage amplifier are almost identical, the highest frequency at which the amplifier can reasonably be operated (f_{MH}) is expressed by the following approximate relationship.

$$f_{MH} = \frac{f_{CH}}{1.1 \sqrt{N}} \qquad \text{Eq. 5.66}$$

where f_{CH} is the maximum frequency at which a single stage can operate, and N = the number of stages.

The amplifier analyzed in the previous problem is one of two identical stages of a two-stage amplifier. Using the results of review problem (R5-30), compute the maximum operating frequency of the multistage unit.   *(R5-33)*

LABORATORY EXPERIMENTS

The following laboratory experiments have been tested and found useful for illustrating the theory in this section.

LABORATORY EXPERIMENT 5-1

OBJECTIVE

To measure the frequency response of a transistor amplifier and determine its bandwidth as a voltage amplifier.

EQUIPMENT AND MATERIALS REQUIRED

1	Transistor, 2N404
3	Capacitors, 10 microfarads
2	Resistance substitution boxes, Heathkit EU-28A or equivalent
1	Signal generator, capable of low-level, adjustable amplitude sinewave output between 10 hertz and 1 megahertz
1	Oscilloscope, dual channel, with sensitivity to 5 mV/cm on the display graticule, with flat frequency response to 10 megahertz or higher
1	Resistor, 2.2 kilohms
1	Resistor, 100 ohms
1	Resistor, 100 kilohms

PROCEDURE

1. Construct a single-stage amplifier using the circuit of figure 5-10, page 188. This circuit uses emitter self-bias for stabilization. Adjust resistor R1 to obtain the best amplifier performance. Leave R_L at infinity.

2. Set the input signal low enough to achieve an undistorted output signal. Use a sinusoidal waveform as the input signal.

3. Starting at the lowest frequency available on the signal generator, vary the frequency to the highest value available on the signal generator. With the load resistance at infinity, record the data necessary to obtain a graph of voltage gain versus frequency. Observe the following rules:
   a. For each frequency level, readjust the amplitude of the input signal to maintain the input signal at a constant amplitude.
   b. Since there is a wide range of frequencies to be covered, take data only where there seems to be a variation in the voltage gain at certain regions of frequency.
   c. Using the oscilloscope, measure the peak-to-peak amplitude of the output and the input voltage signal. Note on the data sheet those places at which it was necessary to change the scales on the scope or readjust the input signal amplitude. These points may account for any discontinuities or irregularities in the plotted graph.

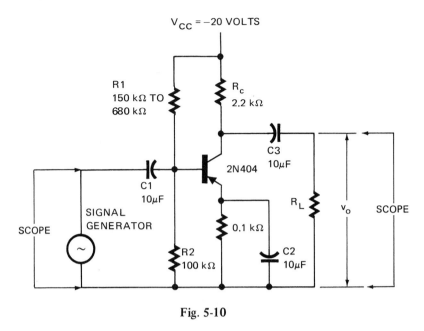

Fig. 5-10

4. Calculate the gain at each frequency in voltage decibels using Equation A.32 of the Appendix.

5. Plot the frequency response on semilogarithmic paper. This is a plot of decibels versus frequency.

6. Mark the places on the frequency response curve where the gain is 3 decibels below the gain of the middle frequency region. These points mark the limits of use of this amplifier with regard to the frequency variation. The range within these limits is known as the bandwidth.

7. Adjust R_L to 2000 ohms and repeat steps 2 through 6.

LABORATORY EXPERIMENT 5-2

OBJECTIVE

To measure the frequency response of the current gain for a transistor amplifier and determine the beta cutoff frequency for this amplifier.

EQUIPMENT AND MATERIALS REQUIRED

1   Transistor, 2N1808
3   Capacitors, 10 microfarads
1   Resistance substitution box, Heathkit EU-28A or equivalent
1   Signal generator, capable of low-level, adjustable amplitude, sine wave output between 10 hertz and 1 megahertz

1 Oscilloscope, dual channel, with sensitivity to 5 mV/cm on the display graticule, with flat frequency response to 10 megahertz or higher
1 Resistor, 2.2 kilohms
1 Resistor, 100 ohms
1 Resistor, 100 kilohms
1 Resistor, 10 ohms, ± 1%
1 Resistor, 10,000 ohms, ± 1%

PROCEDURE

1. Construct a single-stage amplifier using the circuit of figure 5-11. This circuit uses emitter self-bias for stabilization. Adjust R1 to obtain the best amplifier performance.

2. Set the input signal low enough to achieve an undistorted output signal. Use a sinusoidal waveform as the input signal.

3. Record data to make a graph of current gain versus frequency. To do this, measure v_s, v_i, and v_o at various levels of frequency using the high-sensitivity oscilloscope. As the frequency is varied, v_s and v_i are held constant by adjusting the signal generator output. Voltage v_o changes with frequency and is used to calculate the output current. Vary the frequency from the lowest frequency that the signal generator is capable of producing to the highest frequency that it can produce. Take data only where there seems to be a variation in the voltage output at certain frequency levels. There will be a wide range of frequencies over which there is no change in output.

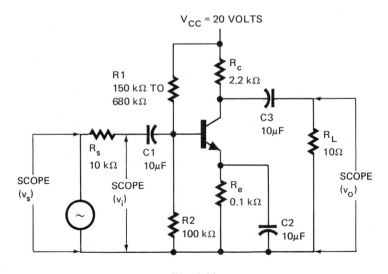

Fig. 5-11

4. Using the oscilloscope, measure the peak-to-peak amplitude of v_o, v_s, and v_i. Note on the data sheet those places where it was necessary to change oscilloscope scales or readjust the input signal amplitude.

5. Make a rough plot of v_o versus frequency. If there are any drastic changes or other irregularities in the plotted output as the frequency changes, repeat the measurements at that frequency. If necessary, recalibrate the equipment and take measurements at closer intervals near those frequency levels at which the irregularities occur.

6. Calculate the current gain. The input current is:

$$i_s = \frac{v_s - v_i}{10,000} \qquad \text{Eq. 5.67}$$

The output current is:

$$i_o = \frac{v_o}{10} \qquad \text{Eq. 5.68}$$

7. Calculate the current gain in decibels at each frequency level using Equation A.33 in the Appendix.

8. Using semilogarithmic paper, plot the current gain versus frequency.

9. Determine and record the frequency above the intermediate range where the current gain drops by 3 decibels. (This is the approximate value of the beta cutoff frequency as defined on pages 159 to 162.)

10. Calculate f_T, f_α, and C_i. Assume that C_o is negligibly small and use the h-parameters for the 2N1808 transistor given in Table 4-3.

LABORATORY EXPERIMENT 5-3

OBJECTIVE

To verify Miller's theorem

EQUIPMENT AND MATERIALS REQUIRED

1   Signal generator, audio frequency, with variable amplitude and frequency
1   Oscilloscope, dual channel
1   Power supply, dc, 0 - 30 volts dc adjustable, 0.5 ampere maximum
2   Decade resistance boxes, Heathkit 1N-17 or equivalent
3   Capacitors, 10 microfarads, 25 volts dc
3   Resistance substitution boxes, Heathkit EU-28A or equivalent
1   Transistor, 2N1808

PROCEDURE

1. Construct the amplifier shown in figure 5-10. It may be necessary to adjust the bias resistors and the voltage inputs to achieve satisfactory operation of the amplifier.

2. Adjust the resistance values as follows:
    a. Set the load resistance at infinity.
    b. Set the decade box in the input circuit to 0 ohms.
    c. Set the resistor R at infinity.

3. Record the values set in steps 1 and 2 and measure and record the voltage gain. Use an input signal with a frequency of 1000 to 2000 hertz. Keep the signal reduced to an amplitude just high enough to obtain an accurate reading with the available equipment.

4. Measure the input impedance. Do this by adding resistance in series with the signal source, using the decade box in the input circuit. When the input voltage is reduced to one-half its value with the decade box set at zero, the input resistance is the value read on the decade box.

5. Measure the output impedance. Do this by adjusting the decade box resistance until the output voltage is one-half its value with the load set at infinity resistance.

6. Adjust R to some arbitrary value (such as 33 kilohms). This value is Z for the equations in step 9. The input signal may change. If the signal does change, reset it to the original level for steps 3 and 4.

7. With R in the circuit, measure the input impedance as in step 4.

8. With R in the circuit, measure the output impedance as in step 5.

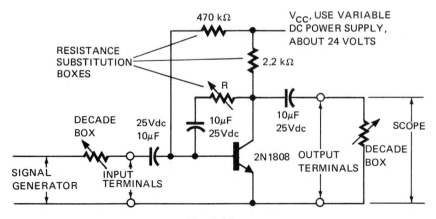

Fig. 5-12

9. Compare the input and output impedances with those obtained from the following equations.

$$Z_1 = \frac{Z}{1 - A_V}$$

$$Z_2 = \frac{ZA_V}{A_V - 1}$$

where   Z   =   R = impedance from the collector to the base
         Z_1 =   equivalent input impedance
         Z_2 =   equivalent output impedance
         A_V =   no load gain

Z_1 and Z_2 are resistances in parallel with the resistances determined in steps 3 and 4. The signal frequency used is such that capacitive reactances are negligible. Any discrepancies in the calculated and measured impedance values are due to transistor nonlinearities and inaccuracies of the dials on the resistance substitution boxes. Show all calculations and list all data.

EXTENDED STUDY TOPICS

1. Using the values given in the circuit shown in figure 2-24, plot the approximate low-frequency response of the current gain of the amplifier. Use $h_{fe} = 85$.

2. Using the high-frequency parameters of a 2N1808 transistor, plot the high-frequency response of the current gain of the amplifier shown in figure 2-24. Use the following high-frequency parameters for the 2N1808 transistor: $f_{hfb} = 8$ megahertz and $C_{ob} = 13$ picofarads.

3. Construct a frequency response graph using data from study topics 1 and 2. This graph is to encompass both the high- and low-frequency dropoff of the current gain. Over what range of frequencies can the response be considered flat?

4. For the high-frequency analysis of a transistor amplifier, under what conditions can the base spreading resistance (r_{bb}') be neglected or included?

5. Explain why the junction capacitances exist in a bipolar transistor.

6. How low a value must be achieved for the reactance of a coupling capacitor before it can be neglected in the computation of the frequency response of a transistor amplifier?

7. What are the ways in which the transistor manufacturer can express the values of C_e and C_c in the transistor specifications?

8. Examine some typical transistor specification sheets. Interpret the terms listed in ways that are familiar from reading this text.

9. Discuss how the high-frequency response characteristics of a transistor affect the phase shift of a common emitter amplifier from the input to the output terminals.

10. What are the h-parameters and the high-frequency parameters that must be supplied as a minimum by a transistor manufacturer to make an approximate assessment of the maximum frequency at which a transistor can be used?

6
Field Effect
Transistor as an
Amplifier

OBJECTIVES

After studying this chapter, the student will be able to:

- define the various types of field effect transistors
- specify a field effect transistor and other high-impedance semiconductor amplifying devices
- analyze a simple common source amplifier

THEORY OF THE JUNCTION FIELD EFFECT TRANSISTOR

The device known as the bipolar transistor uses two types of current carriers for its operation: holes and electrons. The field effect transistor is a *unipolar* device. Its operation depends upon the movement of only one type of current carrier. The carrier can be either holes or electrons, depending upon whether the device has an *n*-type channel or a *p*-type channel. There are many types of field effect transistors which differ in their method of construction and how they are biased for use as amplifying devices. The abbreviation FET is commonly used when referring to the family of field effect transistors. The junction field effect transistor (abbreviated JFET) will be considered first.

Construction of the FET

It will be recalled that, when discussing a reverse biased semiconductor, there was a region within the crystal, at the junction, where there was a lack of free current carriers (either holes or electrons). This was discussed in Chapter 2.

If two of these reverse biased junctions are connected, with the *n*-type materials together, as in figure 6-1(a), or with the *p*-type materials together, as in figure 6-1(b), an FET is formed. The construction is similar to that of a bipolar transistor with a large base section.

The construction of the device starts with a semiconductor crystal known as a *substrate*. This base is covered with an inert material, such as silicon dioxide. The inert material is removed (etched) chemically at specific points. The crystal is then exposed to a gaseous atmosphere and the gas diffuses into the crystal at the places not covered by the inert material. The gas is composed of molecules of a material that supplies either *n*- or *p*-type impurities to the exposed portion of the substrate. For example, if the substrate is a *p*-type material, then the gas is composed of molecules of an *n*-type material. The material into which the gas is diffused forms the center section of the transistor. (For a bipolar transistor,

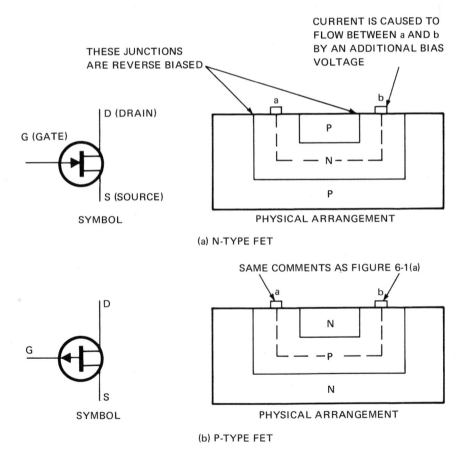

Fig. 6-1 FET configurations

this center section is the base.) The crystal is again covered with an inert materi-al. The surface protection is removed in another area to expose again a previous-ly exposed portion of the crystal. The gaseous atmosphere now introduced con-sists of material having the opposite type of impurity. (For example, if the sub-strate is a *p*-type material, then the first gas introduced must be *n*-type and the second gas introduced must be *p*-type.) Contacts are then added to the device. The resulting transistor is shown in the physical arrangements of either figure 6-1(a) or 6-1(b).

Use figure 6-1(a) as the basis for the following explanation. The center *n*-type material is lightly doped and has a large cross section. The *p*-type materials on either side of the *n*-type material are thin and are heavily doped. (The term, "heavily doped," means that there are a large number of free carriers made avail-able by the addition of a large amount of the right type of impurities.) In normal use, the two *p-n* junctions are reverse biased by an externally applied voltage. If a voltage source is connected across the *n*-type section at the points a and b, then the voltage causes the current carriers to flow between a and b in the *n*-type material. A reverse biased *p-n* junction causes a region that is depleted of free current carriers. The extent of the depletion region depends on the reverse bias potential. The higher the reverse bias potential, the smaller is the area through which the current carriers can flow. Small changes of this reverse bias potential control the current in the center section and make the field effect transistor act like an amplifying device.

Several definitions can be made by referring to figure 6-1. The center sec-tion at terminal a in figure 6-1(a) is known as the *source*. The center section at terminal b is the *drain*. The connections to the outside sections (*p*-type material in this case) are usually common connections and are known as the *gate*. The defi-nitions of the gate, source, and drain are the same for both configurations, figures 6-1(a) and 6-1(b). The gate always represents the outside, reversed biased connections and the drain and source are always the sides of the center section. Figures 6-1(a) and (b) also show the standard symbols for both con-figurations of junction field effect transistors (*n*-type and *p*-type channels).

Bias Connections for the FET

The bias connections of the junction field effect transistor (JFET) are shown in figure 6-2 for a common source connection. Figure 6-2(a) represents the bias connections for the *n*-type FET and figure 6-2(b) represents the bias connections for the *p*-type FET. The voltage required to reverse bias the gate section is designated V_{GG}. V_{DD} is the voltage that sets the bias level of the drain current I_D. The resistor R_d is the drain bias resistor of the JFET. The gate-to-source voltage is V_{GS} and the drain-to-source voltage is V_{DS}. As in previous sections, the capital letters represent the average or dc values of these quantities. Note that the only difference between the bias circuits of the *n*-type and the *p*-type JFETs is the polarity of the applied voltages.

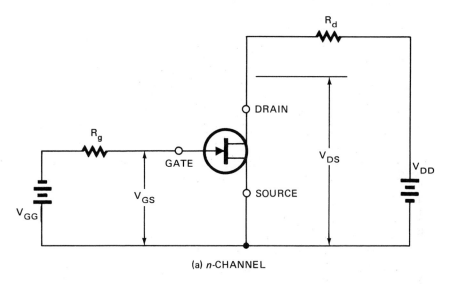

(a) *n*-CHANNEL

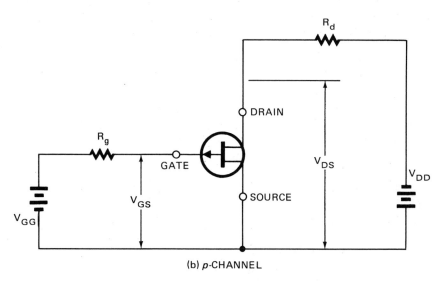

(b) *p*-CHANNEL

Fig. 6-2 FET bias connections

How does the biasing circuit for the n-channel JFET differ from that for the npn transistor? (R6-1)

Discuss the reasons for setting dc bias voltage levels for a JFET or a bipolar transistor. (R6-2)

Characteristic Curves for the JFET

The levels of the bias voltages and current are set by referring to the specifications and characteristic curves of the JFET. Manufacturers list the maximum voltages that can exist across the terminals, as well as maximum currents and power dissipation capabilities. These values must be carefully observed when using the device. Typical characteristic curves for the JFET are shown in figure 6-3. These curves are the common source characteristics for an n-channel junction field effect transistor. The saturation, active, and cutoff regions of this device are defined in the same way that they were defined in Chapter 1, pages 4 to 11. Note that as the gate voltage (V_{GS}) increases more negatively, the drain current is *pinched* off by the narrowing of the channel through which the free carriers pass. As V_{GS} becomes more negative, the gate-to-source junction becomes more reverse biased. An increase of the drain-to-source voltage V_{DS} causes an increase in the reverse biased condition between the p-n junctions. Refer to figure 6-2(a). An increase in V_{DD} (or a decrease of the series resistance R_d) causes the drain end of the channel to be more positive with respect to both the source end and the gate. This condition causes the drain current to level off as shown in figure 6-3. An increase in the voltage from drain to source would result in a current increase, but because of the increase in the reverse bias between the gate and the source, the channel closes and prevents any further increase in the current. Note in figure 6-3 that as the drain-to-source voltage V_{DS} increases to about 27 volts, the action is that of a reverse biased diode reaching an avalanche breakdown condition. With very little change in voltage, a large current can exist through the channel. This is due to the fact that a larger V_{DS} causes the drain terminal to become more heavily positive with respect to the gate; therefore, the gate-to-source junction is more heavily reverse biased than it was before. An increase in the reverse bias voltage causes an avalanche breakdown similar to that of an ordinary semiconductor diode. In an avalanche breakdown, thermally generated carriers acquire sufficient energy to remove additional carriers from their individual valence bonds. An FET should not be operated in this region since it will be permanently damaged.

Is it possible to operate the JFET represented by figure 6-3 at positive values of V_{GS}? Explain the answer given.   *(R6-3)*

Using the given definitions of the cutoff, saturation, and active regions (Chapter 1, pages 4 to 11), identify these regions on figure 6-3.   *(R6-4)*

Since the channel that exists between the drain and source terminals of the FET consists of just one type of semiconductor material, the drain and source terminals are interchangeable. That is, the drain may be used as the source and the source used as the drain in a practical circuit.

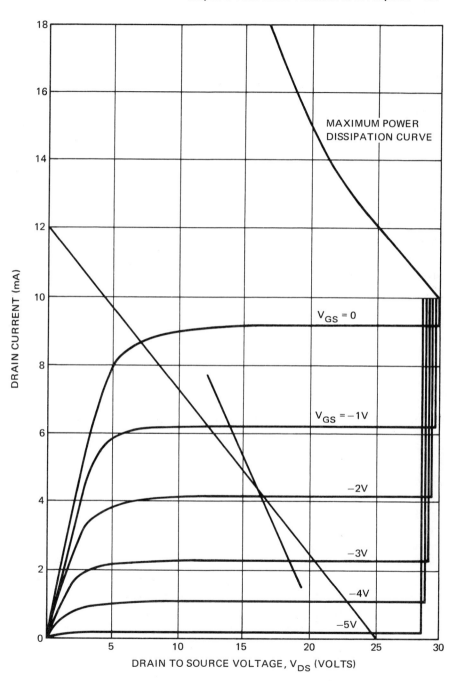

Fig. 6-3 Typical JFET common source characteristic curves

As in the case of the bipolar transistor, the FET can be used in three connection modes. The characteristic curves of the FET in figure 6-3 are drawn for the common source connection. These curves are known as common source curves. The performance of the JFET for the remaining connection arrangements will be examined later.

Refer to the generalized amplifying device connections shown in figure 1-14. What connection is the common source connection analogous to? Assume that the generalized amplifying device is a high-impedance type. Explain fully. *(R6-5)*

How Temperature Changes Affect FETs

The current in the channel of an FET results from the movement of major current carriers and is not due to minor or thermally generated current carriers (as in the case of the current through a reverse biased diode). The action of thermally generated current carriers is a minor factor. The major current carriers are due to the free charges that exist because the channel semiconductor material is doped with impurities. An increase in the temperature causes more thermally generated minority current carriers, but the flow of majority carrier current through the channel section is not affected, except as a result of the decreased mobility of the free charges at a higher temperature. In other words, the free charges move at higher velocities at higher temperatures, with an increase in the average number of collisions between the charges. Because of this increase in collisions, there is a higher resistance to the current flow. In general, then, it can be said that the channel current decreases as the temperature of the FET increases. This action is basically the same mechanism that occurs in any conducting material, including all metal conductors.

It was shown in an earlier chapter that the reverse biased semiconductor diode has a negative temperature coefficient of resistance because of the increased number of free charges that are generated when the temperature increases. (The phrase *negative temperature coefficient of resistance* means that the resistance decreases as the temperature increases.) It is possible that the negative temperature coefficient of the gate-to-source diode will predominate in the determination of the overall temperature coefficient of the device. This dominance is due either to a highly negative gate-to-source voltage or to the physical construction of the FET. In other words, the temperature coefficient can be made positive or negative, depending upon the specific FET and its gate bias voltage. Thus, the FET can have a controlled temperature coefficient of resistance. Although the channel has a positive coefficient, the polarity of the gate bias potential affects the overall coefficient of resistance for the FET. In fact, the bias conditions can be set so that the operation of the FET is relatively insensitive to temperature.

Based on a knowledge of fundamental electronics or by referring to the Appendix of this text, explain the mechanism that causes electrical conductors to increase their electrical resistivity as the temperature increases. How does this mechanism differ from the one for semiconductors? (R6-6)

It appears that a semiconductor diode can have a positive temperature coefficient if it is forward biased and a negative temperature coefficient if it is reverse biased. Justify or dispute this statement based on a knowledge of semiconductor theory. (R6-7)

THEORY OF THE MOSFET

The junction field effect transistor is used in many applications which take advantage of the high resistance it has between the gate and the source in normal operation. This high resistance is a desirable feature because when the FET is used as an amplifier, the input signal is usually applied between the gate and the source. This situation is analogous to the signal applied between the X and Y terminals of the generalized amplifying device described in Chapter 1 or to the signal applied between the base and emitter terminals of a transistor in a common emitter connection.

The high resistance between the gate and the source is due to the reverse bias condition of the FET between these terminals. However, since the material is still a semiconductor crystal, the movement of thermally generated carriers limits the level of input resistance than can be attained.

The MOSFET is an improvement over the JFET in this regard because it provides an insulating material, such as silicon dioxide, between the gate and the source. Thus, a further reduction is made in the number of thermally generated carriers that can affect the gate-to-source resistance. The term MOSFET is an abbreviation of *M*etallic *O*xide *S*emiconductor *F*ield *E*ffect *T*ransistor. This device is sometimes called an insulated gate FET because the gate is insulated from the channel by a thin layer of silicon dioxide.

The following question might be asked: "Doesn't the MOSFET satisfy the conditions for a charged capacitor, since a capacitor is merely a pair of conducting plates separated by a dielectric or insulating material?" The answer is yes. For rapidly changing signals, the behavior of the MOSFET is similar to that of a capacitor that is either charging or discharging. The gate is a conducting material. The channel is also a conducting (semiconducting) material, consisting of a source terminal on one side and a drain terminal on the other side.

The gate and the channel of the MOSFET are separated by a dielectric or insulating layer. In the JFET, the same insulating barrier exists, although the insulating layer is actually a semiconductor material from which all free charges are removed (with the exception of a few thermally generated free charges).

The use of the MOSFET makes it possible to obtain an extremely high input resistance since the input resistance is usually the resistance between the gate and the channel. For the JFET, this resistance can be several megohms, while for the MOSFET, the resistance may be as high as 10^8 megohms.

What are the most desirable levels for the internal capacitance between the terminals of a JFET or MOSFET for a high-frequency amplifier application? Does the internal resistance between the terminals have any effect on the limit of the capacitance values?   *(R6-8)*

Why does the input resistance of the MOSFET have a much higher value than the input resistance for the JFET? Why is the input resistance for the JFET higher than that of a bipolar transistor when normal biasing is used?   *(R6-9)*

The MOSFET is constructed in much the same way as the JFET. However, for the MOSFET, the inert material (silicon dioxide) is not etched the second time. The gate terminal is connected directly to the silicon dioxide layer. There are really only two active sections to the MOSFET crystal: the channel and the substrate. If the substrate is a *p*-type material, then the channel is an *n*-type material.

The Induced Channel MOSFET

The layout of an induced channel MOSFET is shown in figure 6-4. This device has an *n*-type channel and a *p*-type substrate.

If the MOSFET is biased as shown in figure 6-5, the voltage connected to the gate has a positive polarity with respect to the source. As a result, positive charges are placed at the gate side of the silicon dioxide layer and negative charges are collected in the *p*-type substrate on the channel side of the silicon dioxide layer. The *n*-type channel, already containing an excess of negative charges, has more negative charges added between the drain and the source. These additional negative charges are made available by thermal energy levels in the substrate and form an "induced" channel which allows current to flow between the drain and the source. The thickness of the induced channel is determined by the amount of positive voltage applied from the gate to the source. Recall that the terms source and drain are interchangeable. It can just as easily be said that a positive voltage is applied from the gate to the drain. For this

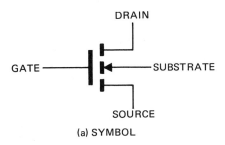

(a) SYMBOL

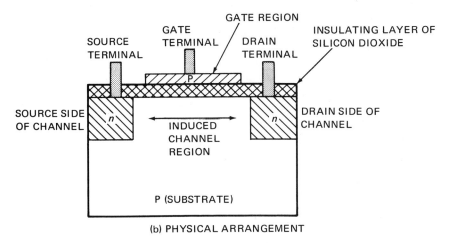

(b) PHYSICAL ARRANGEMENT

Fig. 6-4 Diagram of a metal oxide semiconductor or insulated gate field effect transistor

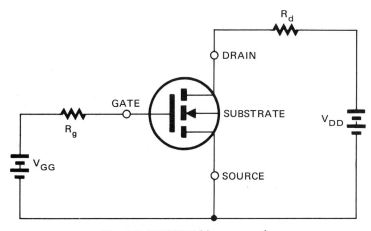

Fig. 6-5 MOSFET bias connections

discussion, however, the source is taken as common. The higher the gate-to-source bias voltage, the greater is the value of the drain current that can exist for a given value of V_{DS}.

The characteristic curves for the induced channel MOSFET, as shown in figure 6-6, are similar to those for the JFET. This type of MOSFET can be operated only at positive gate-to-source voltages. This condition arises from the fact that negative bias voltages only provide for two reverse biased *p-n* junctions between the drain and the source. As a result, conduction cannot take place through the channel. The device whose curves are shown in figure 6-6 has an *n*-type channel. This MOSFET can also be constructed with a *p*-type channel and an *n*-type gate. In this case, all voltage polarities and current directions are reversed; otherwise, the operation of the MOSFET is the same.

The preceding explanation of the induced channel MOSFET is based on an n-channel device. Explain the operation of a MOSFET, based on a p-channel device.   *(R6-10)*

Redraw figure 6-5 to show the correct biasing of a p-channel MOSFET.   *(R6-11)*

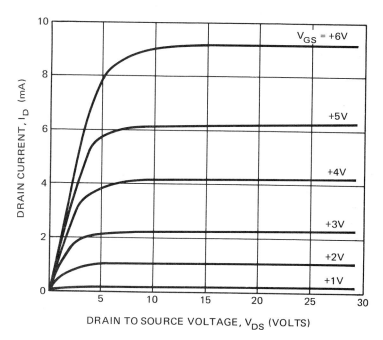

Fig. 6-6 Typical common source curves for a MOSFET with an induced *n*-channel

The MOSFET just described is known as an *enhancement* device. The name, "enhancement," emphasizes the fact that as the bias voltage is increased, the drain current is increased, or enhanced.

Depletion or Diffused Channel-Type MOSFET

It is also possible to construct a MOSFET with a lightly doped *n*-type section in the channel between the heavily doped *n*-type source and drain regions. This is shown in figure 6-7. This device is known as a *depletion*-type MOSFET, or one with a *diffused* channel. This type of MOSFET is made by using the same basic techniques outlined previously. The depletion-type MOSFET can be biased with a negative gate voltage using the same bias polarity connections shown in figure 6-2(a). Because of the negative gate voltage, negative charges accumulate on the gate side of the silicon dioxide layer. At the same time, positive charges build up on the channel side of the insulating layer of silicon dioxide. The greater the negative gate bias voltage applied, the lower is

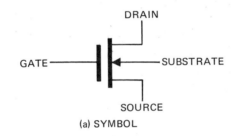

(a) SYMBOL

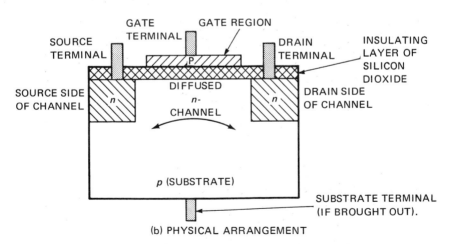

(b) PHYSICAL ARRANGEMENT

Fig. 6-7 Depletion-type MOSFET with a diffused channel consisting of a lightly doped *n*-type region

the drain current. In this case, the drain current is reduced by the positive charges. These positive charges are the majority carriers in the p-type substrate and the minority carriers in the n-type diffused channel between the drain and source regions. The n-type material has an excess of negative charges. However, the positive charges built up by the negative gate bias recombine with the negative charges and thus *deplete* the negative charges available to supply an external circuit.

This transistor can also be operated in the enhancement mode. By applying a positive gate-to-source voltage, more negative charges are attracted to the lightly doped region between the drain and source. The channel is made more conductive than before. The characteristic curves of this MOSFET are shown in figure 6-8. Note that the zero level of the gate bias voltage may occur in or near the center of the active operating region. This means that the gate bias voltage can be eliminated for some linear amplifier circuits using this type of MOSFET. Although the curves in figure 6-8 were developed for an n-channel device, p-channel devices with n-type gates are also available. These devices are operated with opposite voltage polarity and current direction.

Using the curves of figure 6-8, arbitrarily locate a Q-point in the active region such that undistorted linear amplifier performance is obtained for moderately large input signals. What is the advantage of locating the Q-point at $V_{GS} = 0$, if the input signal has a zero dc level?   *(R6-12)*

Discuss the difference between a depletion-type MOSFET and an enhancement-type MOSFET, based on the appearance of the common source characteristic curves. Indicate on figure 6-8 the regions of operation where there is enhancement and depletion.   *(R6-13)*

It should be apparent that a MOSFET can be made to operate in the enhancement mode, in the depletion mode, or in a combination of these modes as a result of its construction, or the applied bias voltage, or both. The options for the use of a MOSFET in a circuit are numerous. It will be shown in the next chapter that there is only one mode for vacuum tubes. (Actually, some tubes constructed for specialized purposes can be built for positive as well as negative input bias voltages.) The vacuum tube still has its uses, but it has been replaced by JFETs and MOSFETs for many applications.

Because the gate insulating material is very thin (with a thickness of as little as 10^{-5} centimeter), the MOSFET can be destroyed easily by transient voltages applied between the gate and the source or the substrate. Dielectric breakdown of this material and possible destruction can result from the presence of static electricity in the vicinity of the MOSFET. This situation can be prevented by the circuit design and careful transistor selection.

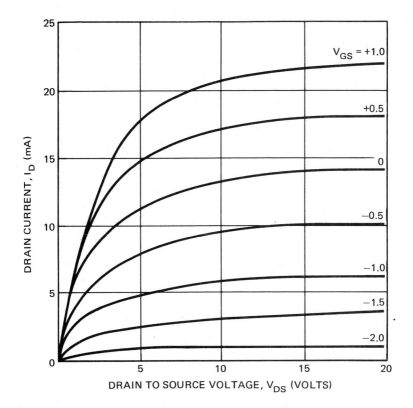

Fig. 6-8   Common source characteristics for an *n*-channel, depletion-type MOSFET

SPECIALIZED FIELD EFFECT TRANSISTOR DEVICES

The development of integrated circuit technology meant that the capability to manufacture JFETs and MOSFETs was now available. Current technology also provides the means by which variations on standard FETs are made for special applications.

As semiconductor technology continued to develop, it became possible to make MOSFET-type devices that were immune to failure by normal handling with static electricity or by the manner in which they are removed from or placed in a circuit. One type of device uses back-to-back Zener diodes connected to the MOSFET using integrated circuit techniques. The circuit diagram and symbol for this device are shown in figure 6-9, page 208. A typical volt-ampere characteristic for Zener diodes is shown in figure 6-10, page 208. The Zener region in figure 6-10 occurs as a result of the strong electric field that is applied in the reverse direction. Because of this strong field, valence electrons

Fig. 6-9   MOSFET containing high-voltage protection for gate insulation

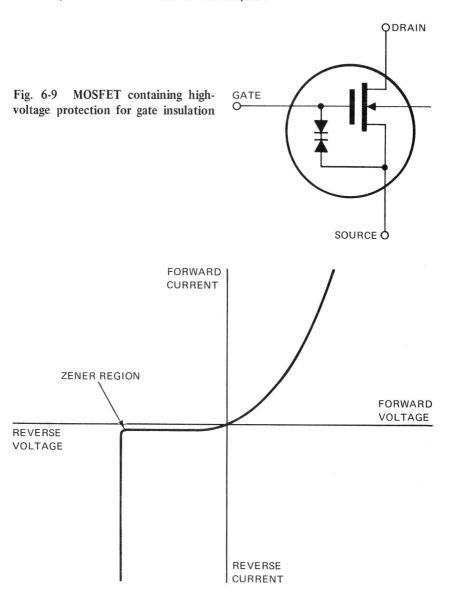

Fig. 6-10 Volt-ampere characteristic for a Zener diode

in the material break out of their bonds and a large number of free charges become available. When these diodes are connected in series and in opposite directions, the MOSFET is protected against large transient voltages of either polarity. The MOSFET is protected because when large voltages are received, the diodes conduct heavily and automatically provide a short between the gate and the source for the MOSFET. The internal resistance of the power source that feeds the short must be high enough to prevent the destruction of the diodes. A disadvantage of this addition to the standard MOSFET configuration is that the input impedance between the gate and source is reduced to the value of an ordinary reverse biased semiconductor diode.

When Zener diodes break down, what limits the flow of current through them? What characteristics must the circuit have to limit the diode current to safe values? *(R6-14)*

There are numerous other MOSFET configurations available for various applications. For example, the substrate of a MOSFET may be internally connected to the source terminal. In another configuration, the source and substrate are brought out to separate external terminals so that the MOSFET becomes a four-terminal device.

Another modification of the MOSFET has two gates with a single channel. A second gate can be made for an ordinary MOSFET by using the substrate as one of the gates. This gate is not an *insulated* gate, but reacts with the channel to form a JFET configuration within the MOSFET. The resulting MOSFET has a single insulated gate with a high impedance between the gate and the source and a second gate with an impedance between the gate and the source which is equivalent to that of a reverse biased semiconductor diode. If a bias voltage is applied to the substrate gate, the common source characteristic curves of the MOSFET can be modified.

The symbols for the various connection modes of this MOSFET are shown in figure 6-11, page 210. The symbol represents a miniature circuit diagram of the device. The actual representation depends upon whether the substrate is brought out separately, connected to the case, connected to the source, or is connected to both the case and the source, as shown by symbols in figure 6-11.

Dual Insulated Gate Field Effect Transistor

A similar configuration is the dual insulated gate field effect transistor (DIGFET). This type of MOSFET has a single channel and two independent gates. These gates are both insulated from each other and from the channel with a layer of silicon dioxide.

The construction of the DIGFET is shown in figure 6-12, page 210. The advantage of this configuration is that either gate can control the amplifying properties without affecting the signal on the other gate.

This device can be used for automatic gain control, as a logical OR gate, or for many other applications because one gate can cut off the flow of drain

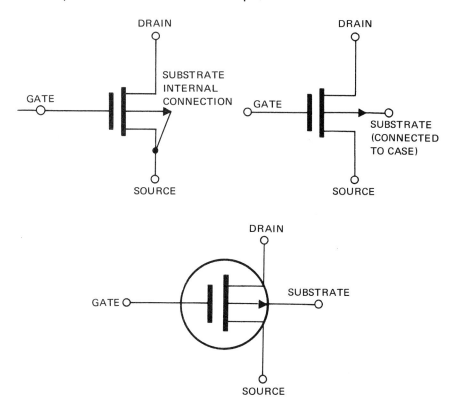

Fig. 6-11   Symbols for MOSFET with substrate connected to case or connected internally

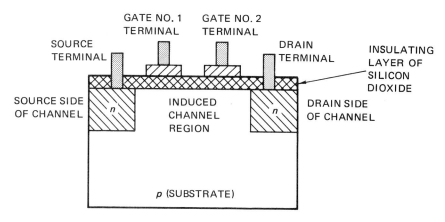

Fig. 6-12   Diagram of a MOSFET with a dual insulated gate. (DIGFET)

current regardless of the signal on the other gate. A typical symbol for the device is shown in figure 6-13. A modified form of this symbol may be used by the device manufacturer if internal connections are involved, such as the substrate connected to the case or the source.

It is more likely that a high static voltage will be applied accidentally between the gate and the source if the MOSFET has two gates. To prevent failure of the device, Zener diode protection frequently is provided. The manufacturer uses integrated circuit techniques to incorporate these diodes in the DIGFET. For each gate of the DIGFET, two back-to-back diodes are internally connected from the gate to the source. A schematic of this arrangement is shown in figure 6-14. A slightly different symbol may be used by the manufacturer to indicate whether the substrate is connected internally or brought out to the case.

Since there are two signals controlling the drain current, the complete characteristic curves become more complex. Whereas each three-terminal device has a family of curves that describes its performance, a four-terminal device theoretically can have an infinite number of families of curves. The DIGFET,

Fig. 6-13   Symbol for a MOSFET with two insulating gates

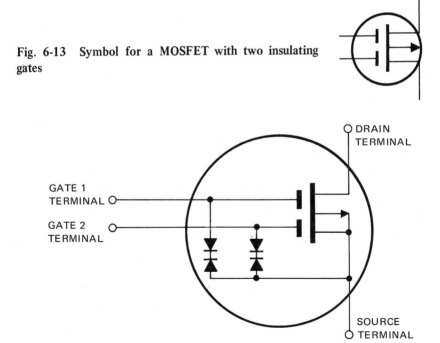

Fig. 6-14   Dual insulated gate MOSFET with internal diodes for gate insulation protection.

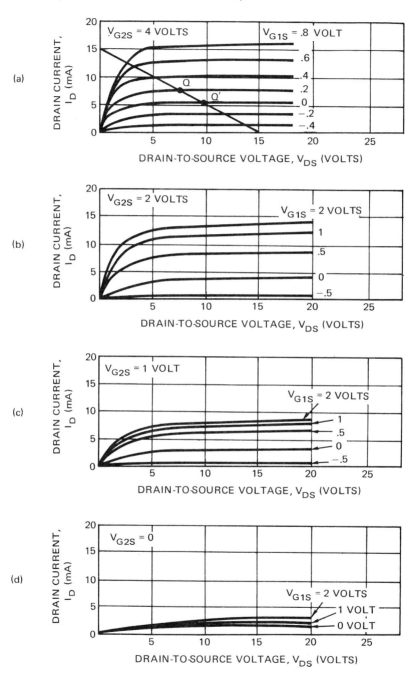

Fig. 6-15 Curve families for a dual insulated gate FET

as well as the MOSFET with the substrate used as a second gate, is described by a series of curve families. Each different gate voltage level has a curve family as shown in figure 6-15.

It should be evident that by using integrated circuit techniques, it is possible to modify a basic device to obtain within one crystal foundation, a wide variety of field effect transistor amplifying devices.

List all of the different combinations, including all possible voltage polarity combinations, that are possible with the use of JFETs and MOSFETs.(R6-15)

A low-frequency input signal is applied to each gate of a dual insulated gate field effect transistor containing Zener diode protection. The signal into one of the gates increases to a much higher amplitude. What effect, if any, is there on the other input signal? Explain the answer fully.   (R6-16)

THE UNIJUNCTION TRANSISTOR

Another type of transistor that is commonly used in oscillators and switching circuits is the *unijunction transistor* (UJT). The unijunction transistor is most closely related to the FET than any other transistor. A diagram of the UJT is shown in figure 6-16. Note that the UJT has two parts, a base region and an

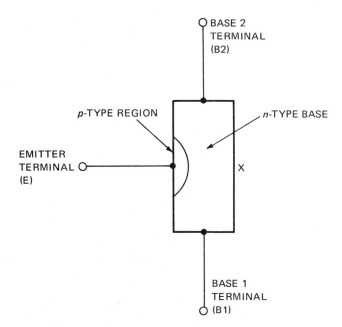

Fig. 6-16 Diagram of unijunction transistor construction

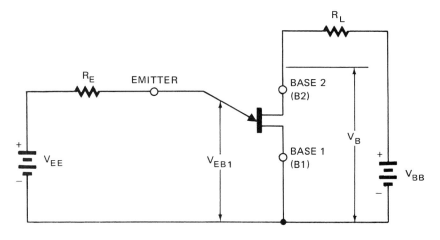

Fig. 6-17 Biasing arrangement for a unijunction transistor

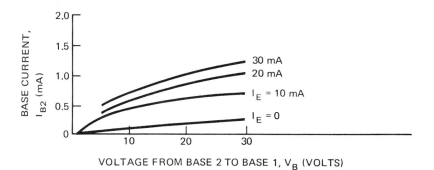

(a) OUTPUT CHARACTERISTIC CURVES

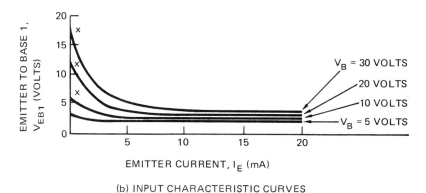

(b) INPUT CHARACTERISTIC CURVES

Fig. 6-18 Characteristic curves for a unijunction transistor

emitter region. The base region has two terminals which are indicated as Base 1 (B1) and Base 2 (B2). If the base region is an n-type material, the emitter is a p-type material. A typical bias circuit using a UJT is shown in figure 6-17. When the voltage V_{EB1} is less than the voltage existing across the base region between point X and terminal B1 in figure 6-16, the p-n junction of the UJT is reverse biased and very little current flows in the base region. If V_{EB1} increases (because of a change in V_{EE}) to the point where it is larger than the voltage from X to B1, the p-n junction becomes forward biased. A large current then flows in the base region, even though V_{EB1} decreases again. The characteristic curves of this transistor are shown in figure 6-18. Only the emitter current for the forward biased condition is shown. Recall that for all semiconductor p-n junctions, a small forward bias voltage is required to start conduction. The path taken by the negative minority carrier current (if plotted) is indicated by dotted lines in figure 6-18(b). Note the negative resistance region in figure 6-18(b) where the current goes up as the voltage goes down. It is this negative resistance feature which makes the unijunction transistor unique and suitable as a switching device in oscillators and timing circuits.

If laboratory facilities are available, a number of interesting experiments can be carried out using a UJT. Obtain a UJT and plot its volt-ampere characteristic curves as shown in figure 6-18. What happens to the meter deflection when each of the X-points shown in figure 6-18 are reached? *(R6-17)*

FET AMPLIFIERS

Figure 6-19, page 216, compares the symbolic representations of the generalized amplifying device, the bipolar transistor, the JFET, and the MOSFET. Each device has three terminals. In addition, each device shares the following characteristics: if a signal is applied to one of the terminals, amplification of the signal can occur as the result of a change in current between the other two terminals.

The analogy between the terminals for the three major types of devices is as follows:

Device	Terminal		
Generalized Amplifying Device	X	Y	Z
Bipolar Transistor	Base	Collector	Emitter
FET	Gate	Drain	Source

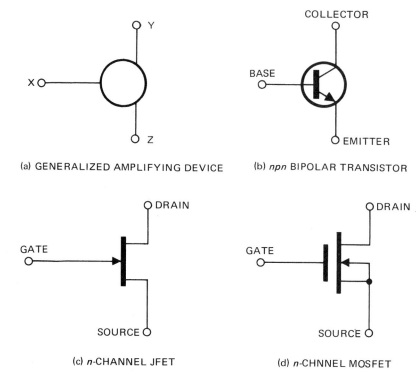

(a) GENERALIZED AMPLIFYING DEVICE

(b) *npn* BIPOLAR TRANSISTOR

(c) *n*-CHANNEL JFET

(d) *n*-CHNNEL MOSFET

Fig. 6-19 Comparison of symbolic diagrams for various amplifying devices

Figure 6-5 is actually the same as Figure 6-2(a) with a different symbol used for the amplifying device and a different polarity shown for the gate bias voltage. Compare the bias connections shown in figure 6-2(a) with those of figure 1-12. Also compare figure 6-5 with figure 2-6. The main difference between the circuit in figure 6-2(a) and the circuit in figure 2-6 is the fact that V_{GG} (figure 6-5) has a polarity opposite to that of V_{BB} (figure 2-6). One difference between the amplifying devices shown in these figures is the input impedance levels; that is, the input impedance is very high for the JFET and MOSFET, but is very low for the bipolar transistor. The JFET and MOSFET are similar in this respect to the high-impedance generalized amplifying device analyzed in Chapter 1.

The circuits of figures 6-2 and 6-5 represent simple FET amplifiers. A more common FET amplifier circuit is the ac coupled amplifier shown in figure 6-20. This amplifier will be used as a model circuit for which circuit values will be established later.

The FET amplifier can be constructed in three different configurations. The following table compares the various configurations of FET-type amplifiers with those of the devices previously analyzed.

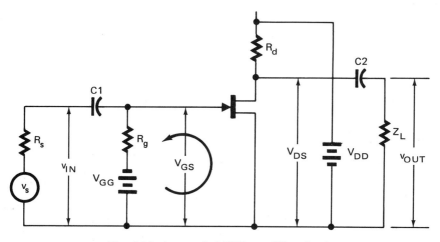

Fig. 6-20 Ac coupled FET amplifier circuit

Device	Configuration		
Generalized Amplifying Device	Common X	Common Y	Common Z
Bipolar Transistor	Common Base	Common Collector	Common Emitter
FET	Common Gate	Common Drain	Common Source

As in the cases of the other amplifying devices studied, the configuration of the FET amplifier is determined by the way in which the input and output terminals are defined. In general then, the device is the same, but the names of the input and output terminals are different.

There are several advantages to be gained from the use of the field effect transistor as an amplifying device: (1) it is capable of a very high input impedance, (2) it has good thermal stability, (3) it is relatively free from unwanted noise, and (4) the drain current is zero at $V_{DS} = 0$. The fact that it has a high input impedance means that the current through the gate-to-source junction can be neglected in the analysis of the FET amplifier.

CONSTRUCTION OF LOAD LINES FOR COMMON SOURCE AMPLIFIERS

A load line can be established for the characteristic curves of the JFET or MOSFET, just as was done for the previously described amplifying devices. The method of establishing circuit values for the JFET or depletion mode MOSFET is the same for both and this method is covered first. Assume that the amplifier circuit shown in figure 6-20 is to be analyzed. The basic procedure is the same as that for the high-impedance generalized amplifying device. For $v_s = 0$, write Kirchhoff's Voltage Law for the drain-to-source circuit:

$$V_{DD} - I_D R_d - V_{DS} = 0 \qquad \text{Eq. 6.1}$$

Using Equation 6.1, one end of the dc load line is established when $I_D = 0$, because $V_{DS} = V_{DD}$. The other end of the load line is located when $V_{DS} = 0$ because

$$I_D = \frac{V_{DD}}{R_d} \qquad \text{Eq. 6.2}$$

The resulting dc load line can be drawn on the characteristic curves shown in figure 6-3.

BIASING FET CIRCUITS

To insure the proper operation of a JFET, all of the values of the gate-to-source voltage V_{GS} must be of such a polarity as to reverse bias the *p-n* junction existing between the gate and source terminals. If this condition does not exist, then there is a gate current and the input signal is shorted out by the forward biased semiconductor diode condition. The device shown in figure 6-3 is an *n*-channel device and the gate must be maintained at a negative polarity with respect to the source to remain in a reverse biased condition. For a *p*-channel device, the reverse biasing is as shown in figure 6-2(b). In this case, the gate-to-source voltage (V_{GS}) must be positive if a reverse bias condition is to be maintained between the gate and the source. To illustrate the biasing technique, the following analysis determines the values for the *n*-channel JFET, described by figure 6-3, when used as an amplifier. The curves shown in figure 6-3 also can represent a MOSFET operated in the depletion mode.

It can be seen that for the FET of figure 6-3, the maximum value of the drain bias voltage (V_{DD}) must be restricted to about 28 volts. If V_{DD} is not restricted, the drain current can become too high if there is a slight increase in V_{DS}. To allow a margin of safety for this FET, the value of V_{DD} is restricted to 25 volts from a regulated voltage supply.

The maximum power dissipation curve is shown on figure 6-3. This device has a maximum power dissipation of 300 milliwatts. The load line must not be too close to this curve if the device is to survive normal use. For an assumed load resistance of 2090 ohms, the load line crosses the current scale at:

$$I_D = \frac{V_{DD}}{R_D} = \frac{25}{2090} = 12 \text{ milliamperes}$$

The slope of the load line is: $1/R_d$ = $1/2090$ = 0.0005. This load line is drawn as an overlay on the characteristic curves in figure 6-3. To insure that the device operates in the linear region, the Q-point is selected at V_{GS} = -2 volts. The gate circuit can be analyzed by applying Kirchhoff's Voltage Law around Loop 1, as shown in figure 6-20.

$$V_{GG} - V_{GS} - I_G R_g = 0 \qquad \text{Eq. 6.3}$$

Recall that the value of I_G is very small for a JFET. As a result, Equation 6.3 reduces to:

$$V_{GG} \approx V_{GS} \qquad \text{Eq. 6.4}$$

The value of R_g is arbitrary. The criteria for selecting R_g are: (1) it must be high enough to limit the gate current in the event V_{GS} becomes positive and the gate-to-source junction becomes forward biased, and (2) it must be low enough to couple V_{GG} to the gate circuit.

The load line and the Q-point are determined in the same way for both the JFET and the MOSFET. A circuit contains a depletion MOSFET which has the same common source characteristic curves as the JFET whose curves are shown in figure 6-3. Assume V_{DD} = 20 volts and I_D at V_{DS} = 0 is 10 milliamperes. Draw the load line and locate the Q-point if the gate bias voltage (V_{GS}) is -1.5 volts. (R6-18)

BIASING CIRCUITS USING A SINGLE DC POWER SUPPLY

The circuit shown in figure 6-20 contains two separate dc power supplies. Both supplies, however, are not necessary since there are ways of providing a dc bias voltage for the gate from one power source.

Examine the circuit shown in figure 6-21, page 220. The resistor R_s causes the source to be positive with respect to the gate for dc voltages. R_g provides a path for the gate bias current. The value of R_g is large enough to insure that the input impedance of the amplifier is not reduced below tolerable levels, and is small enough to allow for the bias connection. Kirchhoff's Voltage Law can be applied around the gate circuit, Loop 1 in figure 6-21.

$$-(I_G + I_D)R_s - I_G R_g - V_{GS} = 0 \qquad \text{Eq. 6.5}$$

Equation 6.5 is solved for V_{GS}.

$$V_{GS} = -I_G R_g - (I_G + I_D)R_s \qquad \text{Eq. 6.6}$$

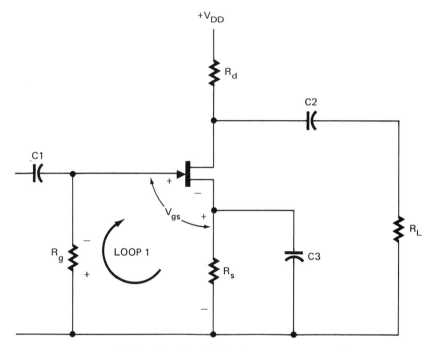

Fig. 6-21 Gate bias for JFET using one power supply

Since I_G is very small as compared to I_D, V_{GS} becomes:

$$V_{GS} \approx -I_G R_g - I_D R_s \qquad \text{Eq. 6.7}$$

Typically, the gate current has a value of 0.1 nanoampere $0.1(10)^{-9}$ ampere. If R_g = 10 megohms, then the voltage drop $I_G R_g$ has a value of only:

$$(10)(10)^6(0.1)(10)^{-9} = (10)^7(10)^{-10} = 0.001 \text{ volt}$$

This voltage is negligible with respect to the required gate bias voltage. Therefore, the required gate bias is:

$$V_{GS} = -I_D R_s \qquad \text{Eq. 6.8}$$

It may be necessary to recalculate the drain-to-source bias level. This value may be different from the value obtained for the circuit shown in figure 6-20. For the transistor whose characteristics are shown in figure 6-3, a value of 500 ohms for R_s is selected. Since I_D at the Q-point is approximately 4 milliamperes, V_{GS} becomes:

$$V_{GS} = -(0.004)(500) = -2 \text{ volts}$$

For the drain circuit, Equation 6.1 is now changed to:

$$V_{DD} - I_D R_d - V_{DS} - (I_D + I_G)R_s = 0 \qquad \text{Eq. 6.9}$$

Since $I_G \ll I_D$, Equation 6.9 reduces to:

$$V_{DD} - I_D(R_d + R_s) - V_{DS} = 0 \qquad \text{Eq. 6.10}$$

At $V_{DS} = 0$,

$$I_D = \frac{V_{DD}}{R_d + R_s} \qquad \text{Eq. 6.11}$$

To operate on the same load line as in the previous example, R_D is selected to be 1590 ohms. I_D is determined from Equation 6.11:

$$I_D = \frac{25}{1590 + 500} = 12 \text{ milliamperes}$$

Therefore, the amplifier operates at the same quiescent conditions as in the previous example.

Refer to the FET characteristic curves shown in figure 6-8. The MOSFET described by these curves has its substrate internally connected to the source. The manufacturer supplies the following information:

- *the maximum allowable drain-to-source voltage is 20 volts;*

- *the minimum gate-to-source voltage is -8 volts (it cannot be more negative than this);*

- *the maximum allowable drain current is 50 milliamperes;*

- *the maximum allowable power dissipation is 330 milliwatts.*

The FET is to be used in an amplifier such as the one shown in figure 6-21 with $R_d = 1000$ ohms, and a load resistance $R_L = 1000$ ohms. Determine the dc load line and a Q-point to insure linear amplification. Use bias voltages that are at safe levels. *(R6-19)*

An FET whose characteristics are shown in figure 6-3 is to be used in a circuit such as the one shown in figure 6-20. $R_d = 1.5$ kilohms, $V_{DD} = 25$ volts, and $V_{GSQ} = -2$ volts. Draw the dc load line and locate the Q-point. *(R6-20)*

A circuit containing a JFET, figure 6-22, can be used for small input signals that vary symmetrically about a zero level. There is no gate bias voltage supplied for this circuit. When the input is positive, the gate-to-source diode conducts and causes the capacitor to charge. The capacitor charge produces a voltage opposing the positive signal input voltage. When the signal goes negative, the gate is reverse biased with respect to the source, as it should be for normal operation. The capacitor holds its charge until the signal input again goes positive, depending on the magnitude of the resistor R_g. The capacitor never becomes fully discharged and thus holds a small negative bias on the gate at all times. This gate bias may be large enough to enable the FET to operate as a linear amplifier. The circuit cannot be used to bias the MOSFET since a MOSFET never allows the gate current to flow for either positive or negative values of V_{GS}.

Construction of the Ac Load Line

When a load is connected to an FET amplifier, such as that shown in the circuits of figures 6-20 and 6-21, an ac load line can be drawn on the common source characteristics. A signal frequency is assumed that allows the reactance of the coupling capacitor (C2) to be neglected. The symbol Z_L is used in place of R_L to indicate that the load may not be a pure resistance. The equivalent load on the drain source circuit of the FET is:

$$\frac{R_d Z_L}{R_d + Z_L}$$

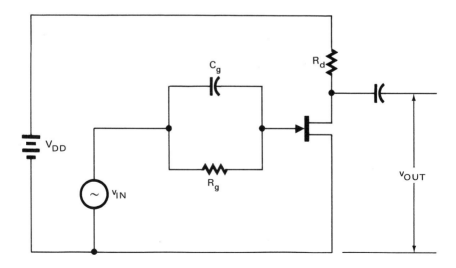

Fig. 6-22 Gate self-bias circuit for JFET

The slope of the ac load line is the reciprocal of the equivalent load:

$$\text{Slope} = \frac{R_d + Z_L}{R_d Z_L} \qquad \text{Eq. 6.12}$$

If the amplifier load is $3000 + j0$, then the equivalent load on the FET is

$$\frac{(3000)(2090)}{3000 + 2090} = 1235 \text{ ohms}$$

Thus, the slope of the ac load line is:

$$\text{Slope} = \frac{1}{1235} = 0.00081$$

The slope can also be defined by:

$$\text{Slope} = \frac{\Delta I_D}{\Delta V_{DS}} \qquad \text{Eq. 6.13}$$

For the case where $\Delta I_D = 1$ milliampere, ΔV_{DS} is:

$$\Delta V_{DS} = \frac{\Delta I_D}{\text{Slope}} = \frac{0.001}{0.00081} = 1.23 \text{ volts}$$

The coordinates $\Delta I_D = 1$ milliampere and $\Delta V_{DS} = 1.23$ volts, as measured from the Q-point, can be used to locate a second point to define the ac load line. The ac load line can now be drawn through this point and the Q-point. This is shown on figure 6-3.

Consider an RC coupled amplifier such as the one shown in figure 6-21. Use the data of review question (R6-19) and draw the ac load line for a load impedance of $R_L = 1000$ ohms. *(R6-21)*

For an RC coupled amplifier such as the one shown in figure 6-20, use the data of review question (R6-20) and draw the ac load line for a load impedance of 1000 ohms resistive. *(R6-22)*

The circuits considered to this point are based on the method of biasing the n-channel JFET or the depletion mode MOSFET. The p-channel FET is biased in the same way with the exception that the voltage polarities are reversed. The question remains as to what type of biasing circuits are to be used for the enhancement-type MOSFET and the MOSFET having both depletion and enhancement levels of gate bias voltage.

Biasing Circuits for the Enhancement-type MOSFET

The drain-to-source circuit for the enhancement MOSFET is biased in essentially the same way as the JFET. The difference is in the gate bias. An enhancement-type, n-channel MOSFET can be biased with a positive gate bias voltage. The resulting circuit is then like the circuit in figure 2-6 for an npn bipolar transistor.

For example, the MOSFET represented by figure 6-6 is to be used in the circuit of figure 6-23. (Note that figures 6-23 and 6-20 are the same except that an enhancement-type MOSFET is used.) When the drain current is zero, $V_{DD} = V_{DS}$. V_{DD} can be set at 25 volts; this value is assumed to be less than the maximum allowable value stated by the manufacturer. At $V_{DS} = 0$, the maximum value of I_D must be less than the value for which the load line crosses the maximum power dissipation curve. This value is assumed to be 10 milliamperes. The value of R_d is then $25/10 = 2.5$ kilohms.

In the previous illustration, if V_{DD} is 20 volts and $R_d = 2.5$ kilohms, what is I_D at $V_{DS} = 0$?   (R6-23)

VOLTAGE GAIN CALCULATIONS

The FET is considered to be a high-impedance amplifying device since a typical gate-to-source resistance is so high. This factor implies that the FET

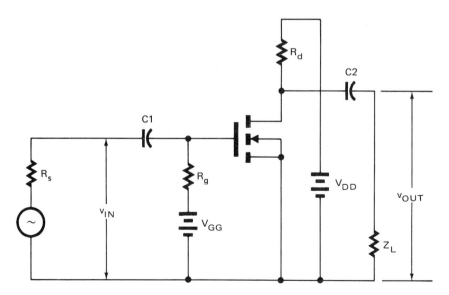

Fig. 6-23 Ac coupled amplifier circuit using enhancement-type MOSFET

amplifier (particularly the MOSFET amplifier) should be considered as a voltage amplifier rather than a current amplifier.

The voltage gain can be calculated for the FET amplifier circuit shown in figure 6-20, using the FET described in figure 6-3. For this circuit, the following values are assumed: V_{DD} = 25 volts and R_d = 2090 ohms. The Q-point values are V_{GSQ} = -2 volts and V_{DSQ} = 16.3 volts. R_L = 3 kilohms. The dc and ac load lines are shown on the curves in figure 6-3. The voltage gain is determined by selecting values of V_{DS} along the ac load line and substituting them in the following equation:

$$A_V = \frac{V_{DS1} - V_{DS2}}{V_{GS1} - V_{GS2}} = \frac{V_{OUT1} - V_{OUT2}}{V_{IN1} - V_{IN2}} \qquad \text{Eq. 6.14}$$

The subscripts 1 and 2 represent initial and final values, respectively. Other subscripts are as indicated in figure 6-20. The reactance of the capacitors is considered negligible. For a ± 1-volt change in V_{GS}, V_{DS} changes by 4.6 volts. Therefore, using Equation 6.14, the voltage gain for this circuit is:

$$A_V = \frac{18.6 - 14.0}{(-3) - (-1)} = -2.3$$

For the RC coupled amplifier considered in review problem (R6-22), assume a signal frequency at which the effect of the coupling capacitors can be considered as negligible. Calculate the voltage gain for an input signal of ± 1 volt about the Q-point. Use values taken from the ac load line drawn for review problem (R6-22). (R6-24)

For the circuit considered in review problem (R6-21) and using values taken from the ac load line drawn for this problem, calculate the voltage gain for an input signal of ± 1 volt about the Q-point. (R6-25)

GATE BIAS REQUIREMENTS

If a depletion-type MOSFET is used and V_{GS} = 0 occurs in approximately the center of the active region of the common source characteristic curves, then a gate bias is not required. Because such a MOSFET also has a high gate-to-source impedance, the FET amplifier may be coupled directly to the input signal source, if there is no dc level to the ac input signal.

Consider an FET amplifier circuit such as the one shown in figure 6-23.
A gate bias voltage V_{GS} is selected to be zero. A depletion-type MOSFET is
used having a gate-to-source impedance of 10 megohms. What amplifier input
impedance can be attained?   *(R6-26)*

If the MOSFET is to be operated primarily in either the depletion mode
or the enhancement mode and not in both, a gate bias of the correct polarity is
required. A number of different circuits can be used to achieve the proper bias
levels using just a single power supply, such as that shown in figure 6-24. Figure
6-24(a) shows a biasing circuit using an enhancement-type MOSFET. In this circuit, the gate voltage polarity with respect to the source is the same as the drain
polarity with respect to the source. An analysis of the bias voltages for this circuit is similar to the analysis of the emitter self-bias circuit of figure 2-17. In
this case, however, the gate current (which is analogous to the base current) can
be assumed to be zero. The gate bias voltage is determined according to the
following equation by using the resistors R_{g1} and R_{g2} as parts of a simple
voltage divider.

$$V_{GS} = \frac{V_{DD}R_{g1}}{R_{g1} + R_{g2}}$$ Eq. 6.15

The drain-to-source circuit can be analyzed using the method developed in this
chapter for the JFET.

The circuit in figure 6-24(b) can be analyzed in the same way as the circuit
of figure 6-21, which used a JFET. The circuit of figure 6-24(b) can only be
used for a depletion-type MOSFET. In this circuit, a gate bias voltage is provided which has a polarity opposite to that of the drain bias voltage. Frequently,
a capacitor is connected across R_s so that the ac gain is not affected by the presence of R_s.

For the circuit in figure 6-24(c), the gate-to-source bias voltage is nearly
equal to the drain-to-source bias voltage. This circuit is used only for an
enhancement-type MOSFET.

The circuit in figure 6-24(d) also is used only for an enhancement-type
MOSFET at low values of V_{DD}. For this circuit, the gate-to-source voltage
(V_{GS}) is nearly equal to V_{DD} (except for very large values of R_g). Although
this circuit does not provide any bias stabilization, this is not as important a factor for the FET as it is for the bipolar transistor.

The gate bias levels for the circuit in figure 6-24(e) can be determined by
writing Kirchhoff's Voltage Law for Loop 1.

$$\frac{V_{DD}R_{g1}}{R_{g1} + R_{g2}} - V_{GS} - I_D R_s = 0$$ Eq. 6.16

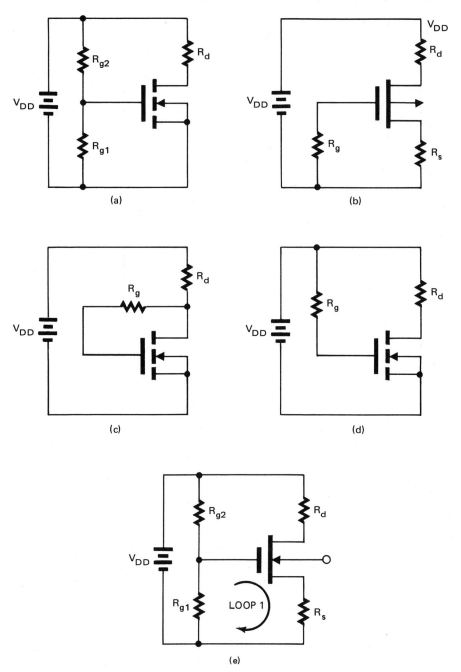

Fig. 6-24 FET amplifier biasing circuits showing different gate biasing techniques

By substituting assumed values, the equation is solved for the unknown circuit constants. This circuit can be used for either depletion mode or enhancement mode MOSFETs and guards against variations in the device parameters. Recall that the emitter self-bias circuit shown in figure 2-17 does the same for the bipolar transistor.

All of the drain-to-source circuits given in figure 6-24 can be solved using the techniques given in Chapter 1 (pages 14 to 17) and in this chapter. To obtain values for the analysis, the manufacturer's data should be studied closely and/or experimental tests should be run to obtain information on the characteristics of the device.

Assume that the biasing circuit of figure 6-24(a) is to be used for an enhancement-type MOSFET. V_{DD} = 15 volts, R_{g2} = 200 kilohms, and R_{g1} = 1.3 megohms. Calculate the gate bias voltage. (R6-27)

In figure 6-24(e), assume that the gate bias voltage (V_{GS}) = 2 volts, and R_s = 800 ohms. R_{g1} and R_{g2} are the same as in (R6-27). What is I_D ? (R6-28)

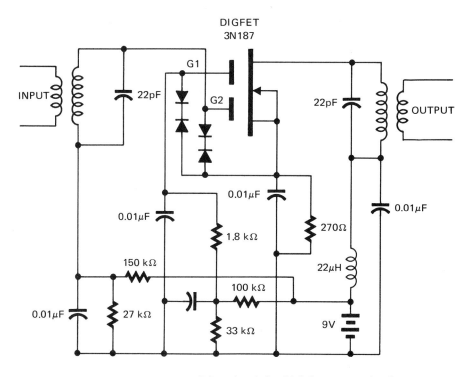

Fig. 6-25   DIGFET amplifier circuit for high-frequency signals

If R_{g2} = 100 kilohms in review problem (R6-28), what is the value of R_{g1} if the same bias conditions are met? (R6-29)

Amplifier Circuits Using Dual Gate MOSFETs

Dual gate MOSFETs are frequently used in high-frequency circuits such as the one shown in figure 6-25. In one application, gate G1 is coupled through a capacitor to ground. This arrangement reduces the feedback capacitance in the device for high-frequency signals. The dual gate feature is also used in communication circuits. For example, one of the gates is used as the means of introducing AGC (automatic gain control) to an amplifier without taking any power away from the output signal. A dual gate MOSFET can also be used in a mixer circuit where signals of two different frequencies are inserted into an amplifier input. Since the gates are virtually isolated from one another, one frequency source has little or no effect on the other frequency source.

If a bias connection is needed for both gates, it is provided as shown in figures 6-26(a) or (b). The bias level of the drain circuit in figure 6-26 is analyzed using the methods for the output circuits of various other configurations. Kirchhoff's Voltage Law is applied, a load line is drawn, and a Q-point is located. Although the gate bias circuits both affect the common source characteristic curves, they do not affect each other. The value of V_G for one gate establishes a set of characteristic curves which defines the device to be used to amplify a signal applied to the other gate.

As an example, consider an FET amplifier similar to the one shown in figure 6-26(b). The FET used is represented by the curves given in figure 6-15. Curve family (a) represents the operation of the device with V_{G2S} = 4 volts. If R_d = 1 kilohm, V_{DD} = 15 volts, and I_D = 15 milliamperes, the resulting load line is as shown in figure 6-15(a). To obtain linear amplification, the Q-point is located at V_{G1S} = 0.2 V. The actual values of R_{g1}, R_{g2}, R_{g3}, and R_{g4} are arbitrary; however, these values must be large enough to prevent a reduction in the amplifier input impedance below desired values. In addition, the R_g values must be in a specific ratio with respect to each other so that the gate voltages are at the proper bias levels. The gate-to-source voltage (V_{G2S}) is determined using Equation 6.14:

$$V_{G2S} = \frac{15R_{g4}}{R_{g2} + R_{g4}} = 4$$

For a value of R_{g2} = 2 megohms,

$$15R_{g4} = 4(2 + R_{g4})$$

(a)

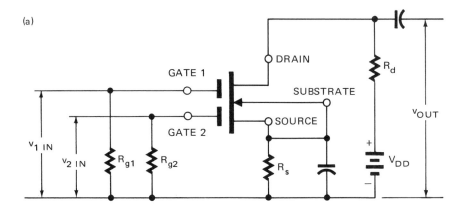

(b)

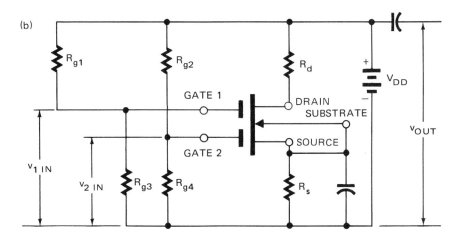

Fig. 6-26 Amplifiers using FETs with dual insulated gates, showing (a) depletion mode bias connections and (b) enhancement mode bias connections

Solve for R_{g4}:

$$R_{g4} = 727 \text{ kilohms}$$

V_{G1S} is established in the same manner:

$$\frac{15R_{g3}}{R_{g1} + R_{g3}} = 0.2$$

For a value of $R_{g1} = 2$ megohms,

$$15R_{g3} = 0.2(2 + R_{g3})$$

Solve for R_{g3}.

$$R_{g3} = 27 \text{ kilohms}$$

This latter value negates one advantage of using a MOSFET due to the fact that the input impedance of the amplifier is effectively only 27 kilohms. A more desirable situation is to select $V_{G1S} = 0$, so that the operation of this device is at point Q' in figure 6-15(a). In this case, the input impedance of the gate 1 input is the input impedance of the bare MOSFET. This value may be as high as 10^{14} ohms.

It was shown previously that a double Zener diode circuit frequently is connected between the gate and source as protection because of the thin silicon dioxide layer that insulates the gate. This arrangement further reduces the input impedance (to that of a reverse biased semiconductor diode). The extremely high input impedance of the MOSFET is not fully utilized, but the protection achieved by this arrangement may be worth the sacrifice.

Refer to the curves of figure 6-15 and the amplifier circuit of figure 6-26(b). Assume $V_{DD} = 20$ volts, $R_D = 1.33$ kilohms, $R_s = 0$, $R_{g1} = 3.9$ megohms, $R_{g3} = 100$ kilohms, $R_{g2} = 4.5$ megohms, and $R_{g4} = 500$ kilohms.

a. Select the curve family that should be used to analyze the amplifier performance.

b. Draw the dc load line and locate the Q-point for the curve family selected.   *(R6-30)*

A MOSFET is being used in an amplifier circuit such as the one shown in figure 6-24(a). The MOSFET has an induced channel and common source curves such as those in figure 6-6. $V_{DD} = 25$ volts, $R_d = 2.5$ kilohms, $R_{g2} = 1.05$ megohms, and $R_{g1} = 200$ kilohms. Using figure 6-6, draw the dc load line and find the values of V_{GS}, V_{DS}, and I_D at the Q-point.   *(R6-31)*

A MOSFET having a diffused channel has common source curves such as those shown in figure 6-8. The MOSFET is used in a circuit like the one in figure 6-21. $V_{DD} = 20$ volts, $R_d = 1$ kilohm, $R_s = 0$, $R_g = 1$ megohm, and $R_L = 2$ kilohms. The capacitive reactances are negligible at the signal frequency.

a. Using figure 6-8, draw the dc load line and identify the location of the Q-point.

b. Draw the ac load line.   *(R6-32)*

LABORATORY EXPERIMENTS

The following laboratory experiments were tested and found useful for illustrating the theory in this chapter.

LABORATORY EXPERIMENT 6-1

OBJECTIVE

To measure the characteristic curves of a junction field effect transistor

EQUIPMENT AND MATERIALS REQUIRED

1   Junction field effect transistor, n-channel, RCA 40468A
1   Curve tracer, Tektronix 575

PROCEDURE

1. Obtain a JFET from the instructor. The specifications for the RCA 40468A device are as follows:

    Maximum Values

    V_{DS} =   20 volts
    I_D =   25 milliamperes
    Power   =   375 milliwatts

2. Insert the JFET into the curve tracer. The switch that applies power to the JFET from the curve tracer should be off. The connection diagram is shown in figure 6-27.

3. Set the curve tracer controls as follows:

    Collector sweep polarity: (+) npn
    Base step generator: (-) pnp
    Base series resistor: down to 1 ohm
    Base step selector: down to 0.2 volt per step
    Steps per family: fully clockwise
    Peak voltage, collector sweep: less than 20 volts
    Dissipation limiting resistor, collector sweep: greater than 2 kilohms

BOTTOM VIEW

Fig. 6-27 Connection diagram for the RCA 40468A field effect transistor

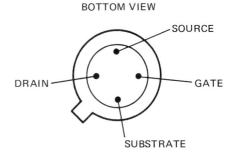

4.   Switch the power from the curve tracer to the JFET. Adjust the controls to obtain an easily measurable set of characteristic curves.

5.   Record all data, including the equipment used. Plot the curves on linear graph paper.

LABORATORY EXPERIMENT 6-2

OBJECTIVES

To measure the volt-ampere curves for a MOSFET and to operate the MOSFET in a simple common source amplifier.

EQUIPMENT AND MATERIALS REQUIRED

1   MOSFET, single gate, diffused n-channel, 3N142 or equivalent

2   Power supplies, dc, adjustable from 0-20 volts dc, 1/2 ampere, regulated

2   Resistance substitution boxes, Heathkit EU-28A or equivalent

1   Volt-ohm-milliammeter, preferably of the high-impedance, FET type operated from batteries (rather than 115 volts ac)

1   Zener diode, 1N5236 or similar

1   Audio signal generator, 0-1 volt peak-to-peak, 0-15,000 hertz frequency, or better

1   Oscilloscope, time-based horizontal axis

PROCEDURE

1.   Obtain a MOSFET from the instructor. The MOSFET probably is packaged with a small metal spring clip around the terminals. This clip is provided to prevent the accidental destruction of the gate insulation by stray electrostatically developed voltages. It should remain in place until the MOSFET is connected in the experimental circuit. The maximum values that this MOSFET can withstand are as follows:

$$V_{DS} : \quad +20 \text{ volts}$$
$$V_{GS} : \quad +1 \text{ to } -8 \text{ volts}$$
$$I_D : \quad 50 \text{ milliamperes}$$
$$\text{Power Dissipation} : \quad 330 \text{ milliwatts}$$

The connection diagram for this MOSFET is shown in figure 6-28, page 234.

2.   Connect the MOSFET in the circuit shown in figure 6-29, page 234. Remove the protective clip from the MOSFET after the Zener diode is connected in the orientation shown. The Zener diode has a Zener voltage of 7.5 volts. When connected in the orientation shown, the Zener diode protects the gate by limiting positive voltages to less than 0.7 volt and negative voltages to less than 7.5 volts. Before applying power to the circuit, adjust the power supply output voltages to 0 (as indicated on the volt-ohm-milliammeter).

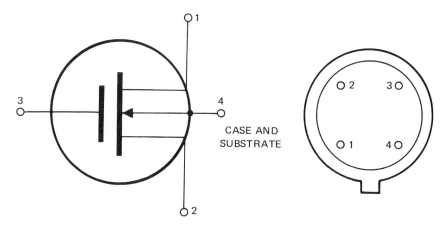

Fig. 6-28 Connection diagram for 3N142 metallic oxide semiconductor FET

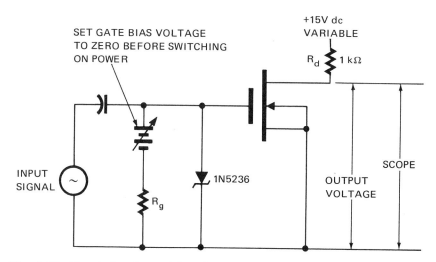

Fig. 6-29 Circuit for determining the common source curves for the 3N142 MOSFET

3. Adjust V_{DD} from 0 to 15 volts in steps of two or three volts each. Measure the drain current and V_{DS} at each step.

4. Adjust the gate bias supply voltage (V_{GG}) to a value between +0.2 and +0.7 volt and repeat step 3. R_g may be set at 0 ohms or any small value.

5. Adjust the gate bias supply voltage to a value between -1 and -7 volts and repeat step 3.

6. Record all applied voltages and drain currents and plot common source characteristic curves that are similar to those of figure 6-8.

7. Adjust R_g to a reasonably high value below 500 kilohms. Set V_{DD} to about 15 volts dc.

8. Apply a sinusoidal input signal of less than 1 volt peak-to-peak at 1000 hertz. Readjust the drain resistor and the gate bias, if necessary, to achieve an undistorted output.

9. Record values of the resistors, voltages, and the drain current at the Q-point. Use the scope to measure the input and the output ac voltages. Calculate the voltage gain.

LABORATORY EXPERIMENT 6-3

OBJECTIVE

To build and operate a common source FET amplifier

EQUIPMENT AND MATERIALS REQUIRED

1	JFET, *n*-channel, 2N4224 or equivalent
3	Resistance substitution boxes, Heathkit EU-28A or equivalent
3	Capacitors, 10 microfarads, 25 volts
1	Power supply, dc, 0-25 volts dc, 1/2 ampere, regulated
1	Oscilloscope, dual channel, time-based horizontal deflection scale

PROCEDURE

1. Obtain a 2N4224 FET and determine its characteristics, using the procedure of Experiment 6-1. The maximum values that this transistor can withstand are as follows:

$$
\begin{aligned}
V_{DS} &: \quad 30 \text{ volts dc} \\
V_{DG} &: \quad 30 \text{ volts dc} \\
V_{GS} &: \quad 30 \text{ volts dc} \\
I_D &: \quad 20 \text{ milliamperes dc} \\
\text{Power Dissipation} &: \quad 300 \text{ milliwatts}
\end{aligned}
$$

The connection diagram for this device is shown in figure 6-30, page 236.

2. Construct a common source amplifier by referring to figure 6-31, page 236. This circuit uses a source resistor to provide the gate bias. The series resistance of $(R_d + R_s)$ determines the dc load line. The voltage across R_s determines the gate bias voltage. R_g couples the gate bias voltage to the gate terminal, without loading the signal source. The capacitors are selected for negligible reactance at a signal frequency of 2 kilohertz.

3. Using figure 6-31 as a guide, determine the load line and locate the Q-point on the experimental characteristic curves.

BOTTOM VIEW

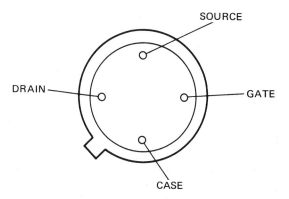

Fig. 6-30 Connection diagram for 2N4224 junction field effect transistor

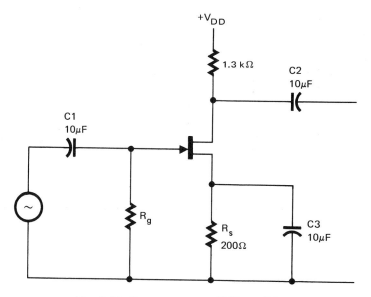

Fig. 6-31 Common source FET amplifier

4. Measure the voltage gain for no load and again for a load resistor of 1200 ohms connected to the output coupling capacitor. Use the maximum sinusoidal signal that can be tolerated and still achieve minimum distortion. Use a frequency of about 2 kilohertz.

5. Adjust the circuit values and voltages to obtain the maximum possible voltage gain. Note that the maximum gain that can be obtained is

considerably lower than that for a bipolar transistor amplifier. Compare this gain with the maximum possible voltage gain as calculated from the characteristic curves.

6. Explain why the dc voltage across R_s is also the gate bias voltage.

7. For a load of 1200 ohms, calculate and draw the ac load line on the characteristic curves.

8. Verify the voltage gain measurements by calculation, using the characteristic curves.

LABORATORY EXPERIMENT 6-4

OBJECTIVE

To build and experiment with an amplifier and a frequency mixer using a dual insulated gate FET.

EQUIPMENT AND MATERIALS REQUIRED

1   Field effect transistor, dual insulated gate, type 3N140 or equivalent
1   Oscilloscope, dual channel, with a time-based horizontal scale
1   Signal generator, 0-1 volt peak-to-peak, 0-10 kilohertz frequency
3   Power supplies, regulated, dc, 0-20 volts dc, 1/2 ampere
2   Resistors, 10 kilohms, ± 10%
2   Capacitors, 10 microfarads, 25 volts
2   Resistance substitution boxes, Heathkit EU-28A or equivalent

PROCEDURE

1. Obtain a 3N140 DIGFET. The maximum values that this transistor can withstand are as follows:

$$V_{DD} \quad : \quad 20 \text{ volts}$$
$$V_{GS1} \quad : \quad -8 \text{ to } +1 \text{ volt dc}$$
$$V_{GS2} \quad : \quad -8 \text{ to } +0.4 \text{ volt dc}$$
$$V_{DG} \quad : \quad 20 \text{ volts}$$

Maximum power dissipation : 400 milliwatts
Drain Current : 50 milliamperes pulse

The connection diagram is shown in figure 6-32, page 238.

CAUTION:

The silicon dioxide insulated layer is very thin and can be punctured easily by either static electricity or by inadvertently applied voltages. While constructing this circuit, the following precautions must be observed:

(1) do not remove the wire clip from the terminals of the DIGFET until the device is installed and the circuit is connected.

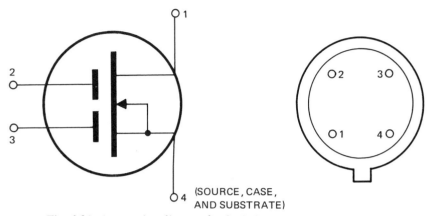

Fig. 6-32 Connection diagram for 3N140 dual insulated gate FET

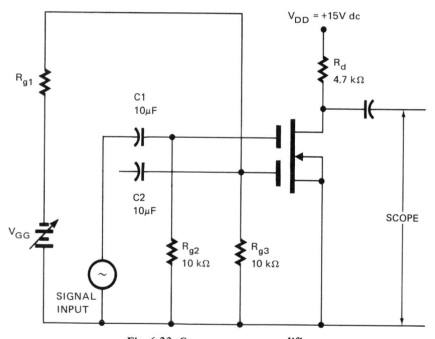

Fig. 6-33 Common source amplifier

(2) do not make any circuit changes without first removing all power from the circuit. Before applying power, set the voltage adjustments to zero. Bring these adjustments to the desired levels after power is restored.

2. Construct an amplifier circuit, figure 6-33. The recorded values should be satisfactory; however, it may be necessary to adjust the values to obtain

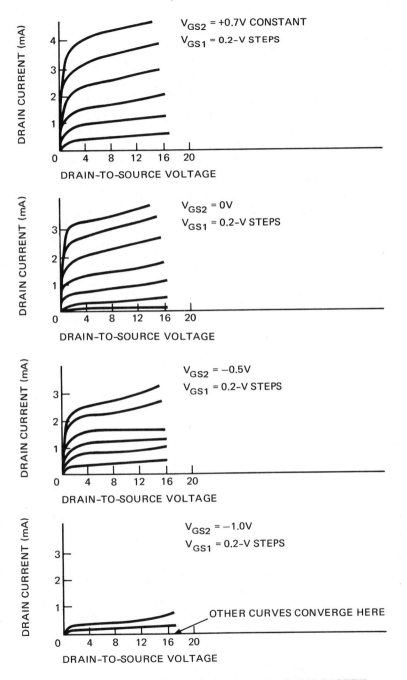

Fig. 6-34 Family of common source curves for 3N140 DIGFET

the maximum gain and the minimum distortion. R_g is any large resistance that can be used to couple a gate bias to one of the gates. Select capacitors having large values. Operate the circuit between 1 kilohertz and 10 kilohertz, using a sinusoidal input signal.

3. Draw a dc load line on one of the characteristic curves of figure 6-34, page 239. Then locate the Q-point (approximately) for the final circuit.

4. Use one gate for the signal and the other gate for the bias. Experiment with the sensitivity of the circuit to changes in the gate bias voltages, both positive and negative, to each gate. Also determine the sensitivity of the circuit to changes in the drain bias resistance and the drain bias voltages. Record the effects of the changes. Sketch the waveforms produced and record the voltages obtained. Compute the voltage gain for several levels of these quantities. Compute the maximum gain that was achieved as a result of circuit adjustments. Use caution in making the adjustments and remain within the maximum values recorded.

5. After becoming familiar with the circuit performance, apply a separate oscillator signal to each gate through coupling capacitors, (C1 and C2 in figure 6-35). Experiment with the effects of applying signals of differing frequencies to the gates. The DIGFET is now operating as a *mixer*. Record the waveforms and the peak voltages produced for several signal levels of differing frequency.

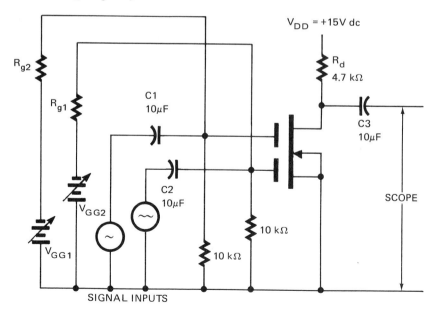

Fig. 6-35 Mixer amplifier

EXTENDED STUDY TOPICS

1. The theory of the JFET was explained in terms of an n-channel unit. Explain how the JFET operates using a p-channel unit as an example. Use diagrams as necessary and indicate a more negative voltage level as "lower" or "less" than a more positive value. (In other words, –2 volts is "less" than +1 volt.)

2. An examination of the transistor data in review problem (R6-19) shows that the maximum drain current times the maximum drain-to-source voltage is 1000 milliwatts of electric energy. However, the maximum power dissipation is given as 330 milliwatts. Explain the apparent discrepancy between these two "maximum level" specifications.

3. Why is it possible to operate a depletion-type MOSFET at negative or positive values of the gate bias voltage, while an enhancement-type MOSFET is restricted to just one polarity of the gate bias voltage? Use a p-channel MOSFET as an example in the explanation.

4. What feature of the MOSFET allows it to be used in an amplifier circuit with direct coupling to its gate, without disturbing the bias voltage levels?

5. Without referring to the text, draw the symbols for the following devices: n-channel JFET, p-channel JFET, n-channel MOSFET with induced channel, and n-channel MOSFET with diffused channel.

6. Explain the analogy between the gate, drain, and source terminals of the FET and the terminals of other amplifying devices covered in the text to this point.

7. Examine the characteristic curves of the unijunction transistor. Explain the feature shown in the curves that makes this device ideal for switching applications and practically useless for linear amplification.

8. Why is the FET sometimes known as a *unipolar* transistor?

9. What FET devices are biased with the same basic circuits used for biasing the bipolar transistor? What are the differences in the analysis of the circuit and in the specifying of circuit parameters?

10. Is the analysis used for setting the bias levels and selecting amplifier components for the JFET the same as that for the depletion-type MOSFET? What differences between the devices must be considered?

7
Vacuum
Tube Amplifiers

OBJECTIVES

After studying this chapter, the student will be able to:

- explain the theory of the vacuum triode
- compare the characteristics of other vacuum tube amplifying devices with those of the triode
- analyze simple triode, tetrode, and pentode amplifiers

THEORY OF VACUUM TUBES

The vacuum tube was the first three-terminal electronic amplifying device produced commercially. This device marked the beginning of the electronic industry. In its most elementary form, the vacuum tube consists of two electrodes placed in an evacuated metal or glass tube. One electrode is called the *cathode* and supplies free electrons. These electrons are directed toward the second electrode which is known as the *plate* or *anode*. There are three mechanisms by which the electrons can leave the cathode and move to the anode:

(1) the mechanical design and materials which make up the electrodes,

(2) the temperature of the cathode,

(3) the voltage applied between the anode and the cathode.

The material used to make the cathode produces a large number of free electrons when it is heated to very high temperatures and yet deteriorates very slowly. Commonly used materials are tungsten or tungsten alloys, or a rare metal with an oxide coating.

Both the anode and the cathode of the vacuum diode are usually cylindrical in shape. Figure 7-1 shows that the electrodes are arranged concentrically with the cathode in the center. An increase in the cathode temperature causes an increase in the kinetic energy of the electrons near the surface of the cathode. The electrons are excited to the point that they are no longer bound to the atoms of the cathode material and they become free charges. As such, they are easily influenced by an electric field. If a voltage is applied between the cathode and anode, figure 7-1, the free electrons are accelerated toward the anode. The electrons then flow around the external circuit to form a continuous electric current. Typical circuit diagram symbols for a vacuum diode are shown in figure 7-2.

The cathode can be heated directly by passing a current through the cathode material by way of a separate heating circuit. A typical circuit arrangement

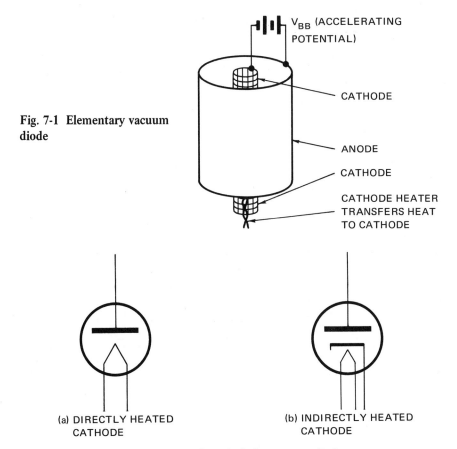

Fig. 7-1 Elementary vacuum diode

V_{BB} (ACCELERATING POTENTIAL)

CATHODE

ANODE

CATHODE

CATHODE HEATER TRANSFERS HEAT TO CATHODE

(a) DIRECTLY HEATED CATHODE

(b) INDIRECTLY HEATED CATHODE

Fig. 7-2 Typical symbols for vacuum diode

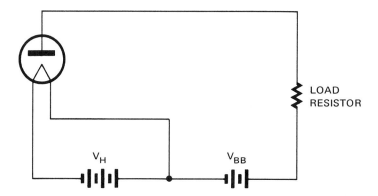

Fig. 7-3   Typical circuit arrangement for a conducting diode with a directly heated cathode

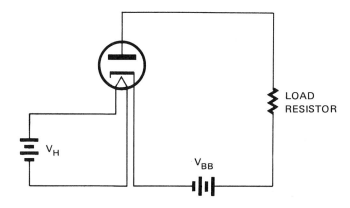

Fig. 7-4   Typical circuit arrangement for a conducting diode having an indirectly heated cathode

for a conducting diode with a directly heated cathode is shown in figure 7-3. An indirect method of heating the cathode is to place an electric heater close to the cathode. Such a heater is designed to conduct heat efficiently to the cathode. The method of indirect heating is preferred because less interference is caused to the external electrical signals that are used during the operation of the tube. A typical circuit arrangement for an indirectly heated diode is shown in figure 7-4.

The directly heated cathode also distorts tube performance as a result of the variation in electrical potential along the surface of the cathode.

The cathode may be operated at temperatures as high as 2300 degrees Kelvin or it may be operated cold. A cold cathode usually cannot supply

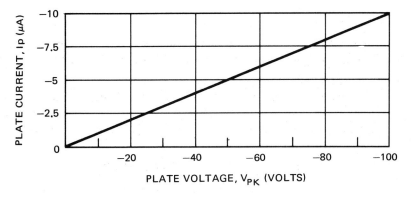

(a) VACUUM DIODE REVERSE BIASED (NOT CONDUCTING)

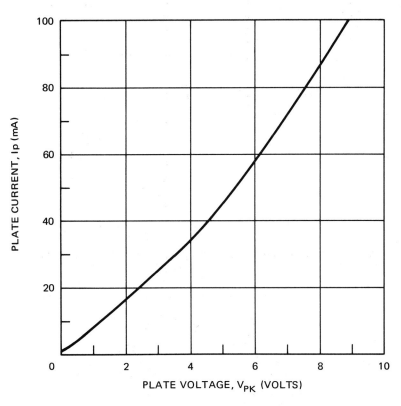

(b) VACUUM DIODE FORWARD BIASED (CONDUCTING)

Fig. 7-5 Volt-ampere characteristic for a vacuum diode

sufficient quantities of electrons. Thus, the cathodes of most vacuum tubes are heated to high temperatures.

If the voltage V_{BB} in figures 7-3 and 7-4 is reduced, the free electrons emitted by the cathode will not be accelerated enough to reach the anode. If the voltage V_{BB} is reversed in polarity, then almost none of the free electrons will travel to the anode. Since the anode is not designed to be a source of free electrons, the current is zero in this case, or nearly so.

Volt-Ampere Curve of a Vacuum Diode

The volt-ampere characteristic of the vacuum diode is shown in figure 7-5, page 245. Its behavior is similar to that of a semiconductor diode. In other words, current flows in the diode only when the anode is positive with respect to the cathode. This means that the diode is positively biased, on, or conducting. If the anode is negative with respect to the cathode (negatively biased), then for all practical purposes, there is no current.

For a forward biased vacuum diode, the current is limited by either of these two ways:

1. At a given temperature, a limited number of electrons at a given speed are available from the cathode, regardless of the value of the accelerating voltage. This is shown in figure 7-6. To increase the electron current above the value available at temperature T_1, the cathode must be heated to a new, higher temperature, T_2.

2. The number of electrons arriving at the anode is limited, regardless of the cathode temperature. The electrons emitted from the cathode form a large negative charge near the surface of the cathode. These electrons tend to repel each other; they do not travel very far into the space between the anode and the cathode unless they are accelerated by an anode voltage. At a given anode voltage, however, only a limited number of electrons reach the anode, figure 7-7. To increase the current to a value greater than that obtainable at voltage V_{PK1}, the voltage must be increased to V_{PK2}.

Since vacuum diodes allow current to flow in one direction only, they are used to rectify alternating current. However, the construction of the vacuum diode can be altered so that it has a wider range of applications. Recall that a base region added to a semiconductor diode results in a three-terminal amplifying device. A vacuum diode can also be modified to include a third terminal which will allow it to function as an amplifying device.

The Triode

By placing a third electrode between the anode and cathode and connecting this electrode to a voltage which is more negative than the cathode, it is possible to control the current flowing between the cathode and the anode of a

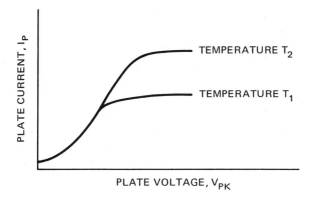

Fig. 7-6   Volt-ampere characteristic of a vacuum diode, showing how current is limited by temperature

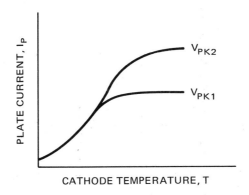

Fig. 7-7   Volt-ampere characteristic of a vacuum diode, showing how current is limited at low voltages

forward biased vacuum diode. This third electrode is called a *grid*. As long as the grid voltage is negative with respect to the cathode, the grid carries little or no current. The grid reduces the accelerating potential between the cathode and anode, resulting in a lower velocity for the free charges. Because the free charges pass a given point at a lower speed, there is a lower level of electric current. Since the grid carries so little current, very little power is consumed in controlling the level of the plate or anode current. The grid is placed closer to the cathode than to the anode so that the plate current is more sensitive to the grid voltage than it is to the anode voltage. When a grid is added to a vacuum diode, the device is called a *vacuum triode*. The symbol for a vacuum triode is shown in figure 7-8, page 246.

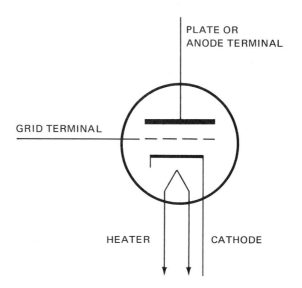

Fig. 7-8 Symbol for a vacuum triode

Characteristic Curves of the Triode

A set of characteristic curves can be drawn for a vacuum triode amplifying device, just as they were drawn for the types of amplifying devices described in previous chapters.   A family of common cathode characteristic curves for a vacuum triode is shown in figure 7-9.   These curves define the output or plate performance of the triode for various grid or input voltages.   Note that this family of curves is the same as the curve shown in figure 7-5(b), but repeated as many times as there are grid voltage levels used.

The linear, or active, region of the curves in figure 7-9 is the region in which the triode is operated as a linear amplifier.   If the device is operating in this region and an input signal is applied between the grid and the cathode, then an output signal from the plate to the cathode can amplify and yet reproduce accurately the waveform of the input signal.   The saturation and cutoff regions of operation for this device are used only when the triode is selected for switching applications or where certain forms of nonlinear power amplification are desired.   These regions of operation are the same as those defined in Chapter 1, pages 4 through 11, for the generalized amplifying device.

Using the curves given in figure 7-9, identify the three regions of operation for the vacuum triode, when the device is used as a common cathode amplifier.
(R7-1)

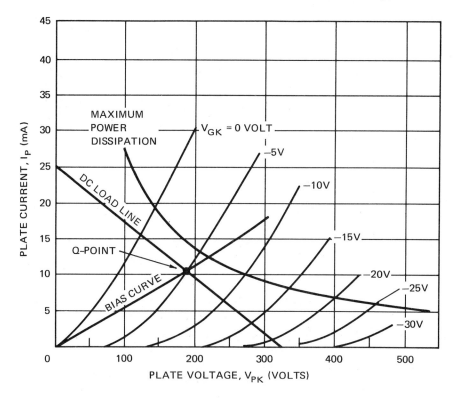

Fig. 7-9 Common cathode characteristic curves for a vacuum triode

How the Triode is Biased

The circuit shown in figure 7-10, page 250, is used to bias the triode so that it operates in the linear or active region of its output characteristic curves. Although two power supplies are used in this elementary amplifier circuit, alternative schemes are available that permit one of the voltage supplies to be eliminated.

In this circuit, the grid voltage must be negative with respect to the cathode. If the grid voltage is positive, then grid current flows because the grid acts as a target for electrons which otherwise flow to the plate. The grid voltage is shown as V_{GK} in figure 7-10. When V_{GK} is positive with respect to the cathode, this voltage accelerates the electrons emitted by the cathode. Some vacuum triode tubes are designed to withstand a positive grid voltage so that they can be used in power amplifier applications where the grid voltage is allowed to go positive. Do *not* assume, however, that a vacuum tube can withstand positive grid voltages. In general, the grid is a wire screen placed between the anode and the cathode. This arrangement permits the passage of electrons between the

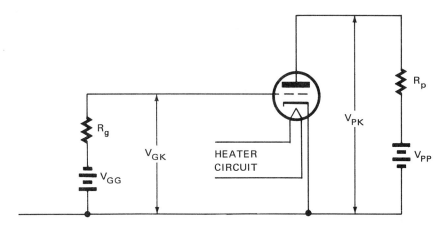

Fig. 7-10 Bias circuit for a vacuum triode

anode and the cathode but establishes an electric field in this location to influence the flow of plate current. Because of its light construction, the grid does not have the capacity to dissipate the I^2R losses that originate within it. Vacuum tube manufacturers generally specify the maximum positive grid voltage that can be applied without damaging the grid. The anode and cathode, however, are designed to withstand the full load current resulting when the device is used in a vacuum tube amplifier.

A tiny amount of grid current can flow even when the grid voltage is negative. This current is due to the ionization of gas molecules in the tube by the electrons emitted from the cathode. The tube around the anode and cathode is evacuated but a perfect vacuum is not achieved. Some gas molecules always remain in the tube atmosphere because it is impossible to remove all of the gas molecules from both the atmosphere and the material of which the container (tube) is made. Other contributions to a grid current (under negative grid bias conditions) are: (1) the movement of secondary emission electrons, (2) the grid acting like a cathode and losing some electrons to the plate, and (3) leakage through the insulation between the grid and the other electrodes. Although some grid current is present, it is very small and is difficult to measure in high-quality tubes.

A curve representing the maximum power that the vacuum tube can dissipate is shown in figure 7-9 as an overlay on the characteristic curves. The tube must always operate below this curve to insure that it is operating within its specifications and has an adequate life.

Refer to the plate characteristic curves shown in figure 7-11. The manufacturer lists the maximum power dissipation for this tube as 1 watt per triode

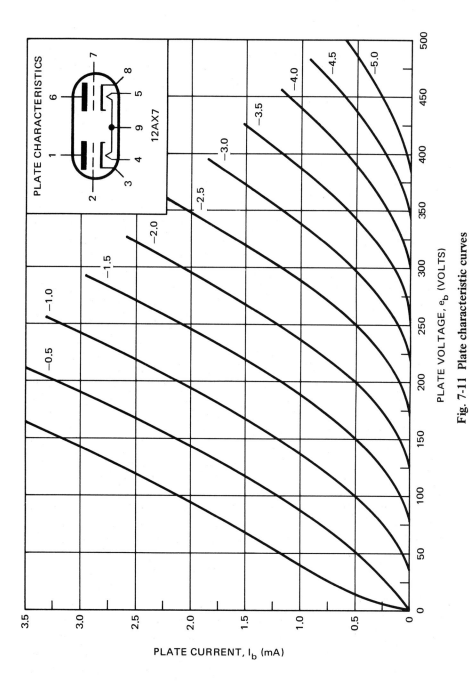

Fig. 7-11 Plate characteristic curves

section. Calculate the location of the points required to plot the curve representing the limits of the maximum operating range of the tube. Graph this curve as an overlay on figure 7-11.   *(R7-2)*

What is the function of the grid in a vacuum triode? How is this grid related to the base of a bipolar transistor? To the gate of an FET?   *(R7-3)*

The following letter subscripts are used to distinguish quantities associated with vacuum tubes:

p and b are used to represent the plate, k is used to represent the cathode, and g is used to represent the grid. If there is more than one grid in a tube, then number scripts are added to the letter subscripts. Upper case letter subscripts mean dc or average values and lower case letter subscripts mean instantaneous or ac values. The lowest number subscript is assigned to the grid nearest the cathode. Examples of the use of this nomenclature are given in the Appendix.

SPECIALIZED VACUUM TUBE AMPLIFYING DEVICES

Transistors have replaced vacuum tubes in so many applications that it is difficult to find a simple triode tube that is still manufactured in significant quantities. The majority of vacuum triodes now manufactured are for special applications that cannot use semiconductor amplifying devices for various reasons. As a result, there are many variations on the simple triode tube. For example, one type of modified tube contains a number of separate devices within the same envelope. Two or more triodes can be mounted within a single sealed container. The 12AX7 tube, whose characteristics are shown in figure 7-11, is such a tube. Another type of tube contains a pair of diodes in the same housing with a triode; both diodes have the same cathode or a number of cathodes share the same heater. By combining several tube functions in one envelope, space is saved. In addition, the characteristics of several triodes can be matched so that they can age together. This factor is important for some types of amplifier circuits.

Applications for which vacuum tubes are still used are: high voltages, wide signal voltage variations, high ambient temperatures or other extreme ambient conditions, high transient overloads, and extremely high signal frequencies.

The Tetrode Tube

Many tubes have more than one grid. A vacuum tube with two grids is known as a tetrode because there are four (tetra) electrodes. This configuration reduces the capacitance in the tube between the grid and the plate. At high frequencies, this capacitance can feed some of the plate circuit energy back into the grid circuit, resulting in uncontrollable surges of current (oscillations) in the amplifier output. A high-frequency equivalent circuit of the triode is shown in figure 7-12. The following capacitances are present: C_{gk} is the internal capacitance from the grid to the cathode, C_{pk} is the internal capacitance from the plate to the cathode, and C_{gp} is the internal capacitance from the grid to the plate. These capacitances shorten the frequency response range of the triode amplifier. The capacitance from the grid to the plate is a particular problem. The addition of a second grid breaks the grid-to-plate capacitance into two capacitances in series, figure 7-13, page 254. This grid, called the screen grid, is placed between the first grid and the plate. This arrangement reduces the effect of the tube capacitances and allows the tube to be used for high-frequency signals. Other special construction features can be incorporated in the tube to reduce the capacitances. A typical bias circuit for the tetrode is shown in figure 7-14, page 254. The screen grid is biased at a positive potential with respect to the cathode. However, this bias voltage is less than the plate voltage V_{PK}. This arrangement reduces the effect of the screen grid in preventing the flow of electrons to the plate. By connecting the screen grid to common through capacitor C_s, capacitive coupling between the plate and the control grid is reduced. The value of C_s is selected so that its reactance at signal frequencies is very low.

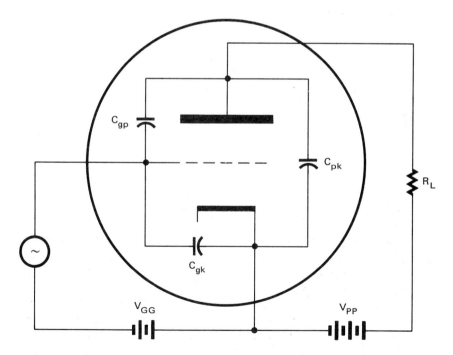

Fig. 7-12 Triode circuit showing interelectrode capacitances

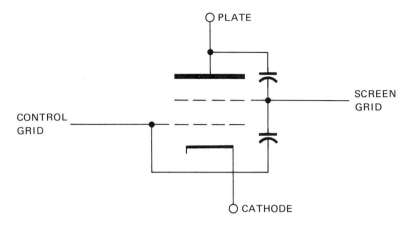

Fig. 7-13 Grid-plate capacitance C_{gp} broken into two capacitances by the addition of a screen grid

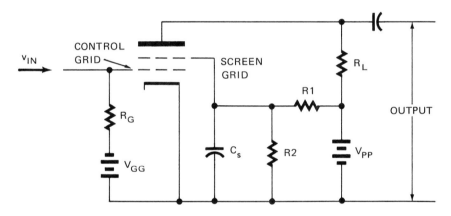

Fig. 7-14 Bias circuit for a tetrode

Typical characteristic curves for the tetrode are shown in figure 7-15. These curves were plotted using values measured in a circuit similar to the one shown in figure 7-14. Note that at values of V_{PK} above 150 volts, there is very little change in the plate current as V_{PK} changes. Small changes in the plate voltage (V_{PK}) do not greatly affect the operation. This is due to the fact that it is the screen grid potential, rather than the plate voltage, that is mainly responsible for accelerating the electrons near the cathode. This gives the tube greater amplifying ability when used as a voltage amplifier. In figure 7-15, note the distortion of the curves for values of V_{PK} between 0 and 150 volts. This distortion results because the level of V_{PK} is too low for the value of the screen

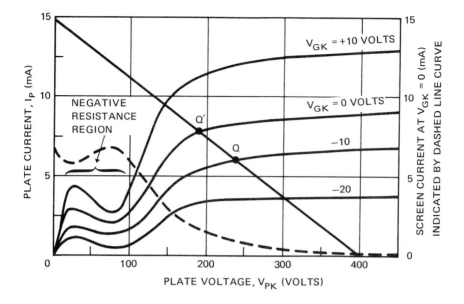

Fig. 7-15 Common cathode characteristic curves for the tetrode

grid-to-cathode voltage (V_{G2K}). The electrons at these low values of V_{PK} either do not reach the plate or they achieve high velocities at a point too close to the cathode. If the electrons pass through the screen grid, they collide with the plate at a velocity high enough to cause additional electrons to be emitted from the plate. These electrons are called *secondary emission electrons*.

The screen grid can also become a target for the cathode electrons, resulting in a screen current variation that is shown as the dashed line curve in figure 7-15.

The nonlinear region in the plate characteristic curves at low values of V_{PK} is a negative resistance region. That is, as the voltage increases, the current decreases (refer to the bracketed area in figure 7-15). When this tube is used as an amplifying device in a circuit, it can cause uncontrollable oscillations in the output at low levels of V_{PK}. Such oscillations are due to a positive feedback condition (see Chapter 11), which severely limits the application of this type of tube.

The beam power tube is constructed so that it considerably reduces the number of secondary electrons emitted from the plate. The screen and control grid of this tube are arranged so that the electron stream is concentrated into thin slices or *beams* between the wire mesh that makes up the grids. As a result, the electron density between the screen grid and the plate is increased to the point that a second space charge area is located near the plate. This area causes a reduction in the accelerating potential near the plate. Thus, fewer secondary electrons are emitted from the plate, or are turned back to the plate

after emission. A typical set of characteristic curves for the beam power tetrode
is shown in figure 7-16. The screen grid for the beam power tetrode is biased in
the same way as the grid for the basic tetrode. It is connected the same because
the difference between the tubes is only in the internal construction.

The Pentode

Still another way to reduce the nonlinear effects of the secondary emission
electrons at low plate voltages is to insert a third grid between the screen grid
and the plate. This grid is known as the *suppressor grid* because it suppresses
the secondary emission electrodes. A diagram showing the relative position of
the different grids is shown in figure 7-17. This vacuum tube containing five
(penta) electrodes is known as a *pentode.*

The pentode is biased so that the suppressor grid voltage is negative with
respect to the anode. This voltage repels and returns the secondary emission
electrons to the plate or anode. This is easily accomplished because these electrons are at comparatively low velocities. A typical biasing circuit for a pentode
is shown in figure 7-18. The suppressor grid is made like a screen and allows the
passage of most of the high-energy electrons. At the same time, the suppressor
grid provides an electric field that causes the secondary emission electrons to
return to the plate. The suppressor grid is at cathode potential, and thus slows
down the primary electrons; however, they still reach the plate. The effect

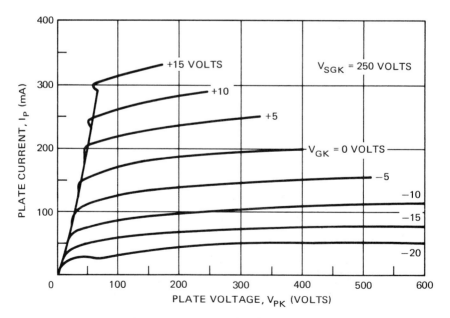

Fig. 7-16 Characteristic curves for the beam power tetrode

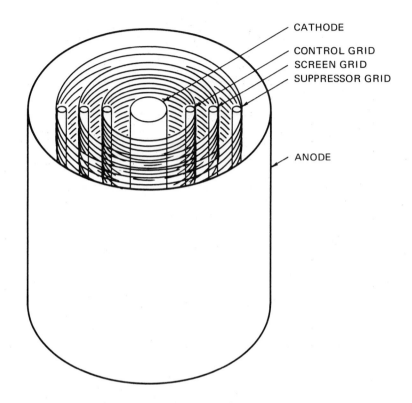

Fig. 7-17 Relative positions of the grids in a pentode

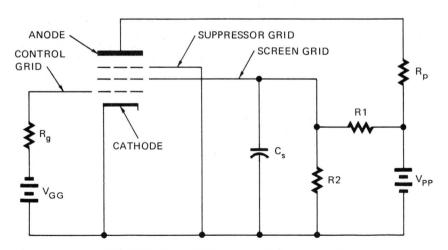

Fig. 7-18 Typical biasing circuit for a pentode

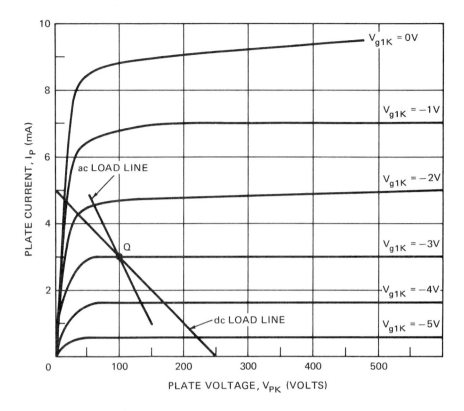

Fig. 7-19 Plate characteristics for a pentode, with screen grid voltage V_{G2K} = 100 volts

obtained by adding the suppressor grid to the tube is shown by the family of characteristic curves for the pentode in figure 7-19. Note that the nonlinearity or dip in the characteristic curves of the tetrode is eliminated in the pentode curves. Also, the effect of the plate voltage V_{PK} on the plate current is further reduced. As a result, regardless of the value of V_{PK}, the curves show that the plate current remains nearly constant. Thus, for a common cathode circuit, the output signal depends almost entirely on the input signal, which is the control grid voltage V_{G1K}. This condition further improves the amplification potential of the vacuum tube when it is used as the amplifying device in a voltage amplifier circuit. Note the similarity of the pentode plate characteristics to the FET common source characteristic curves shown in figure 6-3. The similarity of the curves helps explain why FETs are now used in place of the pentode in many amplifier circuits.

It was shown previously that for most amplifying devices, a large input signal causes an output signal that is not a faithful reproduction of the input signal. (Review Chapter 3 and refer to figures 3-1 through 3-4 which illustrate the type of distortion that results when a large input signal is fed to a transistor amplifier.) At low values of plate current, the tube is operating near its cutoff region. The amplification in this region is different from that which occurs near the center of the active region of operation of the characteristic curves. The difference in amplification results in a distortion of the output signal when the pentode is used as a voltage amplifier. Such distortion is objectionable if linear amplification is required.

Other Types of Vacuum Tubes

To reduce the distortion occurring when the pentode is used in certain applications, a *variable mu* tube is used. This type of tube is also known as a *remote cutoff tube.* "Mu" is the Greek letter μ and refers to the ideal voltage gain of the tube. The variable mu tube has a specially designed control grid. This grid is a screen having variable spacing throughout its length. The grid wires are more closely spaced at the top and bottom of the tube than they are at the center. This arrangement causes a more constant value of μ as the operation of the tube approaches the cutoff region of the characteristic curves.

Another type of beam power tube has an additional suppressor grid to counteract the effects of the secondary electrons. This tube is called the *beam power pentode.*

Multigrid tubes having more than five electrodes are also available. The schematic of a vacuum tube with five grids is shown in figure 7-20. A tube of this type is the 6BE6 pentagrid converter, which is used for frequency mixing and conversion circuits.

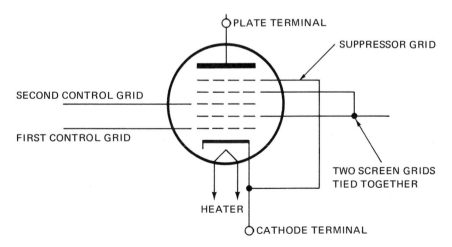

Fig. 7-20 Schematic for pentagrid converter tube

There are many other vacuum tube configurations available. These tubes are used for high-voltage, high-frequency, and other special applications. For example, television transmission and reception systems will continue to require special types of vacuum tubes. As special applications are developed, more tubes are designed and built to meet their specific needs. Many of these tubes are amplifying devices that can be built into the amplifier circuits discussed previously. Each of these special tubes has special biasing requirements. The tube manufacturer will provide these requirements in the specifications for the device; or, they can be developed using the techniques outlined in this text. The student should be aware of the fact that no new vacuum tube or semiconductor device can be applied without a sufficient knowledge of its characteristics under special conditions and its behavior in a circuit. It may be necessary for the circuit designer to purchase one or more samples of a special device and have laboratory experiments run to examine its characteristics and limitations at first hand, before the decision is made to incorporate this device into a proposed circuit.

In vacuum tube technology, what is the definition of μ (mu)?   (R 7-4)

Why is it undesirable to operate multigrid tubes at low values of plate voltage, when they are used in linear amplifiers?   (R 7-5)

VACUUM TUBE AMPLIFIERS

The vacuum tube amplifier generally has a high-impedance input. This condition is due to the fact that the control grid is usually used as the input terminal, and there is a very high impedance between the control grid and any other electrode in the tube. The vacuum tube can be considered to be a high-impedance form of the generalized amplifying device described in Chapter 1. The biasing techniques for vacuum tube amplifiers resemble those used for the FET, since the FET is also a high-impedance device.

A comparison can be made between the vacuum tube and the other amplifying devices described previously in this text, using the following table.

Device	Terminal		
Generalized Amplifying Device	X	Y	Z
Bipolar Transistor	Base	Collector	Emitter
FET	Gate	Drain	Source
Vacuum Tube	Grid	Plate	Cathode

The grid in this table is assumed to be the control grid for the pentode or tetrode tube. The screen grid and the suppressor grid are design features that are added to compensate for basic faults in the triode tube. The screen grid and the suppressor grid generally are connected to overcome these basic faults without becoming active parts of the actual amplifier circuit. They can be used as a part of the amplifier when a time varying signal is applied to both the control grid and the suppressor or screen grids. For the present, only simple amplifier circuits are considered. That is, only the control grid is subjected to a time varying signal.

Ac Coupled Amplifier Configurations

A typical ac coupled amplifier using a vacuum triode is shown in figure 7-21. The ac coupled vacuum tube amplifier shown is capacitor coupled, but transformer coupling may also be used. The configuration of this amplifier is that of a common cathode amplifier. Two other amplifier configurations for the vacuum tube triode are shown in a table that lists the three configurations that can be used for each of the amplifying devices that have been covered.

Device	Configuration		
Generalized Amplifying Device	Common X	Common Y	Common Z
Bipolar Transistor	Common Base	Common Collector	Common Emitter
FET	Common Gate	Common Drain	Common Source
Vacuum Tube	Common Grid	Common Plate	Common Cathode

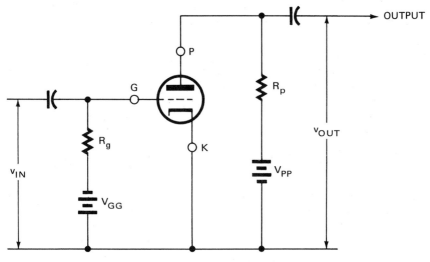

Fig. 7-21 Ac coupled vacuum tube amplifier

The most common amplifier configuration for the vacuum triode is the common cathode. The common plate configuration for a vacuum triode amplifier is known as a *cathode follower*. Recall that all of the configurations shown in the table are determined by the way in which the input and output terminals are defined. It is not necessary to have special designs of vacuum tubes or semiconductor devices for these configurations. Any vacuum triode can be used in a common plate, common grid, or common cathode configuration.

The biasing circuits described previously for the JFET or the depletion-type MOSFET are similar to those required for the vacuum tube. Actually, since the vacuum tube was the first amplifying device available, the biasing circuits for the JFET and depletion-type MOSFET were developed originally for the vacuum tube. By comparing figures 7-10 and 6-2(a), it can be seen that the biasing circuits are the same. Even the values of the resistors used are similar. However, the voltage levels of the vacuum tube are higher by a factor of about 10, causing differences in the circuit. In addition, it is necessary to provide a heater circuit for the cathode in a vacuum tube. It should be obvious that the vacuum tube has the following disadvantages: (1) additional circuitry for cathode heating is necessary, (2) high-voltage bias circuits are required, and (3) the capability of positive as well as negative biasing is not available. With vacuum tubes, only electrons are available for current flow; in semiconductors, both electrons and holes serve as current carriers.

In future applications, JFETs and depletion-type MOSFETs will be substituted for vacuum tubes in various circuits. In addition, bipolar transistor configurations have been adapted to circuits that were originally developed for vacuum tubes. Dual gate MOSFETs and single-package integrated circuit amplifiers are now available to replace pentodes, beam power tetrodes, and other multigrid vacuum tubes in numerous amplifier circuits. Many commercial communication and electronic instrumentation circuits using vacuum tubes have been redesigned or are being redesigned to use FETs and bipolar transistors. For many circuits, the FET meets most of the requirements that vacuum tubes meet. Vacuum tubes are still used in high-voltage circuits, high-frequency circuits, and circuits operating in special environments, such as nuclear reactors. In addition, it is too expensive to completely replace many kinds of equipment by transistorized and/or integrated circuits. In these cases, vacuum tubes are still needed for replacement parts and for minor improvements in equipment now in use in industry.

Describe the changes to be made in an amplifier circuit, such as the one shown in figure 7-21, if the vacuum tube is to be replaced with an FET. What changes must be made in this circuit to replace the vacuum triode with a bipolar transistor and use the circuit for approximately the same type of application?
(R7-6)

BIASING CIRCUITS, LOAD LINES

The circuit shown in figure 7-10 for the triode is a basic bias circuit for a vacuum tube amplifier. The values of the components and the voltages used are determined according to the procedures described in previous chapters.

Consider the circuit shown in figure 7-10 where the tube used has the characteristics represented by the curves in figure 7-9. To find the load line for this tube, apply Kirchhoff's Voltage Law around the plate circuit. The resulting expression is:

$$V_{PP} - I_P R_p - V_{PK} = 0 \qquad \text{Eq. 7.1}$$

By setting $I_P = 0$ so that $V_{PP} = V_{PK}$, one end of the load line is located. When $V_{PK} = 0$, then the other end of the load line is located at:

$$I_p = \frac{V_{PP}}{R_p} \qquad \text{Eq. 7.2}$$

V_{PP} is selected so that the maximum allowable plate-to-cathode voltage is not exceeded for the particular tube being used, even when the tube is operated in the cutoff region of its characteristic curves. The value of R_p is selected so that operation is below the maximum power dissipation curve. This establishes the dc load line with a slope of $1/R_p$.

To achieve linear amplification, a Q-point is used such that the maximum input signal does not cause objectionable distortion of the output signal or cause a positive grid voltage. Thus, V_{GK} is determined at the Q-point. Kirchhoff's Voltage Law is then written for the grid circuit:

$$V_{GG} - I_G R_g - V_{GK} = 0 \qquad \text{Eq. 7.3}$$

Assume the triode is to be operated at a negative grid voltage. A positive grid voltage is seldom used because (1) the grid cannot withstand significant current flow and (2) a positive grid voltage presents a low input impedance to the signal source, resulting in an objectionable load on the source. For a negative grid bias, $I_G \approx 0$ and $V_{GG} = V_{GK}$.

The triode whose characteristics are shown in figure 7-9 is to be used in an amplifier circuit. The maximum plate-to-cathode voltage recommended by the manufacturer is 350 volts. It is possible that the tube may be subjected to cutoff conditions in normal or momentary use. At cutoff conditions, the plate current is zero. Then, according to Equation 7.1, $V_{PP} = V_{PK}$ and V_{PP} must be less than 350 volts. As a safe value, a V_{PP} of 325 volts is assumed.

The load is selected so that the tube operates below the maximum power dissipation curve. The maximum power dissipation of the tube is 275 milliwatts. The locus of points for 275 milliwatts describes the maximum power dissipation curve. This curve is shown as an overlay on the plate characteristic curves in figure 7-9. If R_p is assumed to be 13 kilohms, does the load line fall into the safe operating area? The plate current is calculated using Equation 7.2:

$$I_p = \frac{V_{PP}}{R_p} = \frac{325}{13} = 25 \text{ milliamperes}$$

The dc load line can be drawn and is shown on figure 7-9. It is evident from the curves that the tube can be operated safely for R_p = 13 kilohms. If the Q-point is at V_{GK} = -5 volts, the triode will operate in the linear region for an input signal of ± 5 volts. Since the grid current is essentially zero, $V_{GK} = V_{GG}$ and the grid bias voltage is -5 volts. The value of R_g is arbitrary and is determined by (1) using the value to limit the current at momentarily positive grid voltages, (2) allowing it to transmit a time varying input signal to the grid with minimum signal attenuation, and (3) keeping its value low enough to insure that the grid bias level is not made less negative than is desired. The tube manufacturer frequently specifies a maximum value of R_g so as to prevent too great an increase above the desired grid bias level and causing undesirably high plate currents under normal operating conditions. A typical value of R_g may be 500 kilohms or less.

Using the triode plate characteristic curves shown in figure 7-11, construct a dc load line where V_{PP} = 400 volts and R_p = 115 kilohms. What is the value of I_P at V_{PK} = 0?   (R7-7)

For the load line constructed in review problem (R7-7), does the triode operate safely below the maximum power dissipation curve of 1 watt? Refer to the solution of review probem (R7-2).   (R7-8)

Refer to the results of review problem (R7-7) and assume a grid bias voltage (V_{GKQ}) = -1.5 volts. What is the maximum peak-to-peak input signal that can be accepted if this tube is used in the common cathode circuit shown in figure 7-10? What is the value of V_{GG}? Locate the Q-point on figure 7-11.   (R7-9)

Voltage Gain Calculations

The voltage gain can be calculated for the triode considered in the previous example. The voltage gain for a vacuum triode is given by the following equation.

$$A_V = \frac{V_{PK1} - V_{PK2}}{V_{GK1} - V_{GK2}} \qquad \text{Eq. 7.4}$$

The subscripts 1 and 2 represent the initial and final values, respectively, of V_{GK} and V_{PK}. Referring to figure 7-9, an input signal of ± 5 volts is assumed about the Q-point grid voltage setting of –5 volts. The change in V_{PK} is read from the load line in figure 7-9. V_{GK1} = 0 volts and V_{GK2} = -10 volts. This means that V_{PK1} = 123 volts and V_{PK2} = 240 volts. The voltage gain is:

$$A_V = \frac{123 - 240}{0 - (-10)} = \frac{-117}{10} = -11.7$$

Note that the voltage gain is negative. A negative sign means that as the input voltage goes more negative, the output voltage is driven more positive.

The triode with the characteristics of figure 7-11 is used in the circuit of figure 7-10. The input voltage signal (v_{gk}) is ±1 volt. The dc load line and the Q-point are the same as the ones determined in review problems (R7-7) and (R7-9). Compute the voltage gain for this amplifier. (R7-10)

Bias Circuits Using a Single Dc Power Supply

The circuit of figure 7-10 provides the proper biasing conditions, but requires two separate power supplies. Such an arrangement is both inconvenient and expensive. The circuit can be designed so that a single power supply serves both the grid circuit and the plate circuit. One such circuit is shown in figure 7-22. For this circuit, the drop across the cathode resistor (R_k) causes the cathode to be more positive than the grid. The operating point and load line for this circuit are obtained using the same basic methods outlined previously. Take the grid circuit first. Assume a triode whose characteristics are shown in figure 7-9. This triode is to be operated on the same load line and at approximately

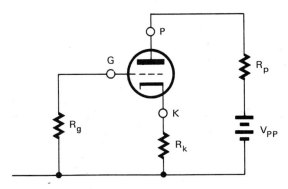

Fig. 7-22 Vacuum tube biasing circuit using cathode resistor bias

the same Q-point as those established for the biasing circuit shown in figure 7-10. The plate current at the Q-point (I_{PQ}) is 10.5 mA. The grid bias voltage at the Q-point (V_{GKQ}) is -5 volts. Applying Kirchhoff's Voltage Law to the grid circuit,

$$-I_P R_k - I_G R_k - V_{GK} = 0 \qquad \text{Eq. 7.5}$$

For a grid current (I_G) of 0, Equation 7.5 becomes:

$$V_{GK} = -I_P R_k \qquad \text{Eq. 7.6}$$

Substituting the proposed values at the Q-point, Equation 7.6 can be solved for R_k.

$$R_k = \frac{V_{GKQ}}{-I_{PQ}} = \frac{-5}{-10.5} = 0.476 \text{ kilohms}$$

The value of R_p can no longer be the value selected for the original circuit shown in figure 7-10. The total resistance external to the plate is $R_p + R_k$. The value of I_P at $V_{PK} = 0$ is given by:

$$I_P = \frac{V_{PP}}{R_p + R_k} \qquad \text{Eq. 7.7}$$

To operate at the original Q-point, the sum of R_p and R_k must be 13 kilohms. Therefore, R_p = 13 - 0.476 = 12.52 kilohms.

If R_k is known, and the Q-point is to be established, a bias curve must be constructed from Equation 7.6. To graph this curve, values of V_{GK} are assumed and values of I_p are calculated. The bias curve is drawn through these points on the characteristic curves for the tube. For the present example, assume that R_k = 476 ohms and that the tube behaves according to the curves shown in figure 7-9. If V_{GK} = 0, this means that I_P = 0 from Equation 7.6. If V_{GK} is assumed to be -10 volts, then I_p is:

$$I_p = \frac{V_{GK}}{R_k} = \frac{-10}{0.476} = -21 \text{ milliamperes}$$

A third value for V_{GK} is assumed at -5 volts. Then,

$$I_p = \frac{-5}{0.476} = 10.5 \text{ milliamperes}$$

For the points just determined, a bias line is drawn, as shown in figure 7-9. The point at which this line crosses the load line is the Q-point established for this circuit.

The value of R_g is arbitrary and is selected according to the criteria established previously. The tube manufacturer recommends a value for R_g of not more than 500 kilohms. As a result, the amplifier will have an input impedance of no more than 500 kilohms.

The triode represented by the curves in figure 7-11 is to be used in a biasing circuit, such as the one shown in figure 7-22. What are the necessary values of R_k and R_p to insure that the triode operates along the same load line and at the same operating point established in review problems (R7-7) and (R7-9)?
(R7-11)

A second bias circuit using just one power supply is the grid leak bias circuit. This circuit was first described in Chapter 6 (pages 217 to 231) and is represented by figure 6-22. The circuit was originally used with vacuum triodes as shown in figure 7-23. R_g is a resistor of relatively large magnitude. When the input signal voltage is on the first positive half cycle, grid current flows and charges the capacitor. During the negative half cycle of the input voltage, the grid current drops to zero. Since the capacitor still has a charge, the grid is left more negative than the cathode. Theoretically, the grid current will not flow again. In addition, the grid will never be positive again with respect to the cathode, except for the period required to replace the capacitor charge that dissipates during the negative half-cycle. This biasing method is useful for small signal excursions that are symmetrical about a zero-volt level. If the input signal is

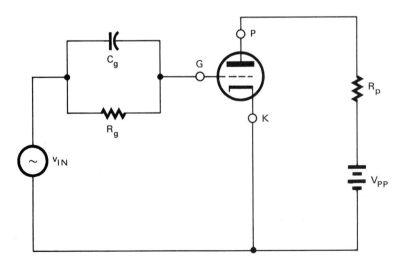

Fig. 7-23 Vacuum tube biasing circuit using grid leak bias

sinusoidal, figure 7-24(a), then the resulting waveform of the grid voltage is shown in figure 7-24(b), using a different voltage scale. The value of the resistor R_g is small enough that the time varying input signal is transmitted to the grid without too much attenuation. However, R_g is large enough to insure that the capacitor holds most of its charge during the negative half-cycle. The usual value of R_g is between 500 kilohms and 1 megohm. This type of bias circuit is used for the first stage of a high-frequency amplifier. Such an amplifier is

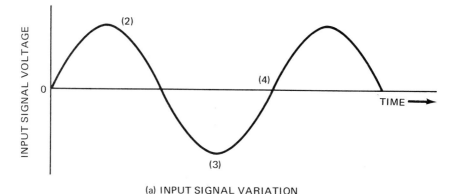

(a) INPUT SIGNAL VARIATION

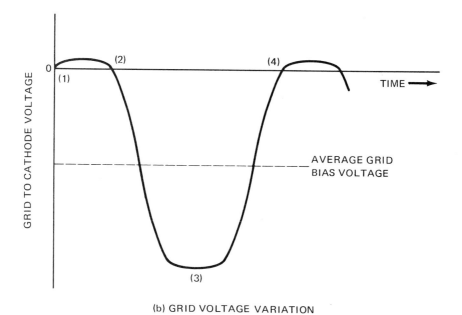

(b) GRID VOLTAGE VARIATION

Fig. 7-24 Grid leak bias waveform

designed to tolerate (1) the slight distortion due to the conducting and non-conducting halves of the input signal and (2) the power losses due to the grid current flowing repeatedly during parts of the input signal waveform.

To analyze this biasing circuit, assume that when the grid is positive with respect to the cathode, it acts like a forward biased diode. When the grid is negative with respect to the cathode, it acts like a reverse biased diode. Refer to figure 7-24 and examine the grid voltage as the signal voltage varies in a sinusoidal manner. Whenever there is a positive signal, grid current flows and V_{GK} is close to zero. These conditions exist from points (1) to (2) on figure 7-24(b). As the signal voltage drops from (2) to (3), the grid acts like a reverse biased diode and the voltage across the charged capacitor adds to the signal voltage. If the capacitor does not lose any charge, then the grid voltage at point (3) is twice the maximum value of the negative peak. The signal voltage then increases to point (4) where it again becomes greater than the capacitor voltage, assuming that some charge was lost from C_g. The "grid" diode is conducting again and continues to conduct until the capacitor is fully charged again. The cycle then repeats itself. Note that the grid bias voltage for this circuit is the average level of the voltage in figure 7-24(b) and is also the maximum value of the input signal. The disadvantage of the bias circuit is that the Q-point changes with the input signal. By varying R_g, some adjustment is possible since C_g can either discharge slowly or rapidly between charging periods. The values of R_g and C_g are determined experimentally.

When using vacuum tubes, it is necessary to refer frequently to the specifications published by tube manufacturers. In general, when these specifications are written, it is assumed that the tube is to be used in circuits similar to the type of circuit in which it normally was used. If a tube is to be changed in a given circuit, then the tube selected must operate well within its maximum limits for all of the conditions under which it may be used. To evaluate the device, a performance test should be conducted in a laboratory.

Analysis of Vacuum Tube Amplifiers

When biasing tetrodes and pentodes, the required analysis is similar to that for triodes, except that the suppressor grid current and the screen grid current must be considered. For the tetrode, it is necessary to study the specifications carefully to judge what level of screen grid supply voltage is to be used and what level of signal is to be applied.

The published specifications are closely followed when the screen grid bias voltage (V_{G2K}) is set. This action insures that the screen grid does not draw a greater amount of power than it can dissipate. The construction of the screen grid does not permit it to dissipate the amount of power that can be dissipated by the plate, without an excessive temperature rise and the possible destruction of the tube. A significant amount of screen grid current is present due to the fact that the screen is normally biased with a voltage that is positive with respect to the cathode. As a result, the screen acts like an anode and draws electrons from the cathode. This condition is especially true for low values of the plate voltage (V_{PK}), since there is only one target for the cathode electrons and the screen current can become excessively high. For this reason, the plate voltage (V_{PK}) should not be removed from the tube terminals while the screen grid bias voltage is on.

The curves shown in figure 7-15 were plotted from data taken when the screen grid voltage was set at 125 volts. For the circuit shown in figure 7-14, a tetrode is used which has the characteristics of figure 7-15. If the screen grid voltage is set at 125 volts, the screen grid current varies according to the dashed line curve in figure 7-15. The apparent screen grid resistance is given by:

$$R_{g2} = \frac{V_{G2K}}{I_{G2}}$$

Eq. 7.8

where I_{G2} is the screen grid current. The value of R_{g2} varies from a low of $125/7 = 17.9$ kilohms to a high of $125/0.2 = 625$ kilohms, depending upon the value of V_{PK}.

In figure 7-14, a voltage divider is formed by R1 and R2. Frequently, however, the resistor R2 is omitted or is made very large when compared to R1. Then, when V_{PK} is small and a large drop in the plate current occurs, a corresponding increase in the screen grid current to dangerous levels can be prevented or made small by the increased voltage drop across resistor R1. For example, assume that the average screen grid resistance is 500 kilohms. It is desired to find the dc load line when $V_{PP} = 400$ volts and $R_p = 27$ kilohms. The plate current at $V_{PK} = 0$ is:

$$I_p = \frac{400}{27} = 14.8 \text{ milliamperes}$$

The dc load line for this condition is drawn on figure 7-15.

A voltage divider is formed with R1 and the average screen grid resistance. The equivalent circuit for this condition is shown in figure 7-25. The screen grid voltage is determined as follows:

$$V_{G2K} = \frac{V_{PP}R_{g2}}{R_{g2} + R1}$$

Eq. 7.9

If R1 = 1.1 megohms and the values given or determined previously for this example are substituted in Equation 7.9, a value is obtained for V_{G2K}:

$$V_{G2K} = \frac{(400)(500)}{500 + 1100} = 125 \text{ volts}$$

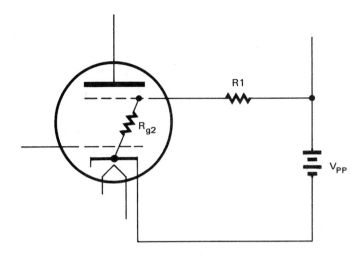

Fig. 7-25 Equivalent circuit showing voltage divider formed by apparent screen grid resistance

At the lowest expected frequency for this circuit, the capacitor C_s must have a low reactance. This will insure that the screen is at zero ac potential and does not restrict the amplifying capability of the circuit.

The only quantity remaining to be determined is the control grid bias voltage. To operate the device in the center of its active region, V_{GK} is about -10 volts. With the grid bias arrangement shown in figure 7-14, the voltage V_{GG} is also at -10 volts. This is due to the fact that the control grid current is nearly zero; therefore, the drop across R_g is negligible. The value of R_g is arbitrary. One disadvantage of this grid bias circuit is the use of two power supplies. The grid bias circuits shown in figures 7-22 and 7-23 can also be used for the tetrode circuit. R_k is again used to develop a voltage to supply the grid by way of R_g.

A tetrode has a set of characteristic curves such as those shown in figure 7-15. These curves were obtained for a screen grid bias voltage of 90 volts. Calculate the screen current for an apparent R_{g2} of 600 kilohms.   *(R7-12)*

What value of V_{GG} is required to establish the Q-point at Q' in figure 7-15, when this tube is used in the circuit of figure 7-14?   *(R7-13)*

What is the slope of the load line on the curves of figure 7-15?   *(R7-14)*

Pentode Circuits — Description and Analysis

The biasing circuit for the pentode is similar to that of the tetrode, with the exception that a suppressor grid is added. The suppressor grid, shown in figure 7-17, is connected so that it is at the same potential as the cathode. (Refer to figure 7-18.) The methods of analysis for the pentode are the same as those for the tetrode.

As an example of the analysis of a pentode circuit, consider the ac coupled pentode amplifier shown in figure 7-26. The series resistor R_s determines the positive dc bias voltage on the screen grid. This resistor acts as a stabilizing influence on the screen grid current because an increase of the screen grid current causes a greater voltage drop across the series resistor R_s and lowers the screen grid voltage (V_{G2K}). The control grid uses a cathode resistor bias rather than a separate dc power supply to provide a negative polarity. Capacitor C_k is connected across the cathode resistor R_k to provide a low-reactance path at the signal frequency. This configuration prevents any decrease in the ac voltage gain due to the presence of resistor R_k. The cathode is indirectly heated by a separate heater circuit, which is not shown in figure 7-26. In general, the heater circuit is rated at 6 or 12 volts ac or dc at a current less than 0.5 ampere. Capacitors C1 and C2 have a low reactance at the signal frequency. They provide dc isolation of the pentode amplifier from any preceding or succeeding stages of amplification.

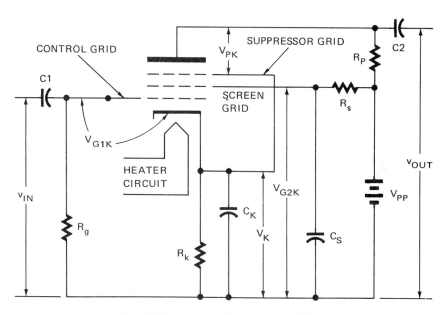

Fig. 7-26 Ac coupled pentode amplifier

It is assumed that this amplifier uses the pentode described by the plate characteristic curves shown in figure 7-19. The characteristics in figure 7-19 were determined for the following conditions:

Suppressor grid voltage to cathode = 0 volts
Screen grid voltage to cathode   = 100 volts
Room Temperature

For this pentode amplifier circuit, assume V_{PP} = 250 volts dc, R_k = 1 kilohm, and R_p = 50 kilohms. V_{PP} locates one end of the load line. At V_{PK} = 0,

$$I_p = \frac{V_{PP}}{R_p + R_k} = \frac{250}{49 + 1} = 5 \text{ milliamperes}$$

These values (V_{PK} = 0 and I_p = 5 milliamperes) locate the other end of the dc load line. According to the tube specifications, for a screen grid-to-cathode voltage (V_{G2K}) of 100 volts, the screen current (I_{G2}) is 0.8 milliampere. The screen grid resistance is given by Equation 7.8:

$$R_{g2} = \frac{V_{G2K}}{I_{G2}} = \frac{100}{0.8} = 125 \text{ kilohms}$$

The plate current and the value of R_k determine the control grid bias voltage. If R_k = 1 kilohm and I_{PQ} = 3 milliamperes, then the control grid bias voltage is:

$$V_{G1KQ} = -I_{PQ}R_k = -(3)(1) = -3 \text{ volts}$$

For these assumed values, figure 7-19 shows the dc load line and the Q-point drawn on the characteristic curves.

The value of R_g is arbitrary, but it must meet the following conditions: it must be large with respect to the signal source and yet very much smaller than the grid-to-cathode resistance when the grid bias is negative. Capacitors C1, C_k, C_s, and C2 must have a low reactance at the lowest expected signal frequency. If the signal frequency is 1000 hertz, the reactance should be at least one-tenth of the equivalent resistance in the circuit. For the cathode resistor bypass capacitor C_k, this means that the reactance must be less than (0.1)(1000) or 100 ohms. If C_k = 2 microfarads, the reactance is determined as follows:

$$X_c = \frac{1}{2\pi f C_k}$$

Eq. 7.10

$$X_c = \frac{1}{(6280)(2)(10^{-6})} = 79.6 \text{ ohms}$$

It can be seen that the value of the capacitor is satisfactory for the frequency of the desired signal.

The Ac Load Line

The ac load line can be drawn if a value is assigned to the ac load impedance Z_L. The slope of the ac load line is given by:

$$\text{Slope} = \frac{R + R_k + Z_L}{(R + R_k)Z_L} \qquad \text{Eq. 7.11}$$

Substituting R_P = 50 kilohms, R_k = 1 kilohm, and Z_L = 50 + j0 kilohms in Equation 7.11, the slope is:

$$\text{Slope} = \frac{50 + 50}{(50)(50)} = \frac{1}{25 \text{ kilohms}} = 4(10)^{-5} \text{ mho}$$

The ac load line must be drawn at this slope through the Q-point. For a small change in V_P, there is a small change in I_P. The slope is $\Delta I_P / \Delta V_{PK}$. The change in current I_P for a change in the voltage V_{PK} from 150 volts to 100 volts is:

$$\frac{\Delta I_P}{\Delta V_{PK}} = 4(10)^{-5}$$

$$\Delta I_p = 4(10)^{-5}(150 - 100) = 2 \text{ milliamperes}$$

The ac load line is then drawn through these points on figure 7-19.

The operation of the ac coupled amplifier is along the ac load line. Using the ac load line just constructed, the voltage gain can be calculated from Equation 7.4:

$$A_V = \frac{V_{PK1} - V_{PK2}}{V_{GK1} - V_{GK2}}$$

Assume the ac input signal varies from -4 volts to -2 volts peak-to-peak. The change in V_{PK} can be read from the curves shown in figure 7-19. The voltage gain is:

$$A_V = \frac{60 - 135}{-2 - (-4)} = -37.5$$

Note that the voltage gain is negative. This value represents a 180 degree phase shift between the input and output signals. Compare the gain of this pentode amplifier with that of the triode amplifier considered previously (the voltage gain for the triode amplifier was about -11). Pentode amplifiers typically have a higher voltage gain capability. For this reason, the more complicated pentode tube is usually preferred over an ordinary triode.

Use the results of review problems (R7-7) and (R7-9). An ac load imped-ance (Z_L) of 100 + j0 kilohms is connected across the output of the amplifier circuit shown in figure 7-21. The circuit conditions are represented by the dc load line and the Q-point drawn for review problems (R7-7) and (R7-9). Draw an ac load line as an overlay on the curves in figure 7-11.   *(R7-15)*

For the amplifier circuit of figure 7-27, V_{PP} = 400 volts, R1 = 500 kilohms, and R2 = 833 kilohms. Determine the screen grid voltage (V_{G2K}).   *(R7-16)*

Use the beam power tube described in figure 7-16 in the circuit of figure 7-27. The value of V_{G2K} calculated in review problem (R7-16) is to be used in this circuit. R_p = 1 kilohm, V_{PP} = 400 volts, and R_k = 0. Draw the dc load line and determine the Q-point for this circuit.   *(R7-17)*

Use the same tube and circuit configuration outlined in review problems (R7-16) and (R7-17). Draw the dc load line from the data developed in review problem (R7-17). The grid resistor R_g is 500 kilohms, according to the manufac-turers' recommendations. Calculate and locate the Q-point if R_k is 35 ohms.   *(R7-18)*

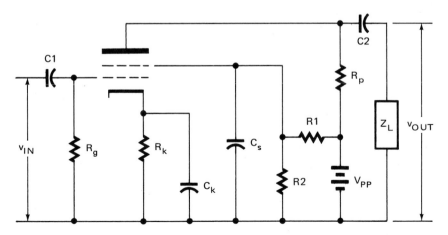

Fig. 7-27 Ac coupled tetrode amplifier

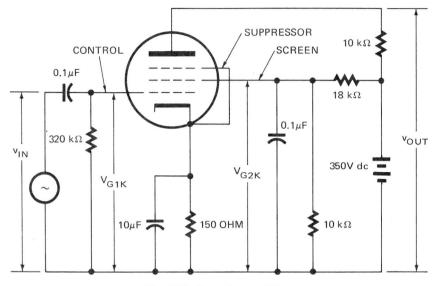

Fig. 7-28 Pentode amplifier

Using the information calculated in review problems (R7-16), (R7-17), and (R7-18), determine satisfactory values of the coupling and bypass capacitors. The signal input frequency is 10,000 hertz. (R7-19)

For the circuit shown in figure 7-28, calculate the screen grid voltage. (R7-20)

Refer to figure 7-28, and assume that the quiescent plate current is 15 milliamperes. Calculate the quiescent value of the control grid voltage. (R7-21)

LABORATORY EXPERIMENTS

The following laboratory experiments were tested and found useful for illustrating the theory in this chapter.

LABORATORY EXPERIMENT 7-1

OBJECTIVE

To determine experimentally the static characteristic curves of a triode.

EQUIPMENT AND MATERIALS REQUIRED

1   Vacuum triode, 12SQ7

1   Power supply, dc, 0-300 volts, regulated, with a 1-milliampere capability

1   Power supply, dc, 0-5 volts, regulated, with a 1-milliampere capability

2   Voltmeters, preferably high-impedance types with an impedance of 500 kilohms per volt or better; full scale settings of 1 volt to 1000 volts dc; one of the voltmeters must be capable of reading the 12.6 volts ac filament voltage

1   Milliammeter, with full scale reading capability of 50 microamperes to 5 milliamperes

1   Transformer, 10:1 transformation ratio, rated at 2 volt-amperes or a variable autotransformer rated at 2 volt-amperes

PROCEDURE

1.   Obtain a triode and check it in a tube tester. The connection diagram for the tube is shown in figure 7-29. The maximum values for the tube are as follows:

Dc plate voltage: 300 volts

Plate dissipation: 0.5 watts

Grid voltage: –2 volts

Plate current: 1 milliampere

Grid voltage: According to the manufacturer, the grid voltage is zero volts. However, a value of +200 millivolts is specified in the experimental procedure to examine the tube behavior at a positive grid voltage. When

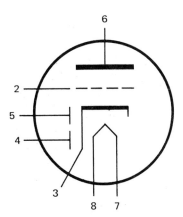

Fig. 7-29   Connection diagram for 12SQ7 triode (also contains two additional diode plates)

tested in the laboratory by the author's students, this voltage level did not damage the tubes.

Heater voltage: 12.6 volts ac or dc
Heater current: 0.15 ampere

2. Connect the triode in a circuit like the one in figure 7-30.

3. Adjust the voltage supplies to a zero voltage output and switch them to "on."

4. Adjust the plate supply voltage in three or four steps to 250 volts dc. Measure the plate current and the plate voltage at each step. Record this data in a table. Immediately plot the data on a piece of graph paper and obtain a curve similar to figures 7-9 and 7-11. By plotting the data while running the experiment, errors can be detected and corrected before the circuit is disconnected.

5. Set the plate voltage back to zero and adjust the grid voltage to -1 volt dc.

6. Repeat step 4.

7. Set the plate voltage back to zero and adjust the grid voltage to -2 volts dc.

8. Repeat step 4.

9. Set the plate voltage back to zero and adjust the grid voltage to +200 millivolts.

10. Repeat step 4.

11. Using a dashed line, plot another curve on the same graph paper used in step 4 to indicate the maximum plate power dissipation.

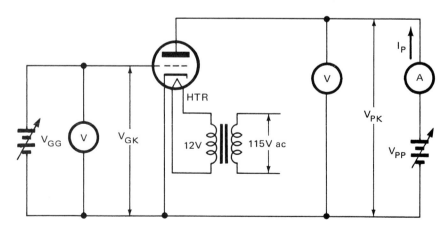

Fig. 7-30 Circuit for measuring the static characteristics of a triode

LABORATORY EXPERIMENT 7-2

OBJECTIVE

To build and operate a common cathode triode amplifier.

EQUIPMENT AND MATERIALS REQUIRED

1   Vacuum triode, 12AX7

1   Power supply, dc, 0-300 volts, regulated, 5-milliampere capability

1   Power supply, dc, 0-5 volts, regulated, 1-milliampere capability

2   Voltmeters, preferably high-impedance types with an impedance of 500 kilohms per volt or better; full scale settings of 1 volt - 1000 volts dc. One of the voltmeters must be capable of reading the ac filament voltage of 6.3 volts or 12.6 volts

1   Oscilloscope, dual channel, with time-based horizontal scale

2   Capacitors, 10 microfarads, 500 volts dc

2   Resistance substitution boxes, Heathkit EU-28A or equivalent

1   Transformer, 10:1 transformation ratio, 2 volt-amperes, or a variable autotransformer rated at 2 volt-amperes

PROCEDURE

1. Check a 12AX7 triode in a tube tester. The characteristic curves for this triode are shown in figure 7-11. The maximum values that this tube can withstand are as follows:

    Plate voltage: 330 volts
    Grid voltage: +0, -55 volts
    Plate dissipation: 1.2 watts
    Heater voltage: 12.6 volts, both heaters in series
                    6.3 volts, one heater, or both heaters in parallel
    Heater current: 0.15 ampere per heater

    The connection diagram for this triode is shown in figure 7-11.

2. Connect the triode in a circuit such as the one shown in figure 7-31, page 280.

3. Apply dc power to the plate and grid circuits and introduce a sinusoidal signal to the grid. Use a 10,000-hertz signal. Adjust the magnitude of the signal so that the output signal has the same wave shape as the input signal. An amplitude of 1/2 volt should be satisfactory.

4. Operate the circuit as a linear voltage amplifier. Measure the voltage output. Calculate the voltage gain of the circuit.

5. Adjust the resistances and the dc voltage values to obtain the best performance of the circuit as a linear voltage amplifier. Try to obtain the highest voltage gain for the largest possible input signal without distortion at the output.

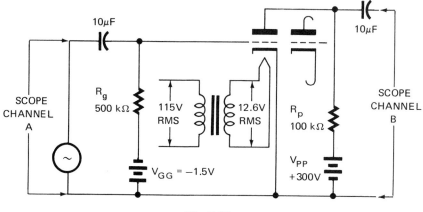

Fig. 7-31

6. For the final circuit constants and voltages, construct a dc load line on the curves shown in figure 7-11. Mark the quiescent point on the load line.

LABORATORY EXPERIMENT 7-3

OBJECTIVE

To build and operate a triode amplifier having cathode resistor bias and an ac load.

EQUIPMENT AND MATERIALS REQUIRED

1 Vacuum triode, 12AX7
1 Power supply, dc, 0-300 volts, regulated, 5-milliampere capability
2 Voltmeters, preferably high-impedance types with an impedance of 500 kilohms per volt or better, and with full scale settings of 1 volt - 1000 volts dc. One of the voltmeters must be capable of reading the filament voltage of 6.3 volts or 12.6 volts
1 Oscilloscope, dual channel, with time-based horizontal scale
2 Capacitors, 10 microfarads, 500 volts dc
2 Resistance substitution boxes, Heathkit EU-28A or equivalent
1 Transformer, 10:1 transformation ratio, 2 volt-amperes or a variable autotransformer rated at 2 volt-amperes

PROCEDURE

1. Check a 12AX7 triode in a tube tester. Information on the maximum values that the tube can withstand is given in the procedure for Laboratory Experiment 7-2.

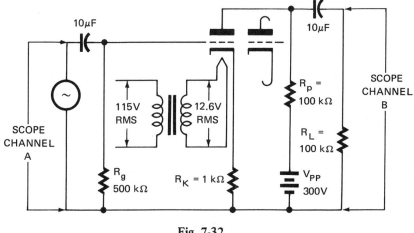

Fig. 7-32

2. Connect the triode in a circuit such as the one shown in figure 7-32.

3. Apply dc power to the plate circuit and introduce a sinusoidal signal to the grid. Use a 10,000-hertz signal. Adjust the magnitude of the signal so that the output signal has the same wave shape as the input signal. An amplitude of 1/2 volt should be satisfactory.

4. Operate the circuit as a linear voltage amplifier. Measure the voltage output. Calculate the voltage gain of the circuit.

5. Adjust the resistances and dc voltage values to obtain the best performance of the circuit as a linear voltage amplifier. Try to obtain the highest voltage gain for the largest possible input signal without distortion at the output.

6. For the final circuit constants and voltages, construct a dc load line on a copy of figure 7-11. Mark the quiescent point on the load line.

7. Construct an ac load line on the curves.

EXTENDED STUDY TOPICS

1. What is a disadvantage of a directly heated cathode?

2. Compare the advantages and the disadvantages of vacuum tube devices and semiconductor devices.

3. What external influences must be applied to a cathode-anode configuration within an evacuated container to make it carry electric current?

4. What is the purpose of the screen grid?

5. What is the purpose of the suppressor grid?

6. What is the reason for the negative resistance (or negative slope) region in the plate characteristics of the tetrode?

7. If a vacuum diode is forward biased, in what two ways can the current be limited in magnitude?

8. Explain why grid current flows in a triode if the grid-to-cathode voltage becomes positive with respect to the cathode.

9. There is some infinitesimally small grid current when the grid-to-cathode voltage is negative with respect to the cathode terminal in a triode. Why is it possible to have this current? Name some contributions to this current.

10. Describe the effects of the interelectrode capacitances within a vacuum tube on its performance as the amplifying device in a high-frequency amplifier. How is this simlar to the effects of the junction capacitance in a transistor on its performance as a high-frequency amplifier?

11. In a vacuum tetrode, why are secondary emission electrons more of a nuisance at low plate voltages?

12. What is the purpose for using a variable mu vacuum tube?

13. What are three circuit configurations in which the vacuum triode can be used?

14. In what ways are the biasing circuits for the vacuum triode similar to those for the FET and the bipolar transistor? In what ways are these circuits dissimilar?

15. Why is a vacuum triode operated at a negative grid bias voltage with respect to the cathode and not a positive grid bias voltage? Is it possible to use a positive grid bias voltage?

16. Compare the pentode curves of figure 7-19 with the FET curves of figure 6-3. Discuss the differences and similarities between these curves.

8
Amplifying Techniques Using High-impedance Amplifying Devices

OBJECTIVES

After studying this chapter, the student will be able to:

- analyze the performance of high-impedance amplifying devices
- list and describe biasing techniques for common drain, common gate, common grid, and common plate amplifier circuits
- describe amplifier circuits using both bipolar transistors and FETs
- calculate small signal parameters for FETs and vacuum tubes

LARGE SIGNALS APPLIED TO COMMON SOURCE OR COMMON CATHODE CIRCUITS

FET and vacuum tube amplifiers can be analyzed using the same techniques since they are both high-impedance amplifying devices and the characteristic curves are similar. The most popular semiconductor device and vacuum tube are the FET and the pentode, respectively. Thus, the analyses covered in this chapter concentrate on the performance of these two devices. FET behavior is used for the examples in this chapter for two reasons: (1) the characteristic curves of the pentode and the FET have the same general shape, and (2) the FET family of devices generally has replaced the vacuum tube family. The curves shown in figure 8-1, page 284, are applicable to JFETs and some types of MOSFETs.

A study was made in Chapter 3 of the distortion resulting from the choice of the Q-point or the magnitude of the input signal, or both. The study was

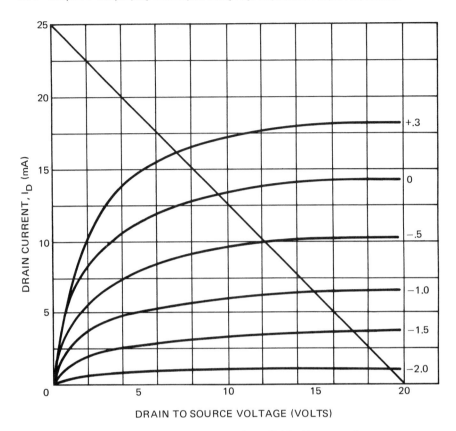

Fig. 8-1 Characteristic curves for a field effect transistor

made using the low-impedance bipolar transistor. A similar study can be made for a high-impedance device. Rather than comparing input and output *currents* as in figure 3-1, this analysis compares the input and output *voltages.* A dynamic transfer characteristic can be drawn for the high-impedance device described by the curves of figure 8-1. The first step in obtaining this characteristic is to connect the device into a typical amplifier circuit, such as the one shown in figure 6-20. A Q-point is established along a load line which is drawn as an overlay on figure 8-1. The transfer characteristic curve shown in figure 8-2 was made by plotting the input voltage (V_{GS}) versus the output current (I_D). For the case where the load is a linear resistance, the curve is equivalent to a plot of the input voltage versus the output voltage. Figure 8-3, page 286, is a plot of the input voltage versus the output voltage using a different Q-point. The high-impedance device being considered has fairly linear characteristic curves as indicated by the transfer characteristic. Although the curve is slighly S-shaped, the output voltage is a fairly accurate representation of the input

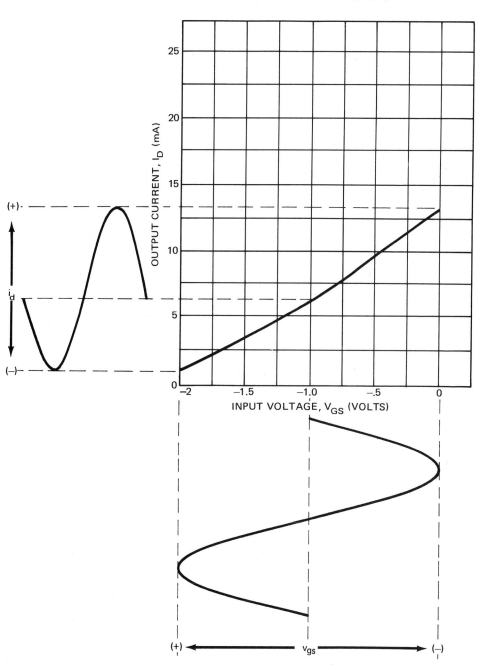

Fig. 8-2 Dynamic transfer characteristic relating input voltage to output current

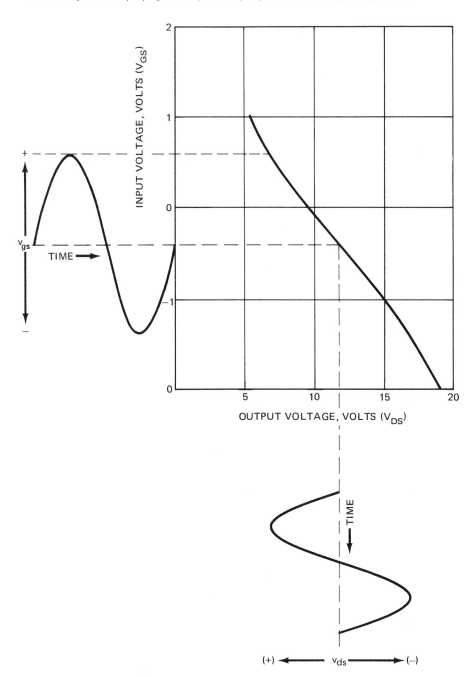

Fig. 8-3 Transfer characteristic for the device described by the curves of figure 8-1

voltage. The voltage gain for this case can be computed graphically by referring to figure 8-3.

$$A_V = \frac{\text{Change in output voltage}}{\text{Change in input voltage}} \qquad \text{Eq. 8.1}$$

$$A_V = \frac{16.7 - 6.8}{-1.4 - (+0.6)} = -4.95$$

This was calculated for a quiescent value of the voltage V_{GS} at 0.4 volts. Since the gain is negative, there is a 180° phase shift through the amplifier. Although the gain value may appear to be low for voltage amplification, it is a common value for many FETs. As in the case of the bipolar transistor, the voltage gain for the high-impedance device is affected by the load resistance. That is, a lighter load results in less slope to the load line. This causes a greater change in the drain voltage for the same change in the gate voltage.

The current gain for a high-impedance device like the FET is very high because the current through the gate is very small when compared to the current through the channel. This condition is especially true for the MOSFET where the current through the gate to the channel is so small that it cannot be measured. However, in pinching off the channel current, the reverse bias *voltage* between the gate and the channel must be relied upon. Thus, the gate current has very little effect on the channel current. As a result, there is a noticeably nonlinear relationship between the gate current and the channel current.

If an FET is to function as a current amplifier, the input signal must come from a pure current source. The impedance of this current source must be much higher than the gate-to-source impedance. This condition is unusual and very unlikely. To achieve any linearity of amplification, the current signal must be extremely small. In theory, the gain of the FET is very high if it is used as a current amplifier. Figure 8-4, page 288, shows a graph of the gate current versus the gate-to-source voltage for a JFET. For an input current of ± 0.2 nanoamperes, the gate voltage (V_{GS}) changes from –0.073 to +0.03 volt. If this JFET is described by the characteristic curves shown in figure 8-1, a change in V_{GS} of this magnitude causes I_D to change from 12.7 milliamperes to 13.5 milliamperes. The resulting current gain is determined as follows:

$$A_I = \frac{\text{Output Current Change}}{\text{Input Current Change}} \qquad \text{Eq. 8.2}$$

$$A_I = \frac{(12.7 - 13.5)(10)^{-3}}{[(-0.2) - (+0.2)](10)^{-9}} = 2(10)^6$$

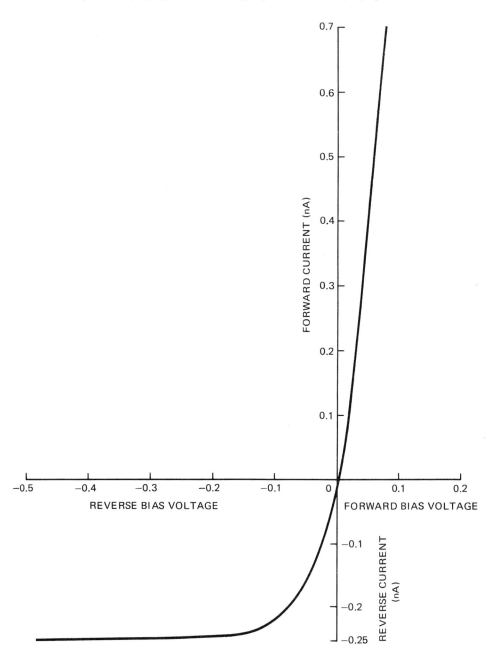

Fig. 8-4 Volt-ampere characteristic of a silicon diode in the region of zero bias voltage

This high value of current gain can be expected for a device with such a high input impedance. Combining the current gain and the voltage gain for this device, the resulting power gain is:

$$A_P = A_V A_I \qquad\qquad \text{Eq. 8.3}$$

$$A_P = (4.95)(2)(10)^6 = 10^7$$

This high power gain is one advantage of using a high-impedance amplifying device.

The JFET described by figure 6-3 is to be used in the RC coupled amplifier circuit shown in figure 6-20. Let $Z_L = \infty$. Using the load line and Q-point drawn on figure 6-3 as the operating condition, draw a dynamic transfer characteristic for the input voltage versus the output current. Space is to be left for a sketch of a sinusoidal input and output signal, such as in figure 8-2.   (R8-1)

Select the drain bias resistor for the amplifier having the load line drawn on figure 8-1.   (R8-2)

For the circuit specified in review problem (R8-1), assume that the applied quiescent gate-to-source voltage is –2 volts and the applied sinusoidal input signal is 4 volts, peak-to-peak. Sketch the output current waveform on the graph prepared in review problem (R8-1). This curve must be extrapolated down to zero current.   (R8-3)

Location of the Q-point for Large Signals

To achieve linear amplification, the Q-point must be located near the center of the active region of the characteristic curves. If the Q-point is located near the cutoff region of the curves, a portion of the input waveform is clipped or distorted.

Refer to the results of review problems (R8-3) and (R8-1). What is the voltage gain for this circuit?   (R8-4)

What is the result when the Q-point is located too high due to a high input bias level? For the JFET or a vacuum tube, a high current flows. For the voltage amplifier, the high current not only distorts the output waveform, but it also disturbs the signal source and may interfere with the proper operation of the signal generator. This situation is more serious for a high-impedance device than it is for a low-impedance device. The gate or grid connection of the high-impedance device is not designed to dissipate the power developed by the high current. The high current results from the fact that there is a forward biased diode junction between the gate-to-source terminals. Although the device may appear to be operating in the safe, active region of the characteristic curves, as reflected in the plate or drain power dissipation, no protection is provided against damage in the gate or grid circuit.

When a MOSFET is used, however, there is no danger of damage resulting from high gate currents when the Q-point is selected at too high a value. The MOSFET always has an insulating layer between the gate and drain circuits. But, too high a voltage on this device can cause a dielectric breakdown; such a breakdown can occur at relatively low voltage values. The manufacturer's recommended maximum voltage values should be noted.

By locating the Q-point well above or well below the center of the active region, a high-impedance generalized amplifying device may be operated in either the saturation or cutoff region for some portion of the input signal. As a result, only nonlinear amplification is possible. The dynamic voltage transfer characteristic shown in figure 8-3 is redrawn in figure 8-5. The gate bias is set at -2 volts (near the cutoff region). Typical input and output waveforms are shown. Note that the shape of the *clipped* output voltage signal is the same as that of the clipped current signal shown in figure 3-4. Voltage, rather than current, waveforms are shown for the high-impedance amplifying device because JFETs, MOSFETs, and vacuum tubes are considered to be voltage amplifying devices rather than current amplifying devices. When operation is close to the cutoff region and the currents are zero (or very close to zero), the device is said to be cutoff. The clipped waveform shown in figure 8-5 occurs because the device is operating below the cutoff level. For this condition, the output current is at its lowest value and the peak output voltage of the amplifier is at its greatest value.

Redraw figure 8-2 showing the gate bias voltage level at -2 volts. Sketch a sinusoidal input waveform and an output waveform to show that it is distorted at a gate bias level of -2 V. *(R8-5)*

If the Q-point of the JFET or vacuum tube is located too close to the saturation region of the characteristic curves, then the opposite condition occurs and the input waveform is clipped. The diode is forwad biased for at least part of the cycle. In the case of the vacuum triode, the grid is no longer effective in controlling the flow of electrons. As a result, the vacuum triode (or pentode, or tetrode) becomes merely a form of vacuum diode. For the JFET, the gate-to-source junction becomes a forward biased *p-n* junction and

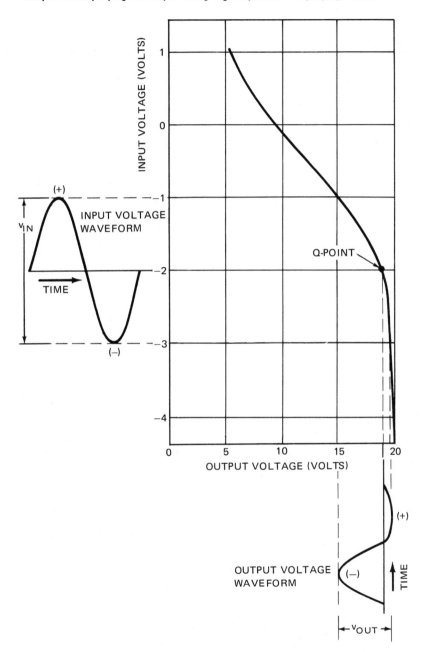

Fig. 8-5   Dynamic transfer characteristic with waveform distortion due to an offcenter Q-point

the channel acts like a resistor made from silicon. The input waveform is clipped because a portion of it operates in the forward biased region. The extent of clipping or distortion that occurs depends almost entirely on the source impedance. This distortion is reflected directly into the output circuit.

If the source impedance of the signal is high, then saturation of the FET or vacuum tube tends to cause the input signal to short out. In this manner, V_{GS} (or V_{GK}) is effectively *clamped* to a level close to zero volts. The equivalent circuit for this condition is shown in figure 8-6. The output current or drain current is the highest and the output voltage is at its lowest level for this condition.

For a large input signal, the output signal is clipped at both the positive and negative peaks, as shown in figure 8-7. This causes a symmetrical square wave signal that is useful in determining an optimum location of the Q-point for a given load condition. If the square wave is symmetrical, having equal positive and negative peaks, as in figure 8-7, the Q-point is centered and located for the maximum amount of input signal that the amplifier can tolerate without distortion. If the input amplitude is reduced just to the level that will eliminate the distortion, this value is the peak input signal that the amplifier can tolerate and still yield linear amplification.

The FET or vacuum tube can be considered to be a voltage controlled switch, similar to the relay described in Chapter 1. For signal input of zero volts, the grid or gate bias can be adjusted so that the "switch" is "on." For a fixed negative voltage input, the "switch" is "off." In other words, the device behaves like a normally closed relay.

Note that this behavior is also similar to that of the bipolar transistor. Because of this similarity, the FET and vacuum tube can also be used in digital logic NOT circuits. The MOSFET has the added advantage that the input impedance is approximately the same high value for either an on or an off signal.

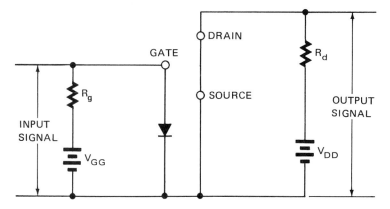

Fig. 8-6 Equivalent circuit for a saturated FET condition, gate forward biased

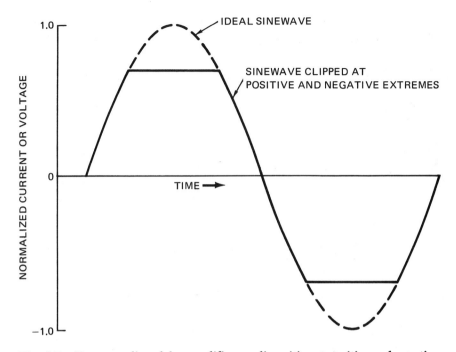

Fig. 8-7 Sinewave clipped by amplifier nonlinearities at positive and negative peaks

ANALYSIS OF A COMPLETE AMPLIFIER

The input resistance, output resistance, current gain, and voltage gain of a complete amplifier can be calculated from the information presented thus far.

Consider the RC coupled amplifier circuit shown in figure 6-21. The capacitors are selected to have negligible reactance at the signal frequency, and thus, can be ignored. The gate current is essentially zero. The equivalent resistance between the gate and the source can be assumed to be infinite, for all practical purposes.

The JFET described in the previous paragraphs has a gate current maximum of 0.25 nanoampere at a gate voltage of -2 volts. Compute the apparent equivalent resistance between the gate and the source terminals for this JFET.
(R8-6)

If the resistance R_g has too high a value, then there are problems in operating the amplifying device. This situation is due to several reasons which depend upon the device itself and the frequency at which the device is operated. The dc voltage drop across R_g must be negligible so that the gate bias can be set properly. The gate-to-source capacitance must be charged through R_g. This means that too high a value of R_g can increase the time constant of the gate circuit beyond acceptable levels. If the value of R_g is in the range between a few hundred kilohms and one megohm, then there is an acceptably high input impedance for the amplifier. The input impedance for the amplifier can be assumed to equal R_g for the condition that the equivalent gate resistance is much higher than R_g.

For any of the described amplifying devices, the characteristic curves are the result of plotting the dc load current versus the dc voltage across the output terminals of the device. The output curves shown in figure 8-1 were obtained by plotting the drain current versus the source voltage. Ohm's law can be used in conjunction with these curves to find the value of the apparent resistance between the drain and the source terminals.

For example, consider the point on the curves in figure 8-1 at which $V_{DS} = 10$ volts and $V_{GS} = -0.5$ volts. When V_{DS} changes from 8 to 12 volts, the drain current I_D changes from 9.2 milliamperes to 9.75 milliamperes. The apparent drain-to-source resistance is determined as follows:

$$r_{ds} = \frac{8 - 12}{9.2 - 9.75} = 7.72 \text{ kilohms}$$

Calculate the apparent drain-to-source resistance (r_{ds}) again, using the data given in the example, but with $V_{GS} = 0$.   *(R8-7)*

The output impedance for the amplifier circuit shown in figure 6-21 is R_d in parallel with the equivalent drain-to-source resistance.

The output curves like those in figure 8-1 can also be used to calculate the current gain for the circuit of figure 6-21. The input current (i_{in}) can be considered to flow through the resistor R_g only. The ac voltage drop across R_g is also equal to the change in V_{GS}. If i_{in} has a value of ± 1 microampere through a resistance (R_g) of 1 megohm, then the value of v_{gs} is ± 1 volt. This value is also the amplifier input voltage.

A circuit has the following values: $R_d = 1500$ ohms and $R_L = 3000$ ohms. If $V_{DSQ} = 12$ volts and $V_{GSQ} = -0.5$ volt in figure 8-1, what is the change in the drain current for a change in V_{GS} of ± 0.5 volt?   *(R8-8)*

The ac drain current divides between the output current and the current through R_d. Using the current divider theorem, calculate the output current, using the results of review problem (R8-8).   *(R8-9)*

Calculate the overall current gain for the amplifier of figure 6-21, using the values i_{in} = 1 microampere and R_g = 1 megohm, and the results of review problem (R8-9). (R8-10)

The value of current gain calculated in review problem (R8-10) is considerably lower than the current gain for the amplifying device by itself. This problem shows that the current gain for such an amplifier is actually determined by necessary and desirable circuit components other than the basic high-impedance amplifying device.

The voltage gain can be determined by using the curves given in figure 8-1. The values are read from the curves, using the ac load line, when a load resistance is attached.

Using the values given in problem (R8-8), compute the voltage gain for the simple amplifier circuit shown in figure 6-21. (R8-11)

Refer to the RC coupled amplifier circuit shown in figure 6-20. The capacitive reactances at the signal frequencies are assumed to be negligibly small. A JFET described by the curves in figure 8-1 is used in the circuit. Let V_{DD} = 20 volts, R_d = 800 ohms, R_L = 1000 ohms, and I_{DQ} = 10 milliamperes. Compute the voltage gain for an input of ± 0.5 volt. (R8-12)

Using a JFET in the circuit shown in figure 6-21, compute the value of R_s needed to achieve the approximate Q-point specified in problem (R8-12). To what value must R_d be changed to achieve operation on the same dc load line? (R8-13)

Compute the current gain of the amplifier described in problem (R8-12). Assume that the gate coupling resistor (R_g) has a value of 0.8 megohm and the gate current is negligible. (R8-14)

Refer to the amplifier circuit shown in figure 7-26. Assume that the pentode used is described by the curves in figure 7-19. The circuit values are as follows: R_g = 500 kilohms, R_k = 1 kilohm, V_{GK2} = 100 volts, V_{PP} = 400 volts, and R_p = 49 kilohms. The capacitors are selected to have negligible reactance at signal frequencies. The ac load R_L is set at 50 kilohms. Calculate the voltage gain for an input voltage of ± 1 volt. (R8-15)

What is the approximate input resistance of the amplifier described in problem (R8-15)? (R8-16)

What is the current gain of the amplifier described in problem (R8-15) if the ac load resistance is 50 kilohms. Assume that the grid currents are negligible. (R8-17)

DEFINITIONS OF TERMS USED IN FET SPECIFICATIONS

Manufacturers of FETs frequently use terms in their specifications that are not defined or explained. These terms generally are defined and used in textbooks covering the subject in some detail. Since these terms may not be defined in the specifications, the technician may not be able to determine if the information is important. For this reason, the following list defines those terms that have a bearing on the maximum operating capabilities of an FET device.

- BV_{GSS} is the avalanche breakdown voltage from the gate to the source when the gate is reverse biased. The drain is connected to the source.

- BV_{GDS} is the avalanche breakdown voltage from the gate to the drain when the gate is reverse biased and the source is the common lead.

- BV_{DGO} is the avalanche breakdown voltage from the drain to the gate when the source is open-circuited.

- BV_{SGO} is the avalanche breakdown voltage from the source to the gate when the drain is open-circuited.

- BV_{DSS} or BV_{DGS} is the avalanche breakdown voltage from the drain to the source when the gate is connected to the source.

- I_{GSS} is the leakage current in the gate.

- I_{DSS} is the drain current with V_{GS} at zero and V_{DS} at a high enough value to insure that any further increase in V_{DS} does not cause any additional increase in the drain current.

FET AMPLIFIER CONFIGURATIONS: THE COMMON DRAIN CIRCUIT

The common source amplifier configurations, such as the one shown in figure 6-20, were analyzed previously to determine the typical gains and impedances that can be expected. Two other FET amplifier configurations are still to be examined: (1) the common drain mode and (2) the common gate mode.

The common drain or source follower amplifier is shown in figure 8-8.

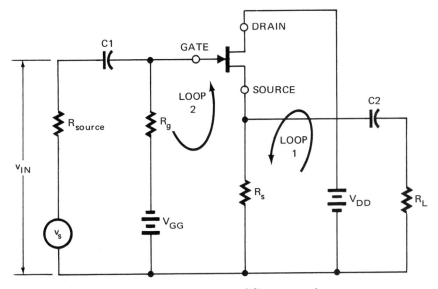

Fig. 8-8 Common drain amplifier connection

The common plate circuit is known as the cathode follower, the common collector circuit is known as the emitter follower, and the common drain circuit is known as the source follower. Explain why. (R8-18)

The method of setting the drain-to-source bias is the same for both the common source amplifier and the common drain connection. Kirchhoff's Voltage Law is used around Loop 1 in figure 8-8 to obtain the following expression:

$$V_{DD} - I_D R_s - V_{DS} = 0 \qquad \text{Eq. 8.4}$$

The expression is the same as Equation 6.1, with the exception that the resistor R_s is used rather than R_d. When $I_D = 0$, $V_{DD} = V_{DS}$, and when $V_{DS} = 0$, $I_D = V_{DD}/R_s$. These points locate both ends of the load line. The load line is drawn at a slope of $-\dfrac{1}{R_s}$. It can be compared with the load line drawn for figure 6-3.

To set the gate bias, Kirchhoff's Voltage Law is applied to Loop 2 in figure 8-8.

$$V_{GG} - (I_G + I_D)R_s - V_{GS} - I_G R_g = 0 \qquad \text{Eq. 8.5}$$

But, $I_G \approx 0$. So, solving for V_{GS}:

$$V_{GS} = V_{GG} - I_D R_s \qquad \text{Eq. 8.6}$$

$I_D R_s$ is set for the proper drain-to-source bias. V_{GG} must be of the opposite polarity with respect to ground to provide for the gate bias. To eliminate a second voltage source for V_{GS}, a number of circuits can be used. Typical circuits are shown in figures 8-9, 8-10, and 8-11.

The following analysis is made of the common drain amplifier shown in figure 8-8. The FET in the circuit is described by the curves given in figure 6-3. Let V_{DD} = 25 volts and R_s = 2 kilohms. The load line can be drawn through the points (I_D, V_{DS}) = (0, 25) and (I_D, V_{DS}) = (12.5, 0). V_{GS} = –2 volts at the Q-point. The Q-point value of –2 volts is the result of V_{GG} and the voltage across the resistor R_s. To eliminate the second power supply, the circuit shown in figure 8-9 is used.

The voltage across R_s is $I_{DQ} R_s$. Substituting I_{DQ} from the Q-point in figure 6-3, the voltage across R_s is (4.2)(2) = 8.4 volts. Note that this voltage is practically at cutoff. V_{GG} must be provided to counteract this condition. R_{g2} must be limited to a minimum value to prevent too low a value of the amplifier input impedance. R_{g2} is assumed to be 500 kilohms. V_{GG} is determined by the voltage divider formed by the resistors R_{g1} and R_{g2}.

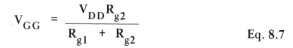

$$V_{GG} = \frac{V_{DD} R_{g2}}{R_{g1} + R_{g2}} \qquad \text{Eq. 8.7}$$

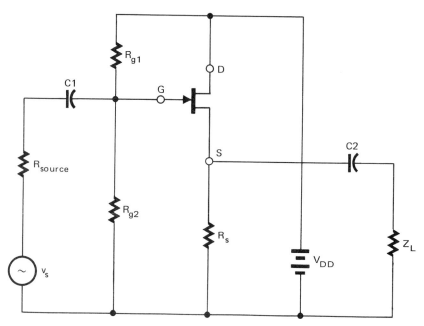

Fig. 8-9 Common drain amplifier connection showing gate biasing arrangement

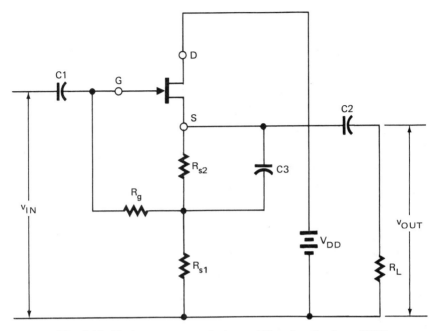

Fig. 8-10 Biasing a common drain amplifier circuit using a JFET

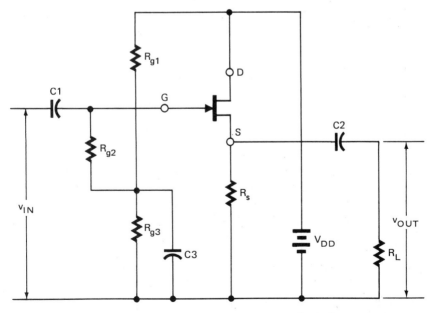

Fig. 8-11 Common drain amplifier showing a method of gate biasing

Known values are substituted in Equation 8.6 to solve for V_{GG}:

$$-2 = V_{GG} - 8.4$$

$$V_{GG} = 6.4 \text{ volts}$$

This value of V_{GG} can now be substituted in Equation 8.7 which then is solved for R_{g1}:

$$6.4 = \frac{(25)(500)}{R_{g1} + 500}$$

$$R_{g1} = 1450 \text{ kilohms}$$

This circuit has an input impedance of less than 500 kilohms (since R_{g1} is actually in parallel with R_{g2}).

Figure 8-10 is a preferred circuit for this application. In this circuit, the gate bias voltage is obtained by adjusting the relative values of R_{s1} and R_{s2}. The drain-to-source circuit is biased by adjusting the sum of R_{s1} and R_{s2}. If the same conditions are maintained as in the previous example, $R_{s1} + R_{s2} = 2$ kilohms and the value of $I_{DQ}(R_{s1} + R_{s2})$ is still 8.4 volts. Since V_{GSQ} is to be -2 volts, then this expression can be used:

$$V_{GSQ} = -I_{DQ}R_{s2} \qquad \text{Eq. 8.8}$$

because R_{s1} and R_{s2} form a voltage divider to supply the proper value of V_{GS}. Since $I_{DQ} = 8.4/(R_{s1} + R_{s2}) = 8.4/2 = 4.2$ milliamperes, this value is substituted in Equation 8.8 to obtain the value of R_{s2}:

$$-2 = -4.2R_{s2}$$

$$R_{s2} = 0.476 \text{ kilohms}$$

Thus,

$$R_{s1} = 2 - 0.476 = 1.524 \text{ kilohms}$$

R_g has little effect on the gate gias if its value is very much less than the input impedance of the FET. A rule of thumb is that the value of R_g should be less than one-tenth the input impedance of the FET.

A common drain amplifier, like the one in figure 8-9, uses a MOSFET whose characteristics are shown in figure 6-8. The drain-to-source voltage at

cutoff is 20 volts. The maximum drain current at $V_{DS} = 0$ is 25 milliamperes.
$V_{GSQ} = -0.5$ *volt.*

   a. Assuming that no load is connected, what is the value of the source resistor?

   b. Select the values of R_{g1} and R_{g2} to set V_{GSQ} at -0.5 volt.   *(R8-19)*

Other types of high-impedance amplifying devices can be used in a common drain circuit configuration. However, some of these devices may have bias voltage polarities different from those considered in the previous analysis. For example, there are enhancement-type MOSFETs, depletion-type MOSFETs, and p-channel FETs of various kinds. Recall that n-channel FETs are used in the examples considered to this point because the methods of analysis are then applicable to vacuum tubes as well. Figures 8-12 and 8-13 show common drain

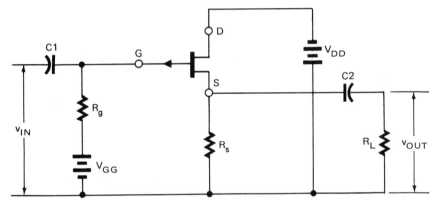

Fig. 8-12 Common drain circuit using a p-channel JFET (bias polarities shown)

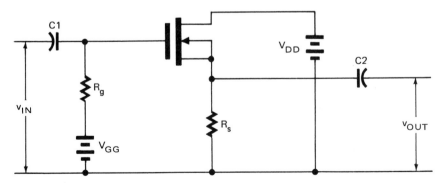

Fig. 8-13   Common drain circuit using an enhancement-type MOSFET (bias polarities shown)

circuits using these other types of devices. Note that the voltage polarities required are indicated. The circuit shown in figure 8-11 can also be used for the enhancement mode MOSFET.

The same MOSFET and conditions of operation given in review problem (R8-19) are applied to the circuit shown in figure 8-10. $R_L = \infty$ and $R_g = 500$ kilohms. Find the values of R_{S1} and R_{S2} necessary to achieve operation at the Q-point used for problem (R8-19). (R8-20)

Figure 8-14 shows a common drain amplifier circuit that utilizes the full high-input resistance capability of the MOSFET. A double Zener diode can be connected at the arrows to shunt the input. This diode device will break down and pass current when the input voltage exceeds the maximum allowable values. The disadvantage of this type of circuit protection is that the input resistance is decreased to a level equivalent to that of a reverse biased semiconductor diode.

THE COMMON GATE CIRCUIT

A common gate circuit for an *n*-channel JFET is shown in figure 8-15. To find the load line for the JFET, apply Kirchhoff's Voltage Law for the dc voltages in Loop 1.

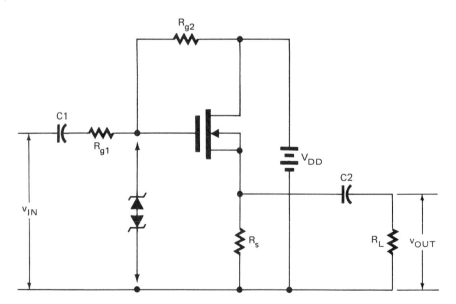

Fig. 8-14   Common drain amplifier using a MOSFET in the enhancement mode, with a high-impedance input. Two Zener diodes can be connected at the arrows to protect the MOSFET from high input voltages.

$$V_{DD} - V_{DS} - I_D(R_s + R_d) = 0 \qquad \text{Eq. 8.9}$$

When $I_D = 0$, $V_{DD} = V_{DS}$, and when $V_{DS} = 0$, $I_D = V_{DD}/(R_s + R_d)$. These points locate the load line which can then be drawn on the output characteristic curves. The resistor R_s includes the internal resistance of the signal source if the signal generator is connected in series with the source, as in figure 8-15. If the signal input is to be ac coupled, then the circuit appears like that of figure 8-16. An ac coupled input means that R_s must be matched to the signal source. The ac coupled circuit in figure 8-16 can be converted by Thévenin's theorem to the circuit shown in figure 8-15. Therefore, the setting of the bias resistors and voltages is accomplished in the same manner for both circuits.

As an example of the analysis of a common gate circuit, assume that a JFET having the characteristic curves of figure 6-3 is to be used in the circuit

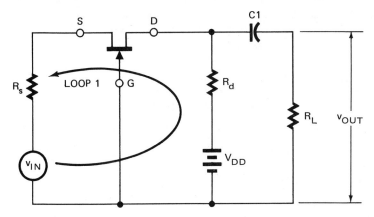

Fig. 8-15 Common gate circuit using an *n*-channel JFET

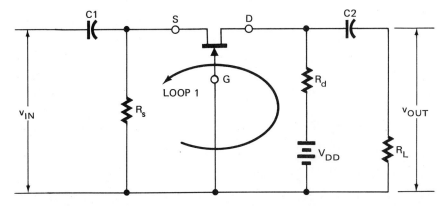

Fig. 8-16 Common gate circuit using *n*-channel JFET and ac coupled input

shown in figure 8-16.   Assume that the JFET is to be operated at the same Q-point as it was previously. V_{GSQ} = -2 volts and I_{DQ} = 4.2 milliamperes. The total of R_s and R_d is 2 kilohms for the same dc load line. What value must R_s be to make V_{GS} = -2 volts? A separate dc supply is *not* to be used for V_{GS}. Therefore, Equation 8.6 is to be used for V_{GG} = 0.

$$V_{GS} = -I_D R_s \qquad \qquad \text{Eq. 8.10}$$

Substituting and solving for R_s,

$$R_s = \frac{-(-2)}{4.2} = 0.476 \text{ kilohms}$$

The value of R_d is 2 - 0.476 = 1.524 kilohms. This value of R_d is the same as that used for setting the bias levels of the common drain amplifier. It is the same because the same device is being used at the same Q-point and load line. The only difference between the circuits is that the input and output terminals are redefined.

R_s is matched to the internal impedance of the signal source. The signal source resistance must be much smaller than R_s if the JFET is to act as a voltage amplifier. One of the disadvantages of the common gate circuit is that the circuit will have a low input resistance and a high output resistance. When R_s includes the source resistance, as in figure 8-15, the input signal is not attenuated. However, the signal source must be closely matched to the amplifier and it has to withstand the dc bias drain current.

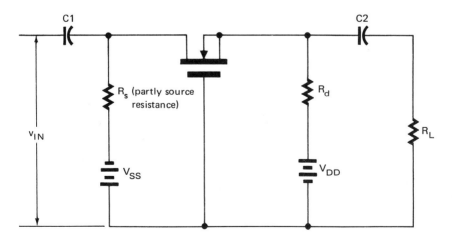

Fig. 8-17   Common gate amplifier using a MOSFET operated in an enhancement mode

A depletion mode MOSFET is biased in the same way as a JFET. If an n-channel MOSFET is to be operated in an enhancement mode as a common gate amplifier, it is biased as shown in figure 8-17. This arrangement is similar to the connection of an npn bipolar transistor in a common base circuit. (Refer to figure 3-9).

The common gate circuit is little used because of its performance characteristics. If a gate bias is required, the use of a second bias power supply cannot be avoided.

The MOSFET described by the curves in figure 6-8 is used in the common drain circuit shown in figure 8-11. V_{DD} = 20 volts, R_{g2} = 500 kilohms, and R_s = 1 kilohm. What values of R_{g1} and R_{g3} must be used to force the gate bias voltage to remain at 0 volts? (R8-21)

VACUUM TUBE CIRCUITS: COMMON PLATE CIRCUIT AND COMMON GRID CIRCUIT

The circuits and techniques used to set the bias levels for the common grid circuit and the common plate circuit are the same as those used for the common gate and common drain circuits of the JFET.

The common grid and common plate circuits are shown in figures 8-18 and 8-19. Note the similarity of these circuits to the common gate and common drain circuits for the JFET (refer to figures 8-8 and 8-15).

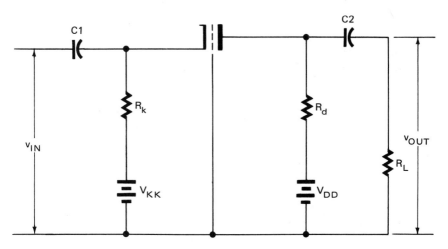

Fig. 8-18 Common grid circuit

The circuits and equations used to set the bias values for the vacuum tube circuits are the same as those for the JFET with the following substitutions:

	For JFET	For Vacuum Tubes
Symbol	GATE — DRAIN / SOURCE	GRID — PLATE / CATHODE
Output Bias Voltage	V_{DS}	V_{PK}
Input Circuit Bias Resistor	R_g, R_s	R_g, R_k
Input Bias Voltage	V_{GS}	V_{GK}
Input Supply Voltage	V_{GG}, V_{SS}	V_{GG}, V_{KK}
Output Supply Voltage	V_{DD}	V_{PP}
Output Current	I_D	I_P
Output Circuit Bias Resistor	R_d	R_p

Table 8-1

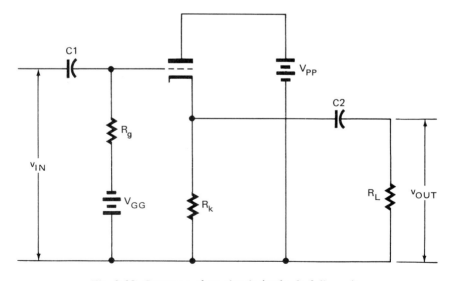

Fig. 8-19 Common plate circuit (cathode follower)

A pentode described by the curves given in figure 7-19 is to be used in a common plate circuit, such as the one shown in figure 8-19. V_{PP} is 400 volts. Additional circuitry not shown in figure 8-22 provides a screen grid voltage of 100 volts. The following assumptions are made: the load resistance $R_L = \infty$ and $R_k = 50$ kilohms. Determine V_{GG} for a control grid bias voltage of -3 volts.
(R8-22)

VACUUM TUBE AND FET SMALL SIGNAL PARAMETERS

Once the voltages and circuit components are selected so that the operating point of an amplifier can be set, the dc voltages and currents lose their significance. Further analysis of the amplifier is conducted by ignoring the bias currents and voltages and concentrating on its performance as an amplifier. This analytic process was started at the beginning of Chapter 8 where the amplifier response to large signals was investigated. In general, the analysis was based on prior knowledge of the Q-point and the desired load line.

The characteristic curves are useful in the study of amplifier performance for large signals and for those occasions where the Q-point is selected at some point close to cutoff or saturation. For linear amplifiers, the operation is restricted to small voltage or current signals. Linear amplifiers frequently are used in communications and for industrial control and instrumentation applications.

To analyze linear amplifiers, the dc levels are ignored and a group of small signal parameters is used to determine the ac equivalent circuit. This circuit permits an analysis of the amplifier performance using standard ac and dc network analysis techniques, including Thévenin's Theorem, Norton's Theorem, Kirchhoff's Laws, and others.

The small signal parameters are nearly constant (within reasonable limits) near the center of the active region of the device, regardless of the signal input or load. This factor is a major reason why these parameters are so useful.

The small signal parameters are first defined for the middle range of input signal frequencies. For ac coupled amplifiers, the signal frequency is assumed to be above the level where the capacitive or inductive coupling between the stages interferes with the amplifier performance. Another assumption is that the signal frequencies are low enough to insure that interelectrode capacitances and any other capacitances and inductances in the circuit have a negligible effect on the amplifier performance.

The small signal parameters for the various vacuum tube configurations (such as the triode and pentode) are the same as those for the FET configurations (such as JFETs and MOSFETs). They are the parameters that are to be used generally for high-impedance amplifying devices. The parameters are as follows:

1. the amplification factor, using the symbol μ.

2.   the dynamic plate or drain resistance (the symbol r_p is used for the plate resistance and the symbol r_d is used for the drain resistance).

3.   the transconductance or mutual conductance, using the symbol g_m.

For a vacuum tube, the amplification factor μ is defined as the ratio of the change in the plate voltage to the change in the grid voltage while the plate current is held constant at the quiescent point. Mathematically, the amplification factor is expressed by:

$$\mu = \frac{v_{pk}}{v_{gk}}, \text{ with } i_p = 0 \qquad \qquad \text{Eq. 8.11}$$

where the lower case symbols v and i indicate peak-to-peak variations in the voltages and currents. The subscripts define the voltages and currents at the circuit points which were determined in Chapter 7 (pages 262 to 276).

For the JFET or MOSFET, the amplification factor μ is defined as the ratio of the change in the drain voltage to the change in the gate voltage while the drain current is held constant. Mathematically, it is expressed by:

$$\mu = \frac{v_{ds}}{v_{gs}}, \text{ with } i_d = 0 \qquad \qquad \text{Eq. 8.12}$$

where the subscripts refer to the circuit points discussed in Chapter 6 (pages 217 to 231).

The plate resistance r_p is defined as the ratio of the change in the plate voltage to the change in the plate current while the grid-to-cathode voltage is held constant. It is expressed symbolically by:

$$r_p = \frac{v_{pk}}{i_p}, \text{ with } v_{gk} = 0 \qquad \qquad \text{Eq. 8.13}$$

The drain resistance is defined in much the same way as the plate resistance, except that the subscripts are changed to FET nomenclature. Thus, the drain resistance is:

$$r_d = \frac{v_{ds}}{i_d}, \text{ with } v_{gs} = 0 \qquad \qquad \text{Eq. 8.14}$$

FET specifications may list r_d values as r_{ds} or r_{DS}. Occasionally, FET specifications may give values for $r_{ds(ON)}$. The symbol "$r_{ds(ON)}$" is *not* r_d in the active region and should not be used in linear amplifier analysis. The parameter $r_{ds(ON)}$ is more useful when the FET is used at low voltages as a voltage-sensitive resistor or switch.

Transconductance for a vacuum tube is defined as the ratio of the change in the plate current to the change in the grid-to-cathode voltage while the plate-to-cathode voltage is held constant. This parameter is expressed as follows:

$$g_m = \frac{i_p}{v_{gk}}, \text{ with } v_{pk} = 0 \qquad \text{Eq. 8.15}$$

Since it is convenient to express a current such as i_p in milliamperes, g_m is expressed in millimhos. In addition, r_p can be expressed in kilohms, for the same reason.

For an FET, the transconductance is defined as the ratio of the change in drain current to the change in the gate-to-source voltage while the drain-to-source voltage is held constant. FET transconductance is expressed by the following equation:

$$g_m = \frac{i_d}{v_{gs}}, \text{ with } v_{ds} = 0 \qquad \text{Eq. 8.16}$$

FET specifications commonly list data for the quantities g_{mo} or y_{fs}. The values given for these symbols are the same as the transconductance g_m. For example, g_{mo} is actually the value measured at $V_{GS} = 0$, but it is approximately the same as g_m. The specifications will indicate that g_{mo} is g_m measured for $I_D = I_{DSS}$, which is I_D at $V_{GS} = 0$.

The interrelationships between these parameters will be analyzed in the following paragraphs. The peak-to-peak variations should be as small as possible if the given equations are to represent the parameters accurately. For small variations, the following equation expresses the relationship between the parameters for the vacuum tube.

$$\mu = g_m r_p \qquad \text{Eq. 8.17}$$

A similar expression applies to the FET:

$$\mu = g_m r_d \qquad \text{Eq. 8.18}$$

This expression is obtained for either vacuum tubes or FETs by combining the equations for the three parameters. For example, substitute Equations 8.14 and 8.16 into Equation 8.18 to obtain μ for the family of FET devices.

$$\mu = \frac{v_{ds}}{i_d} \cdot \frac{i_d}{v_{gs}}$$

$$\mu = \frac{v_{ds}}{v_{gs}}$$

Note that this last expression is Equation 8.12. This relationship between the parameters is strictly true only when the peak-to-peak variations approach zero magnitudes.

The parameters can be obtained by testing the FET or vacuum tube device in a circuit specifically designed for this purpose. Alternatively, if characteristic curves are available, then the parameters can be measured on the curves.

As an example of the use of curves to calculate the parameter μ for a vacuum tube, refer to the plate curves shown in figure 8-20. The Q-point is assumed to be at V_{GKQ} = -1.5 volts, I_{PQ} = 1.5 milliamperes, and V_{PKQ} = 220 volts. The amplification parameter μ is determined by substituting the appropriate values as measured from figure 8-20 in Equation 8.11. These values are measured over an arbitrarily selected increment of ΔV_{GK} = 0.5 volts while I_P is held constant. Thus, μ is expressed as follows:

$$\mu = \frac{v_{pk}}{v_{gk}} = \frac{V_{PK1} - V_{PK2}}{V_{GK1} - V_{GK2}} \qquad \text{Eq. 8.19}$$

$$\mu = \frac{220 - 268}{-1.5 - (-2.0)} = -96$$

V_{PK1}, V_{PK2}, V_{GK1}, and V_{GK2} are identified on figure 8-20.

For this same device, the plate resistance is calculated by substituting values in Equation 8.13. These values are measured on figure 8-20 over an arbitrarily selected increment of ΔV_{PK} = 30 volts about the Q-point for V_{GKQ} = -1.5 volts.

$$r_p = \frac{v_{pk}}{i_p} = \frac{V_{PK1} - V_{PK3}}{I_{P1} - I_{P3}} \qquad \text{Eq. 8.20}$$

$$r_p = \frac{220 - 250}{1.5 - 2.06} = 53.6 \text{ kilohms}$$

V_{PK3}, I_{P1}, and I_{P3} are identified on figure 8-23.

Finally, the transconductance for this vacuum tube is determined by substituting the appropriate values in Equation 8.15. These values are measured over an arbitrarily selected increment of ΔV_{GK} = 0.5 volts for V_{PK} = 220 volts. The transconductance, g_m, is as follows:

$$g_m = \frac{i_p}{v_{gk}} = \frac{I_{P1} - I_{P2}}{V_{GK1} - V_{GK2}} \qquad \text{Eq. 8.21}$$

$$g_m = \frac{1.5 - 0.75}{-1.5 - (-2.0)} = 1.5 \text{ millimhos}$$

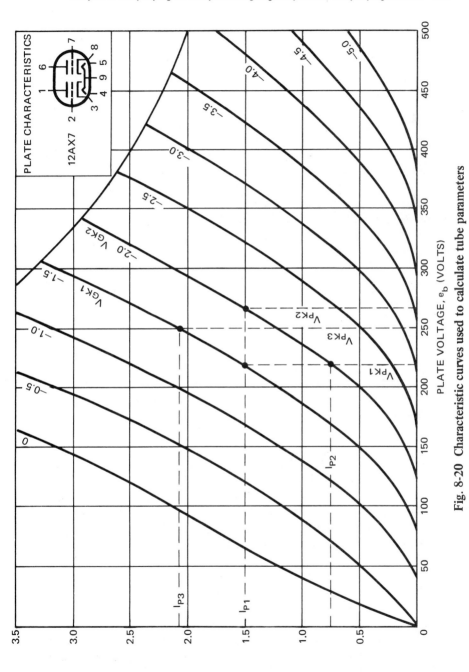

Fig. 8-20 Characteristic curves used to calculate tube parameters

The value of μ can be checked by substituting values of g_m and r_p in Equation 8.17.

$$\mu = g_m r_p = (1.5)(53.6) = 80.4$$

Compare this value for μ with the one obtained previously by taking measurements from the curves in figure 8-20. If smaller increments are selected over which to take the measurements, then there is a closer correspondence between the two values. The value of μ is a negative 80.4, but the origin of the sign was lost during the calculation of g_m and r_p.

Refer to figure 7-9, and calculate values of μ, g_m, and r_p for this vacuum triode about the Q-point determined in Chapter 7 (pages 262 to 276) and recorded on figure 7-9. Select arbitrarily small increments about the Q-point for the measurements.   *(R8-23)*

Check the value of μ obtained in review problem (R8-23) by the use of Equation 8.18.   *(R8-24)*

Recalculate values of μ, g_m, and r_p for the conditions given in review problems (R8-23) and (R8-24) for changes in V_{PK}, I_P, and V_{GK} that are about one-quarter the magnitudes used previously. Recheck the value of μ using Equation 8.17. Compare this value of μ with the value calculated from Equation 8.11.   *(R8-25)*

Using the same techniques, values of μ, g_m, and r_d can be calculated for a JFET or a MOSFET. The graphical calculations for the FET will be shown because the values of the parameters differ from those for the vacuum tube and the characteristic curves have a different shape. Refer to the FET curves shown in figure 8-21. A Q-point is assumed that is specified by $V_{GSQ} = -1$ volt, $V_{DSQ} = 10.4$ volts, and $I_{DQ} = 5.67$ milliamperes. The amplification factor is determined by substituting measured values in Equation 8.12. These values are taken over an arbitrarily selected increment of $v_{gs} = \Delta V_{GS} = 0.5$ volt while I_D is held constant at 5.7 milliamperes.

$$\mu = \frac{v_{ds}}{v_{gs}} = \frac{V_{DS1} - V_{DS2}}{V_{GS1} - V_{GS2}} \qquad \text{Eq. 8.22}$$

$$\mu = \frac{2.6 - 10.4}{-0.5 - (-1.0)} = -15.6$$

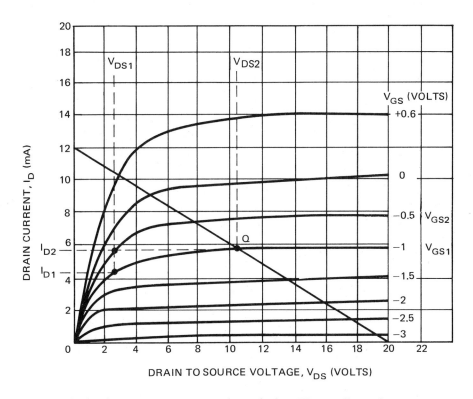

Fig. 8-21 Characteristic curves used to calculate FET small signal parameters

The drain resistance for the FET is determined by substituting measured values in Equation 8.14. The change in the drain-to-source voltage (v_{ds}) used for the calculation of μ is used for this calculation also. The gate-to-source voltage is held at a constant value of –1.0 volt.

$$r_d = \frac{v_{ds}}{i_d} = \frac{V_{DS1} - V_{DS2}}{I_{D1} - I_{D2}} \qquad \text{Eq. 8.23}$$

$$r_d = \frac{2.6 - 10.4}{4.3 - 5.7} = 5.57 \text{ kilohms}$$

Finally, the transconductance for the FET is determined by substituting measured values in Equation 8.16. The change in the drain current (i_d) and the gate-to-source voltage (v_{gs}) used in the previous calculations are used here also. V_{DS} is held constant at 2.6 volts.

$$g_m = \frac{i_d}{v_{gs}} = \frac{I_{D1} - I_{D2}}{V_{GS1} - V_{GS2}} \qquad \text{Eq. 8.24}$$

$$g_m = \frac{4.3 - 5.7}{-1 - (-0.5)} = 2.8 \text{ millimhos}$$

The value of the transconductance (g_m) for the FET compares to the value for the vacuum triode. However, μ and the dynamic resistance for the FET are much lower than the corresponding parameters for the vacuum triode. It is suggested that the calculations required for review problems (R8-26) and (R8-27) be performed and the resulting values compared.

The value of μ can be checked against Equation 8.18 by using the calculated values of g_m and r_d as follows:

$$\mu = g_m r_d = (2.8)(5.57) = 15.6$$

The sign of μ is negative, but the sign was lost during the calculations for g_m and r_d.

The parameters for a MOSFET described by the curves shown in figure 6-6 are to be determined. The Q-point is defined by V_{DSQ} = 15 volts, V_{GSQ} = 4 volts, and I_{DQ} = 4.2 milliamperes. Calculate values for the parameters μ, g_m, and r_d.   (R8-26)

A pentode described by the curves of figure 7-19 has a Q-point defined by V_{G2KQ} = 100 volts, V_{PKQ} = 100 volts, V_{G1KQ} = -2 volts, and I_{PQ} = 4.7 milliamperes. Calculate the values for μ, g_m, and r_p for this device.   (R8-27)

Tabulate the values of μ, g_m, and r_p (r_d) from a JFET and a triode, as calculated from the illustrative example and from a MOSFET and pentode as calculated in review problems (R8-26) and (R8-27). Compare the relative values for each of these devices in an amplifying circuit. Which device has the highest gain, μ? When a judgment is made as to which device is more versatile, what is the significance of r_p and r_d?   (R8-28)

The development of the FET led to the use of *y-parameters*. These parameters are analogous to the h-parameters for the bipolar transistor, and they are defined in a similar way. (Refer to the introductory discussion of h-parameters in Chapter 4.) The y-parameter definitions, for ac signals only, are made as follows:

1. y_i = input admittance with output terminals shorted

For a common source circuit, y_i is defined as the change in the gate current divided by the change in the gate-to-source voltage, with the drain-to-source voltage held constant. In the form of an equation, it is:

$$y_{is} = \frac{i_g}{v_{gs}}, \text{with } v_{ds} = 0$$

Eq. 8.25

The subscript "s" indicates that it is for a common source circuit. The parameter is significant only when the FET is used for high-frequency circuits. It is equivalent to the reciprocal of the small signal gate-to-source resistance at low frequencies.

2. y_o = output admittance with input terminals shorted

For a common source circuit, y_o is defined as the change in the drain current divided by the change in the drain-to-source voltage, with the gate-to-source voltage held constant. In equation form, it is:

$$y_{os} = \frac{i_d}{v_{ds}}, \text{with } v_{gs} = 0$$

Eq. 8.26

Referring to Equation 8.14, it is seen that:

$$y_{os} = \frac{1}{r_d}$$

Eq. 8.27

3. y_f = forward transfer admittance with output terminals shorted

For a common source circuit, y_f is defined as the change in the drain current divided by the change in the gate-to-source voltage, with the drain-to-source voltage held constant. This parameter is the same as the transconductance (g_m) defined by Equation 8.16. For the common source circuit, the forward transfer admittance is expressed by the symbol, y_{fs}.

4. y_r = reverse transfer admittance with the input terminals shorted

For a common source circuit, y_r is the change in the gate current divided by the change in the drain-to-source voltage, with the gate-to-source voltage held constant. In equation form, this is:

$$y_{rs} = \frac{i_g}{v_{ds}}, \text{with } v_{gs} = 0$$

Eq. 8.28

The subscript "s" in the four y-parameters changes to "d" or "g" for the common drain or the common gate circuits. The common source y-parameters are preferred for describing FET small signal characteristics. Generally a circuit can be analyzed sufficiently without resorting to the use of y-parameters. However, some knowledge of y-parameters is required because manufacturers frequently define FET performance in terms of y-parameters. They are very useful in analyzing high-radio-frequency circuits. Further discussion of y-parameters is given in Chapter 9.

AMPLIFIERS USING MIXED TYPES OF DEVICES

Frequently, two types of amplifying devices are connected together to form an amplifier that uses the advantages of one type of device and overcomes the faults caused by the second type of device. One example of such an arrangement is the *cascode amplifier* shown in figure 8-22. This amplifier uses bipolar transistors to produce a high gain and high output impedance. Transistor Q1 is a common emitter stage and transistor Q2 is a common base stage. The use of two bipolar transistors means that the input impedance of the amplifier is low. The reason for this is that the base-to-emitter junction of a bipolar transistor is normally forward biased for operation in the active region. The problem of a low input impedance is solved by the use of an FET for the first stage, as in figure 8-23. The bipolar transistor is used as a common base stage because the low input impedance for this stage does not affect the external circuitry. This circuit is analyzed by considering the common base stage as a separate common

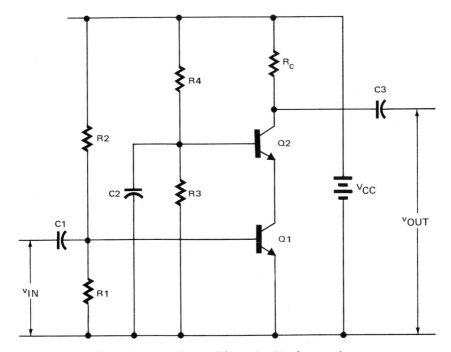

Fig. 8-22 Cascode amplifier using bipolar transistors

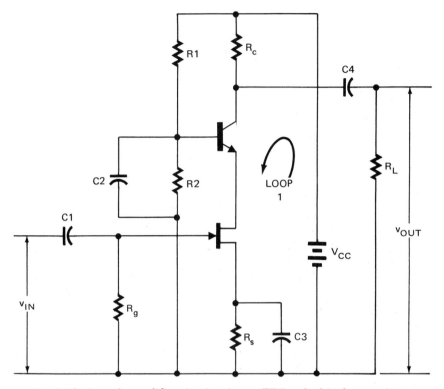

Fig. 8-23 Cascode amplifier circuit using an FET and a bipolar transistor

base amplifier, and then treating the FET stage as a separate common source amplifier. A cascode amplifier using two FETs will have a circuit very similar to figure 8-23, if the first stage is connected as a common source amplifier and the second stage is connected as a common gate amplifier.

As an example of the analysis of a cascode amplifier, assume that the bipolar transistor in figure 2-7 and the JFET in figure 6-3 are used in the circuit shown in figure 8-23. Let V_{CC} = 25 volts, R_c = 0.68 kilohm, R_s = 0.32 kilohm, R1 = 37 kilohms, R2 = 100 kilohms, and R_g = 500 kilohms. To analyze the biasing of this circuit, it is assumed that linear amplification is achieved when both of the transistors operate near the center of their active regions. The easiest way to analyze the biasing is to assume some values for the currents and voltages. The circuit equations are then solved to determine if the assumptions made are correct. To achieve linear amplification, assume that V_{GS} = -1.6 volts and V_{DS} = 12 volts. Using Equation 8.10,

$$I_D = \frac{V_{GS}}{R_s} = \frac{1.6}{0.32} = 5 \text{ milliamperes}$$

If the base current is neglected, then I_D is also the collector current. The biasing of the bipolar transistor can be analyzed by making a Thévenin equivalent circuit of the transistor base-to-emitter circuit, as shown in figure 8-24. The Thévenin equivalent voltage applied to the base is:

$$V_{Th} = \frac{V_{CC}R2}{R1 + R2}$$

Eq. 8.29

The Thévenin equivalent resistance looking toward the input from the base of the transistor is:

$$R_{th} = \frac{R1R2}{R1 + R2}$$

Eq. 8.30

Writing Kirchhoff's Voltage Law around this circuit:

$$\frac{V_{CC}R2}{R1 + R2} - \frac{I_b R1R2}{R1 + R2} - V_{BEQ} - V_{DSQ} - I_D R_s = 0$$

Eq. 8.31

V_{BEQ} is small and can be neglected for approximate calculations. For this analysis, it is assumed that 0.6 volt is a good value to use as V_{BEQ} for a silicon transistor. The values known to this point in the analysis can be substituted in Equation 8.31.

$$\frac{(25)(100)}{37 + 100} - \frac{(100)(37)}{100 + 37} I_B - 0.6 - 12 - 1.6 = 0$$

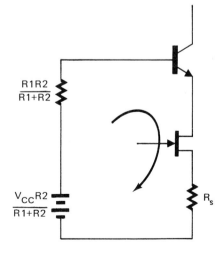

Fig. 8-24 Equivalent circuit used to obtain the gate bias for the circuit of figure 8-22

Thus, I_B = 0.148 milliampere. This value of base current is too high and means that the bipolar transistor is saturated. This condition means that the FET is operating on a steeper dc load line than was assumed. If the FET has a steeper load line, V_{DSQ} is higher for the same values of I_{DQ} and V_{GSQ}. Assume that V_{DSQ} = 14 volts. I_B can be recalculated using Equation 8.31 with V_{DSQ} = 14 volts instead of 12 volts. As a result, I_B = -0.074 milliampere. The total resistance in series with the collector bias voltage includes the apparent drain resistance of the FET. This resistance is equal to:

$$\frac{V_{DSQ}}{I_{DQ}} = \frac{14}{5} = 2.9 \text{ kilohms}$$

The total collector bias resistance locates the dc load line for the bipolar transistor. This resistance is:

$$\frac{V_{DSQ}}{I_{DQ}} + R_c + R_s = 2.9 + 0.32 + 0.68 = 3.9 \text{ kilohms}$$

As shown in figure 8-25, page 320, the load line intersects the I_c axis at:

$$\frac{V_{CC}}{\text{Total collector bias resistance}} = \frac{25}{3.9} = 6.4 \text{ milliamperes}$$

An examination of this load line indicates that a value of I_{BQ} of 0.074 milliampere is still too high. Therefore, a greater value of V_{DSQ} is assumed and I_B is again recalculated using Equation 8.31. For V_{DSQ} = 14.5 volts, I_B changes to 0.055 milliampere. This value is close to a match up with the dc load line. At I_{BQ} = 0.055 milliampere and I_{CQ} = 5 milliamperes, V_{CEQ} is about 6 volts. To check this value, Kirchhoff's Voltage Law is applied around Loop 1 in figure 8-24.

$$V_{CE} = V_{CC} - V_{DS} - I_D(R_s + R_c) \qquad \text{Eq. 8.32}$$

$$V_{CE} = 25 - 14.5 - 5 = 5.5 \text{ volts}$$

The total drain bias resistance that locates the dc load line for the FET includes the apparent collector resistance of the bipolar transistor. The apparent collector resistance is equal to:

$$\frac{V_{CEQ}}{I_{CQ}} = \frac{6}{5} = 1.2 \text{ kilohms}$$

The total drain bias resistance is determined by:

$$\frac{V_{CEQ}}{I_{CQ}} + R_c + R_s = 1.2 + 0.32 + 0.68 = 2.2 \text{ kilohms}$$

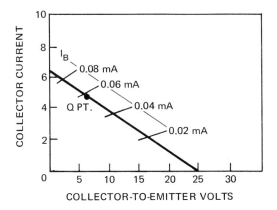

(a) LOAD LINE AND BASE BIAS CURRENTS FOR THE
BIPOLAR TRANSISTOR OF FIGURE 2-7.

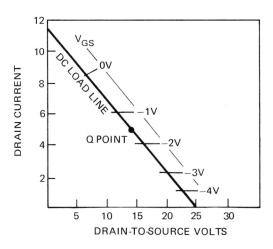

(b) LOAD LINE AND GATE BIAS VOLTAGES FOR THE FET OF FIGURE 6-3.

Fig. 8-25   Load lines and bias values for the FET of figure 6-3 and the bipolar transistor of figure 2-7

The load line intersects the I_D axis at a value of:

$$\frac{V_{CC}}{\text{Total drain bias resistance}} = \frac{25}{2.2} = 11.36 \text{ milliamperes}$$

In this manner, the dc load line is determined for the FET and the Q-point is located at approximately $I_{DQ} = 5$ milliamperes, $V_{DSQ} = 14.2$ volts, and

V_{GSQ} = -1.6 volts. Note that these values are close to the assumed values. Figure 8-25 shows that for this circuit, the FET has a better operating bias level than does the bipolar transistor.

A cascode amplifier circuit can be built using one dual insulated gate FET, as in figure 8-26. Gate 1 serves as the input of the first stage. The source is connected to ground through capacitor C1 for ac signals, thus providing the common source stage. Gate 2 is connected to ground through capacitor C2 for ac signals. The output is taken from the drain to the source, to provide conditions for a common gate stage.

The cascode amplifier is commonly used for high-frequency signals. For high-frequency signals, the inductances and capacitances in the circuit are tuned to the expected signal frequency.

A variation on the emitter coupled differential amplifier shown in figure 3-16 is a circuit using an FET, such as the circuit in figure 8-27, page 322. Such a circuit is useful in those situations where the source of the signal for the left stage has a vastly different resistance than does the source for the right stage. The bias voltage and current levels for this amplifier are analyzed by taking the FET and bipolar sections separately. That is, the bias conditions are determined first for one section and then for the other section.

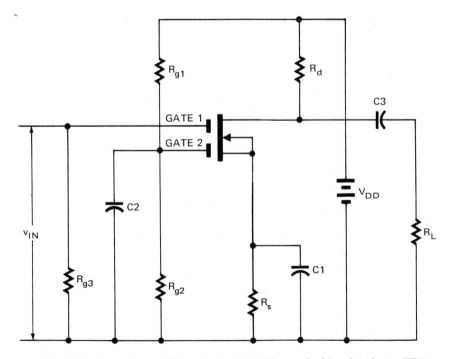

Fig. 8-26 Cascode amplifier constructed with one dual insulated gate FET

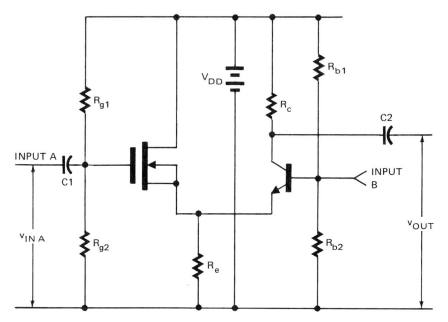

Fig. 8-27 Difference amplifier using one FET and one bipolar transistor

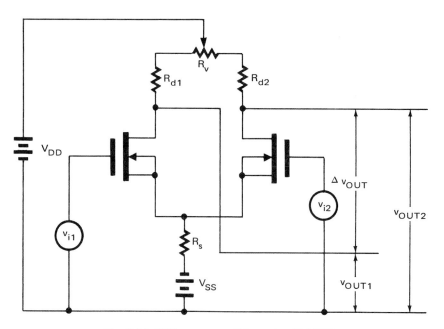

Fig. 8-28 Difference amplifier using MOSFETs

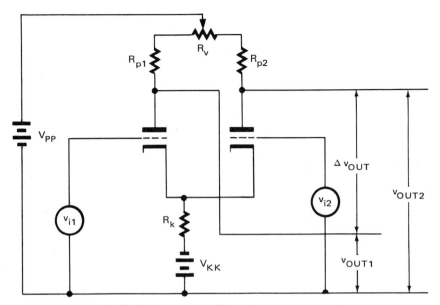

Fig. 8-29 Difference amplifier using vacuum triodes

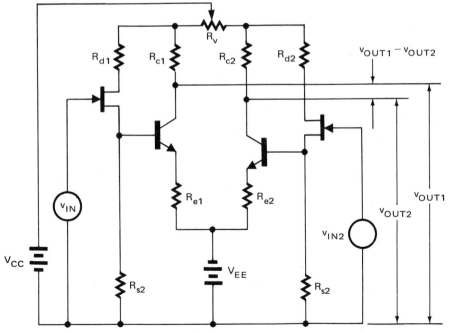

Fig. 8-30 Difference amplifier using FETs and bipolar transistors together

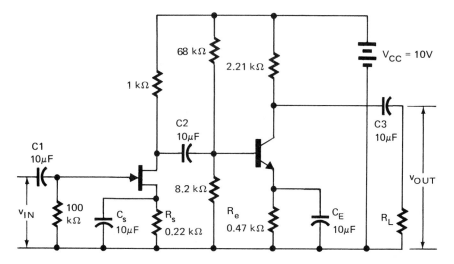

Fig. 8-31 Two-stage amplifier using an FET and a bipolar transistor

The difference amplifier can use two FETs or two vacuum tubes, as in figures 8-28 or 8-29. One of the FETs can be a JFET and the other a MOSFET. One of the vacuum tubes can be a pentode and the other can be a beam power tetrode.

Figure 8-30 is the schematic of a two-stage difference amplifier. An FET is used so that each input signal can see a high impedance. The output signal is the difference between v_{out1} and v_{out2}, or between v_{out1} and common, or between v_{out2} and common.

Another circuit uses both FETs and bipolar transistors. This circuit consists of ordinary common emitter and common source amplifiers connected in cascade for a mixed form of multistage amplifier. Thus, an amplifier such as the one shown in figure 3-15 can be built using an FET for the first stage, as shown in figure 8-31. Capacitors C_s and C_e short out the source and emitter circuit resistors R_s and R_e for ac signals, so that the amplifier has common source and common emitter stages. Resistors R_s and R_e provide the proper bias levels and temperature stabilization for the circuit.

LABORATORY EXPERIMENTS

The following experiments represent the theory covered in this section and have been successfully performed.

LABORATORY EXPERIMENT 8-1
OBJECTIVE

To build and operate a common drain amplifier.

EQUIPMENT AND MATERIALS REQUIRED

1   Transistor, 2N4224

2   Capacitors, 10 microfarads, 50 volts

1   Resistor, 220 kilohms, 1/4 watt

1   Decade resistance box, like the Heathkit DR-1

1   Oscilloscope, dual channel, time based

1   Volt-ohm-milliammeter, high impedance, capable of measuring voltages down to 1 millivolt, ac or dc

2   Power supplies, dc, regulated, 1/2 ampere, 0-30 volts adjustable

1   Audio signal generator, 0-1 volt, 0-20 kilohertz, sinusoidal waveform

PROCEDURE

1.   Obtain a 2N4224 JFET. The connection diagram and maximum allowable voltages are given in Laboratory Experiment 6-2.

2.   Connect the JFET into a circuit such as the one shown in figure 8-32.

3.   Adjust the power supply voltages: V_{DD} = 25 volts and V_{GG} = -2 volts.

4.   Apply a sinusoidal input signal of 20 kilohertz from the signal generator. Bring the magnitude of the signal up to a maximum value which is below the maximum voltages and currents that are allowed for this transistor. Then reduce the signal to a value low enough to insure that linear amplification is achieved.

5.   Measure the values of v_{in} and v_{out} on the oscilloscope. Use both channels simultaneously so that the phase relationships can be observed.

6.   Calculate the voltage gain of this circuit using Equation 8.1:

$$A_V = \frac{v_{out}}{v_{in}}$$

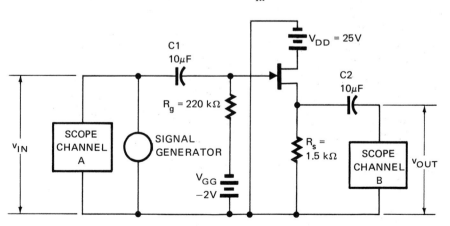

Fig. 8-32 Common drain amplifier

7. Measure the approximate phase shift of the amplifier by observing the time-phase relationship between v_{in} and v_{out} on the oscilloscope.

8. This circuit has the disadvantage of requiring two dc power supplies. To overcome this feature, reconnect the circuit so that it is like figure 8-33. This circuit uses a portion of the drain resistance to develop a voltage across it in such a way that a negative bias is put on the gate of the FET.

9. Adjust the relative values of R_{s2} and R_{s1} so that the Q-point obtained is the same as for the first circuit. The total resistance of $R_{s1} + R_{s2}$ should equal the resistance of 1.5 kilohms that was used as the drain resistor for the circuit shown in figure 8-32. The voltage across R_{s1} is the gate bias for the circuit. Calculate R_{s1} from:

$$R_{s1} = \frac{V_{GS}}{I_D} \qquad \text{Eq. 8.33}$$

10. Demonstrate that this circuit performs as well as the circuit of figure 8-32.

11. Calculate the voltage gain and the phase shift of the circuit as in steps 6 and 7 for the circuit of figure 8-32.

LABORATORY EXPERIMENT 8-2

OBJECTIVE

To measure the small signal parameters of a high-impedance amplifying device

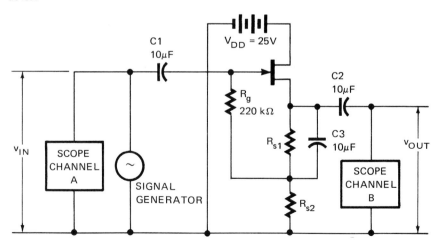

Fig. 8-33 Common drain circuit with self-biased gate

EQUIPMENT AND MATERIALS REQUIRED

1   Transistor, 2N4224
2   Capacitors, 10 microfarads, 50 volts
1   Resistor, 220 kilohms, 1/4 watt
1   Decade resistance box, like the Heathkit DR-1
1   Resistor, 1 ohm, 1%
1   Oscilloscope, dual channel, time based
1   Volt-ohm-milliammeter, high impedance, capable of measuring voltages down to 1 millivolt, ac or dc
2   Power supplies, dc, regulated, 1/2 ampere, 0-30 volts adjustable
1   Audio signal generator, 0-1 volt, 0-20 kilohertz, sinusoidal waveform

PROCEDURE

1. Obtain a 2N4224 JFET. The connection diagram and maximum allowable voltages are given in Laboratory Experiment 6-2.

2. Connect the JFET into the circuit shown in figure 8-34. Use the decade resistance box as the drain bias resistor and adjust its value to 1500 ohms.

3. Adjust the power supply voltages: V_{DD} = 25 volts and V_{GG} = -2 volts.

4. Determine I_D by measuring the voltage across the 1500-ohm bias resistance and dividing this value by 1500 ohms.

5. Measure V_{DS}.

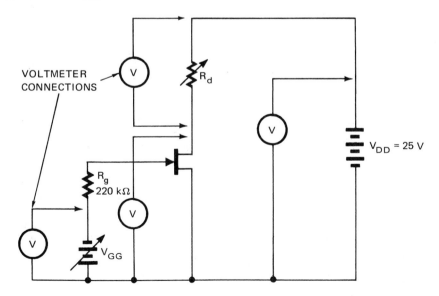

Fig. 8-34 Circuit for measuring the amplification factor

6. Adjust R_d to 2 kilohms. This adjustment will change V_{DS} and, to a lesser extent, I_D.

7. Adjust V_{GS} (by varying V_{GG}) so that I_d is adjusted to the original level.

8. Measure V_{GS} and V_{DS}.

9. Calculate the amplification factor μ using the following equation:

$$\mu = \frac{(V_{DS} @ 1500 \text{ ohms}) - (V_{DS} @ 2000 \text{ ohms})}{(-2) - (V_{GS} @ 2000 \text{ ohms})} \qquad \text{Eq. 8.34}$$

10. Connect the circuit shown in figure 8-35 so that the dynamic drain resistance can be determined.

11. Apply a small 20 kilohertz sinusoidal signal to the drain circuit. The magnitude of the signal is not critical, but it should have a minimum amplitude and yet be large enough to be measured accurately with the oscilloscope.

12. Measure v_{ds} and i_d. To determine i_d, measure the ac voltage across the 1-ohm resistor with the oscilloscope or a high-impedance ac volt-ohm-milliammeter (VOM). (If the VOM is used for one measurement and the scope is used for the other, remember that the VOM measures rms values and the oscilloscope reads the peak-to-peak value directly.)

13. The dynamic drain resistance is given by Equation 8.14:

$$r_d = \frac{v_{ds}}{i_d}$$

Note that the ac short placed across the gate-to-source circuit in figure 8-35 means that v_{gs} is made equal to zero.

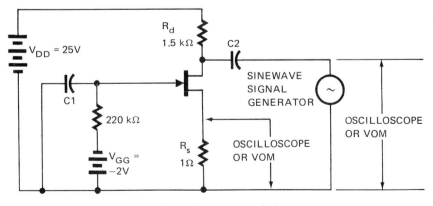

Fig. 8-35 Circuit used to measure drain resistance, r_d

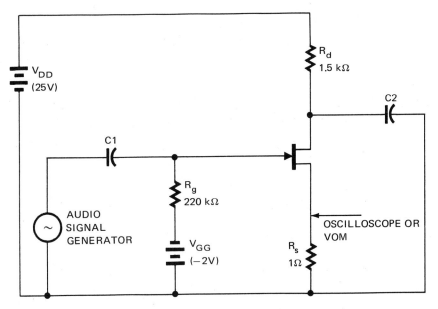

Fig. 8-36 Circuit used to measure transconductance

14. The transconductance is the next value to be measured. To accomplish this, the capacitor C2 is connected as shown in figure 8-36. This causes an ac short across the drain-to-source terminals, so that v_{ds} is made equal to zero. Connect the circuit shown in figure 8-36 to make this measurement.

15. Apply a sinusoidal signal of 20 kilohertz to the gate circuit as shown in figure 8-36. The amplitude should be at a minimum and yet large enough to be measured accurately with the available laboratory equipment.

16. Measure v_{gs} and i_d. Refer to step 12 for the method of measuring i_d.

17. Calculate the transconductance using Equation 8.16.

$$g_m = \frac{i_d}{v_{gs}}$$

18. Using the characteristic curves for the 2N4224 transistor, measure and calculate the three small signal parameters and compare them with the test results.

EXTENDED STUDY TOPICS

1. What conditions must exist if an amplifying device is to produce an undistorted signal?

2. Describe the differences and similarities between the characteristic curves for JFETs, MOSFETs, pentodes, triodes, and bipolar transistors.

3. Sketch a common source amplifier circuit and show the proper voltage polarities for a p-channel MOSFET in the enhancement mode.

4. Explain why a p-channel JFET is *not* operated with the gate negative with respect to the source.

5. Compare the voltage gains obtainable with the FET to the voltage gains obtainable with vacuum tubes and bipolar transistors. Which device produces the least voltage gain?

6. Can a JFET be used in an enhancement mode? Give the reason for this answer.

7. Explain how amplitude distortion is caused in a JFET amplifier for this situation: the magnitude of the input signal causes the gate to be driven at a voltage whose polarity with respect to the source is the same as the drain-to-source bias voltage.

8. When a MOSFET is operating in the enhancement mode, the gate current is almost zero. Note that the gate current is also almost zero when the MOSFET is operated in the depletion mode. However, there is a limit to the input signal magnitude. Explain why this limit exists and why it must be observed.

9. If the output signal from a single-stage amplifier is clipped an equal amount for both the positive and negative peaks, what conclusion can be made about the location of the Q-point?

10. What criteria are used to specify the resistor which is connected from the gate to the source in a self-biased MOSFET amplifier?

11. What is the purpose of connecting a double Zener diode between the gate and the source of a MOSFET?

12. State an advantage of combining FETs and bipolar transistors in a multi-stage amplifier.

13. Can some of the circuits used in the stabilization of bipolar transistor amplifier circuits be used for stabilizing FET amplifier circuits as well? What circuits can be used to stabilize an amplifier using an enhancement mode MOSFET?

14. Write the equations defining μ, g_m, and r_d.

15. Derive the equation $\mu = g_m r_p$.

16. The small signal parameters μ, r_p, r_d, and g_m were all calculated at the Q-point for each device. The devices lose their amplifying ability when they are operated too close to the saturation or cutoff regions. What

happens to the parameters in these two regions? The amplification factor μ is of particular interest. Check its value at saturation and at cutoff for some of the examples in the text.

17. Using semiconductor data sheets, locate an FET and a bipolar transistor that are acceptable for the circuit shown in figure 8-31. It will be necessary to draw the load lines and locate the Q-point on the characteristic curves of the devices. Characteristic curves may be used for devices for which external reference materials are available; or, the characteristic curves given in this text may be used as references.

9
Equivalent Circuits for Field Effect Transistors and Other High-impedance Amplifying Devices

OBJECTIVES

After studying this chapter, the student will be able to:

- set up equivalent circuits of FETs and vacuum tubes for performace analysis
- analyze FET and vacuum tube amplifier circuits at high and low frequencies
- specify and calculate high-frequency parameters for FETs and vacuum tubes

EQUIVALENT CIRCUITS USING DEVICE PARAMETERS

The use of small signal parameters permits ac equivalent circuits to be drawn for FETs and vacuum tubes so that their performance in an amplifier circuit can be analyzed. The equivalent circuits reduce these three-terminal devices to groups of resistances or impedances and voltage sources between the input and output terminals of an amplifier circuit. FETs and vacuum tubes form a family of devices that can be described by the same basic ac equivalent circuit. The following discussion assumes that the ac signals applied to the amplifier have a frequency such that reactances can be omitted from the equivalent circuit. However, to make the generalization that reactance can be present under some conditions, it is necessary at times to consider that the circuit components have impedance instead of resistance. The symbol "Z" is used to represent impedance.

As an example, consider the simple FET common source amplifier shown in figure 6-20. Omitting the capacitors and the dc voltage supplies, the ac equivalent circuit is shown in figure 9-1. Z_L is the load impedance connected across

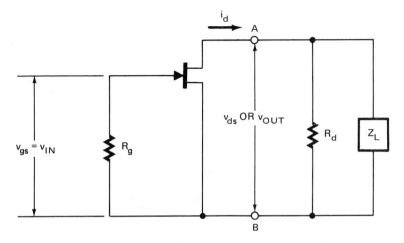

Fig. 9-1 Exact ac circuit for the amplifier shown in figure 6-20

the amplifier output terminals. The impedance is the parallel combination of R_d and Z_L and is expressed by:

$$Z_D = \frac{R_d Z_L}{R_d + Z_L}$$
Eq. 9.1

A Thévenin equivalent circuit can be formed between points A and B looking into the transistor drain-to-source circuit. The Thévenin equivalent voltage for any network is the open circuit voltage appearing across the terminals used to form the equivalent circuit. The open circuit output voltage across terminals A-B is v_{ds} with $i_d = 0$ (I_D constant). If μ is defined as the ratio of voltages with a constant drain current, then μ is the *no load* amplification. Thus, the drain-to-source voltage variation must be the same as the open circuit Thévenin equivalent voltage. The amplification factors can be defined in terms of the gate-to-source voltage and the amplification. Using Equation 8.12:

$$v_{ds} = -\mu v_{gs}$$
Eq. 9.2

Since the resistance between the gate and the drain (or source) is so high when compared to the dynamic drain resistance (r_d), the Thévenin equivalent resistance reduces to r_d. The complete ac equivalent circuit is shown in figure 9-2, page 334. Note that the voltage source μv_{gs} is shown negative with respect to common to observe the minus sign in Equation 9.2.

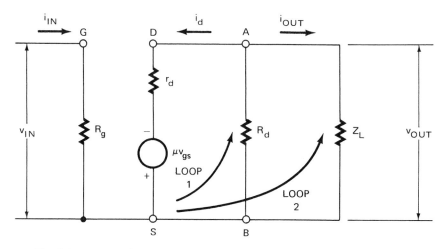

Fig. 9-2 Ac equivalent circuit for the FET amplifier shown in figure 6-20

Voltage Gain Calculation Using the Equivalent Circuit

The equivalent circuit can be used to determine the voltage gain in Loop 1.

$$A_V = \frac{v_{out}}{v_{in}}$$

Eq. 9.3

Kirchhoff's Voltage Law can be applied around the output circuit to obtain the following expression.

$$-\mu v_{gs} + i_d r_d + i_d Z_D = 0$$

Eq. 9.4

Collect the terms in the previous expression and solve for i_d to obtain:

$$i_d = \frac{\mu v_{gs}}{r_d + Z_D}$$

Eq. 9.5

The input and output voltages are given by the following expressions:

$$v_{in} = v_{gs}$$

Eq. 9.6

$$v_{out} = -i_d Z_D$$

Eq. 9.7

Equations 9.5, 9.6, and 9.7 can be combined and substituted into Equation 9.3 to obtain the following expression for the voltage gain:

$$A_V = \frac{-\mu Z_D}{r_d + Z_D}$$

Eq. 9.8

For a value of Z_D very much larger than r_d, the voltage gain is μ. If Z_D is very small (or a short), the voltage gain is very small or close to zero.

As an example of the effect of Z_D on the voltage gain, a typical situation will be considered. For the transistor shown in figure 8-21, μ was calculated to be 15.6. Let $V_{DD} = 20$ volts. The operation of this amplifier is to be along a load line determined by $R_d = 1.67$ kilohms. This load line is drawn on figure 8-21. Let the resistive load be 3 kilohms. According to Equation 9.1,

$$Z_D = \frac{(1.67)(3)}{1.67 + 3} = 1.073 \text{ kilohms}$$

The voltage gain can be computed using Equation 9.8. Note that it is not necessary to refer to the characteristic curves or the ac load line. In Chapter 8 (pages 307 to 316) the value of r_d was found to be 5.57 kilohms.

Thus, $$A_V = \frac{-\mu Z_D}{r_d + Z_D}$$

$$A_V = \frac{-(15.6)(1.073)}{5.57 + 1.073} = -2.52$$

It is evident that the voltage gain can be improved if the load is reduced.

Current Gain, Using the Equivalent Circuit

An equation can be written for the current gain using the equivalent circuit. The current gain is:

$$A_I = \frac{i_{out}}{i_{in}} \qquad \text{Eq. 9.9}$$

Use the current divider theorem to obtain an expression for i_{out}.

$$i_{out} = \frac{i_d R_d}{Z_L + R_d} \qquad \text{Eq. 9.10}$$

Kirchhoff's Voltage Law can be applied around Loop 2 to obtain:

$$-\mu v_{gs} + i_{out} Z_L + i_d r_d = 0 \qquad \text{Eq. 9.11}$$

Equations 9.10 and 9.11 can be combined and solved for i_{out}.

$$i_{out} = \frac{\mu v_{gs}}{Z_L + \dfrac{r_d(R_d + Z_L)}{R_d}} \qquad \text{Eq. 9.12}$$

Referring to figure 9-2 and Equation 9.6, it is evident that the following expression can be written for v_{gs} if the gate current is assumed to be negligible.

$$v_{gs} = v_{in} = i_{in}R_g \qquad \text{Eq. 9.13}$$

Substitute Equation 9.13 into Equation 9.12 to obtain:

$$i_{out} = \frac{\mu i_{in}R_g}{Z_L + \dfrac{r_d(R_d + Z_L)}{R_d}} \qquad \text{Eq. 9.14}$$

Equation 9.14 can now be substituted in Equation 9.9 to give an expression for the current gain:

$$A_I = \frac{\mu R_g}{Z_L + \dfrac{r_d(R_d + Z_L)}{R_d}} \qquad \text{Eq. 9.15}$$

If $Z_L = \infty$ (the condition existing at no load), the current gain is zero. If the amplifier output terminals are shorted, then $Z_L = 0$ and the current gain is:

$$A_I = \frac{\mu R_g}{r_d} \qquad \text{Eq. 9.16}$$

The following example demonstrates a method of calculating the current gain. A typical value of R_g is 300 kilohms. The short circuit current gain is to be calculated for $\mu = 18.2$ and $r_d = 6.33$ kilohms.

$$A_I = \frac{(18.2)(300)}{6.33} = 862.6$$

The current gain can also be calculated for the previously used resistive load of 3 kilohms, with R_g again set at 300 kilohms and $R_d = 1.67$ kilohms. Equation 9.15 is used as follows:

$$A_I = \frac{(18.2)(300)}{3 + \dfrac{(6.33)(3 + 1.67)}{1.67}} = 263.8$$

The input impedance of the FET amplifier is usually considered to be R_g. This is true because the usual value of R_g is very much smaller than the gate-to-source resistance of an FET. Therefore,

$$R_i = R_g \qquad \text{Eq. 9.17}$$

The output impedance is the parallel combination of r_d and R_d and is expressed as follows:

$$Z_o = \frac{r_d R_d}{r_d + R_d} \qquad \text{Eq. 9.18}$$

Z_o is also the Thévenin equivalent impedance of the amplifier when it is viewed from its output terminals. For the amplifier considered in the previous example, $R_i = 300$ kilohms and the output impedance is:

$$Z_o = \frac{r_d R_d}{r_d + R_d} = \frac{(6.33)(1.67)}{6.33 + 1.67} = 1.32 \text{ kilohms}$$

The no load voltage output for the amplifier is limited to the value determined by the following equation.

$$v_{outNL} = \frac{-\mu v_{gs} R_d}{r_d + R_d} \qquad \text{Eq. 9.19}$$

This value is also the Thévenin equivalent voltage of the amplifier when it is viewed from the output terminals.

A MOSFET is to be used in a common source amplifier such as the one shown in figure 6-23. R_g is 1 megohm. R_d is 2.5 kilohms, as determined by the procedures outlined in Chapter 6. Let $\mu = 15$ and $r_d = 6$ kilohms. Draw the ac equivalent circuit. Calculate the voltage gain, the current gain, and the input and output impedances for the case that $Z_L = \infty$ and $R_s = 0$. (R9-1)

Calculate the voltage gain and current gain for the amplifier described in review problem (R9-1) for (a) $R_L = 0$, and (b) $R_L = 3$ kilohms. (R9-2)

For a common source amplifier, such as in figure 6-20, $R_d = 2$ kilohms, $Z_L = 1$ kilohm resistive, $g_m = 3$ millimhos, and $r_d = 3$ kilohms. Compute the voltage gain of this amplifier. (R9-3)

Calculate the voltage gain for the circuit of review problem (R9-3) for $Z_L = 10$ kilohms resistive. (R9-4)

Calculate the current gain for the circuit of review problem (R9-3) if $R_g = 180$ kilohms. (R9-5)

The preceding analysis and examples apply to vacuum tubes as well. To use the expressions for vacuum tube circuits, replace the subscripts in the equations with the proper nomenclature for vacuum tubes (refer to Table 8-1).

The vacuum tube amplifier described in figure 7-21 is to be analyzed. The following values will apply: $\mu = 96$, $r_p = 53.5$ *kilohms,* $R_g = 500$ *kilohms, and* $R_p = 121$ *kilohms. Draw the ac equivalent circuit. Calculate the voltage gain, the current gain, and the input and output impedances. Assume that the input signal* (v_{in}) *is from a perfect voltage source and that a load is **not** connected to the amplifier.* *(R9-6)*

Calculate the voltage gain and the current gain for the amplifier described in review problem (R9-6) for (a) $R_L = 0$, *and (b)* $R_L = 100$ *kilohms.* *(R9-7)*

EQUIVALENT CIRCUITS FOR COMMON PLATE (OR DRAIN) AMPLIFIERS

The small signal parameters can be used in the analysis of amplifiers of the source follower or cathode follower type. In the following examples, the FET is used. However, the same analysis can be made using vacuum tubes.

The amplifier used in this example is the source follower or common drain amplifier shown in figure 8-8. The ac equivalent circuit for this amplifier is shown in figure 9-3. Compare this circuit with that of figure 9-2 and note that the voltage source (μv_{gs}) is *positive* with respect to common. This polarity implies that the voltage output of the source follower amplifier is in phase with the voltage input (or nearly so). The equations show that this is the case. Z_D is redefined as:

$$Z_D = \frac{R_s Z_L}{R_s + Z_L} \qquad \text{Eq. 9.20}$$

Voltage Gain for the Source Follower

Kirchhoff's Voltage Law applied around the output circuit (Loop 1) will yield an equation which is the same as Equation 9.4. The expression for i_d is the same as Equation 9.5:

$$i_d = \frac{\mu v_{gs}}{Z_D + r_d}$$

The quantity v_{gs} can be defined in terms of the input and output voltages. Kirchhoff's Voltage Law applied around Loop 2 yields:

$$v_{gs} = v_{in} - v_{out} \qquad \text{Eq. 9.21}$$

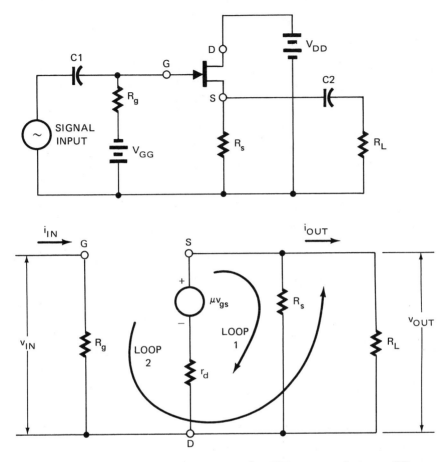

Fig. 9-3 Actual and ac equivalent circuit for FET common drain amplifier

Equation 9.7 defines v_{out} although this term now has the opposite polarity.

$$v_{out} = i_d Z_D \qquad \text{Eq. 9.22}$$

An expression for i_d is obtained by substituting Equation 9.22 into Equation 9.21 and combining the result with Equation 9.5.

$$i_d = \frac{\mu v_{in} - \mu i_d Z_D}{Z_D + r_d} \qquad \text{Eq. 9.23}$$

Equation 9.23 can be solved for v_{in} to yield:

$$v_{in} = \frac{i_d(Z_D + \mu Z_D + r_d)}{\mu} \qquad \text{Eq. 9.24}$$

The voltage gain is found by substituting Equations 9.22 and 9.24 into Equation 9.3.

$$A_V = \frac{\mu Z_D}{(1 + \mu)Z_D + r_d} \qquad \text{Eq. 9.25}$$

Z_D may be large or small, but since μ is a large number, then in most cases, $(1 + \mu)Z_D \gg r_d$. Thus, an approximate value for the voltage gain is given by:

$$A_V = \frac{\mu Z_D}{(1 + \mu)Z_D} \approx \frac{\mu}{1 + \mu} \qquad \text{Eq. 9.26}$$

This expression means that $A_V \approx 1$. Such a value is typical for the source follower and cathode follower amplifiers. (Note that the common collector amplifier also has a voltage gain of approximately one!)

The following example shows how the voltage gain is calculated using typical values in Equation 9.25. Using $\mu = 15.6$, $r_d = 5.57$ kilohms, and $Z_D = 1.03$ kilohms resistive,

$$A_V = \frac{(15.6)(1.03)}{(1 + 15.6)(1.03) + 5.57} = 0.718$$

For this example, μ is small and the amplifier is heavily loaded. Thus, the calculated value of the voltage gain for this example is not as close to unity as it usually is.

Current Gain Calculations

The output current for this amplifier is given by Equation 9.10 with the substitution of R_s for R_d.

$$i_{out} = \frac{i_d R_s}{R_s + Z_L} \qquad \text{Eq. 9.27}$$

An expression for i_d is determined from Equation 9.24.

$$i_d = \frac{\mu v_{in}}{Z_D + \mu Z_D + r_d} \qquad \text{Eq. 9.28}$$

Substituting Equation 9.20 for Z_D into Equation 9.28 yields:

$$i_d = \frac{\mu v_{in}}{\dfrac{(1 + \mu)R_s Z_L}{R_s + Z_L} + r_d} \qquad \text{Eq. 9.29}$$

Equation 9.13 ($v_{in} = i_{in}R_g$) can be applied to this circuit as well. By combining Equations 9.13 and 9.29 and substituting the result into Equation 9.27, an expression for i_{out} is obtained.

$$i_{out} = \frac{\mu i_{in} R_g R_s}{(1 + \mu)R_s Z_L + r_d(R_s + Z_L)} \qquad \text{Eq. 9.30}$$

The current gain is found by substituting Equation 9.30 into Equation 9.9.

$$A_I = \frac{\mu R_g R_s}{(1 + \mu)R_s Z_L + r_d(R_s + Z_L)} \qquad \text{Eq. 9.31}$$

At no load, $Z_L \approx \infty$ and the current gain is zero. If the output terminals are shorted, then $Z_L \approx 0$ and the current gain equation reduces to Equation 9.16. As a result, the short circuit current gain of the common drain amplifier is the same as that of the common source amplifier.

The input impedance of the circuit is R_g if the value of R_g is significantly less than the resistance between the gate and the source terminals of the FET.

Figure 9-3 indicates that the output impedance of the amplifier is the Thévenin equivalent impedance of the simple network that includes r_d, R_s and the voltage source μv_{gs}. This output impedance is given by:

$$Z_o = \frac{R_s r_d}{r_d + R_s} \qquad \text{Eq. 9.32}$$

As an example of the analysis of a typical source follower, the current gain, and the input and output resistances can be calculated using figure 8-11. This common drain circuit uses a self-bias circuit for the gate. For the expected signal frequencies, the value of capacitor C3 is such that the resistor R_{g3} is regarded as shorted out for the ac equivalent circuit. The ac equivalent circuit of this amplifier is the same as the circuit shown in figure 9-3. Thus, the equations in the preceding paragraphs are applicable. Assume that $\mu = 15.6$ and $r_d = 5.57$ kilohms, based on the curves given in figure 8-21. Let $R_{g3} = 10$ kilohms, $R_{g1} = 290$ kilohms, $R_{g2} = 500$ kilohms, and $R_s = 1.67$ kilohms, V_{GSQ} is formed by a combination of the dc voltage drop across the resistance R_s and the voltage formed by the voltage dividers R_{g1} and R_{g3}. R_{g2} is connected to common through the capacitor C3. This capacitor also provides an ac short to resistors R_{g1} and R_{g3}. Therefore, for an infinite gate resistance (r_{iss}), the input resistance is:

$$R_i = R_{g2} = 500 \text{ kilohms}$$

Assume an ac load impedance (Z_L) of $3 + j0 = R_L = 3$ kilohms. The current gain can be determined from Equation 9.31:

$$A_I = \frac{(15.6)(500)(1.67)}{(1 + 15.6)(1.67)(3) + (5.57)(1.67 + 3)}$$

$$A_I = 119.3$$

Equation 9.32 is then used to determine the output impedance:

$$Z_0 = \frac{(1.67)(5.57)}{1.67 + 5.57} = 1.28 \text{ kilohms}$$

The common drain amplifier shown in figure 8-11 is to be analyzed. The Q-point of the FET is at $V_{GSQ} = -1$ volt, $I_{DQ} = 5.6$ milliamperes, and $V_{DSQ} = 10$ volts. The dc load line is established and has a slope of $1/2000$ originating at $V_{DD} = 20$ volts. The amplifier is connected to a maximum ac load of $5 + j0$ kilohms. The capacitors have a negligible reactance at the signal frequency. The voltage drop across the source resistor (R_s) is too large for the required gate-to-source bias. Therefore, a voltage divider is formed using R_{g1} and R_{g3} to bring the gate bias to -1 volt. The following values apply to the circuit: $R_{g2} = 320$ kilohms, $R_{g3} = 10$ kilohms, $\mu = 15.6$, and $r_d = 5.57$ kilohms.

    a. What value of R_{g1} is required to insure the proper gate bias voltage?
    b. Calculate the voltage gain.
    c. Calculate the current gain of the amplifier.
    d. Calculate the input and output impedances of the amplifier.   *(R9-8)*

Calculate the voltage gain of the circuit shown in figure 8-9. Let $\mu = 15.6$, $r_d = 5.57$ kilohms, $R_s = 1.67$ kilohms, $R_{g1} = 18.7$ kilohms, $R_{g2} = 10$ kilohms and $Z_L = \infty$.   *(R9-9)*

Compute the input and output impedances for the amplifier described in review problem (R9-9).   *(R9-10)*

A common drain amplifier, such as the one shown in figure 8-13, is to be used. Assume the following values: $R_s = 3$ kilohms, $\mu = 5$, $Z_L = 3$ kilohms resistive, $R_g = 300$ kilohms, and $r_d = 6$ kilohms. Compute the current gain and the voltage gain for this amplifier.   *(R9-11)*

EQUIVALENT CIRCUITS FOR COMMON GATE
(OR COMMON GRID) AMPLIFIERS

The ac equivalent circuit of the circuits shown in figures 8-15 and 8-16 is given in figure 9-4. The input side of the circuit in figure 8-15 is the same as the Thévenin equivalent circuit of the input side shown in figure 8-16. A ac analysis of the circuit in figure 8-15 can also be made applicable to the circuit in figure 8-16. To do this, v_{in} and R_s (figure 8-15) are considered as the Thévenin equivalent voltage and the Thévenin equivalent resistance, respectively, of the input connections to the gate and source terminals of the circuit in figure 8-16. (The Appendix contains an explanation of Thévenin's Theorem.) In other words, figure 8-16 must be converted to the form shown in figure 8-15 before an ac equivalent circuit like figure 9-4 can be made from it.

To compute the voltage gain, the output voltage is defined in terms of the input voltage and the circuit parameters. Kirchhoff's Voltage Law is applied around Loop 1 in figure 9-4 to yield:

$$v_{in} - i_d R_s - i_d r_d - \mu v_{gs} - i_d Z_d = 0 \qquad \text{Eq. 9.33}$$

Recall that Z_D is defined by Equation 9.1. The relationship between v_{gs} and v_{in} is expressed by:

$$v_{gs} = -(v_{in} - i_d R_s) \qquad \text{Eq. 9.34}$$

(Refer to figure 9-4.) Substitute Equation 9.34 into Equation 9.33 and solve for v_{in}:

$$v_{in} = \frac{i_d[(1 + \mu)R_s + r_d + R_D]}{1 + \mu} \qquad \text{Eq. 9.35}$$

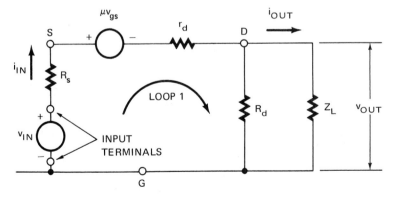

Fig. 9-4 Ac equivalent circuit for a common gate amplifier

The output voltage is:

$$v_{out} = i_d Z_D \qquad \text{Eq. 9.36}$$

Now substitute Equations 9.35 and 9.36 into Equation 9.3 to obtain the voltage gain.

$$A_V = \frac{(\mu + 1)Z_D}{(\mu + 1)R_s + r_d + Z_D} \qquad \text{Eq. 9.37}$$

In some instances, $r_d + Z_D$ in the denominator of Equation 9.37 may be negligible when compared to $(\mu + 1)R_s$. Thus, the voltage gain for the dc coupled common gate amplifier is given approximately as:

$$A_V \approx \frac{Z_D}{R_s} = \frac{R_d Z_L}{(R_d + Z_L)R_s} \qquad \text{Eq. 9.38}$$

At no load,

$$A_V \approx \frac{R_d}{R_s} \qquad \text{Eq. 9.39}$$

Current Gain and Impedance Calculations

The current gain for this amplifier is determined by the current division between R_d and Z_L. The current divider theorem gives the following expression for i_{out}.

$$i_{out} = \frac{i_d R_d}{Z_L + R_d} = \frac{i_{in} R_d}{Z_L + R_d} \qquad \text{Eq. 9.40}$$

The current gain, therefore, as defined originally in Equation 9.9 becomes:

$$A_I = \frac{i_{out}}{i_{in}} = \frac{R_d}{Z_L + R_d} \qquad \text{Eq. 9.41}$$

The input impedance for the amplifier can now be calculated, using figure 9-5 (which is derived from figure 9-4). The input impedance is:

$$Z_i = R_s + r_d + \frac{R_d Z_L}{R_d + Z_L} \qquad \text{Eq. 9.42}$$

The output impedance is R_d with nothing connected to the input terminals. If the source is connected and has a negligible impedance, then the output impedance becomes:

$$Z_o = \frac{R_d(r_d + R_s)}{R_d + r_d + R_s} \qquad \text{Eq. 9.43}$$

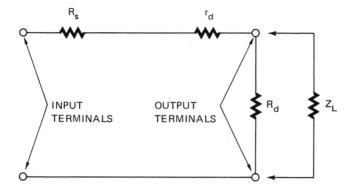

Fig. 9-5 Equivalent circuit used in the computation of the network resistance of a dc coupled common gate amplifier

As an example of a common gate amplifier, consider an FET circuit such as the one shown in figure 8-15. This circuit has the following constants: r_d = 5.57 kilohms, R_d = 1.67 kilohms, μ = 15.6, and the total series input resistance is 167 ohms. The voltage gain, current gain, and the input and output impedances are to be calculated using the equivalent circuit shown in figure 9-4. The internal resistance of the source is assumed to be 50 ohms. The total external resistor required to set the gate bias is R_s = 167 - 50 = 117 ohms. The maximum voltage gain occurs when a negligible amount of load is connected (Z_L = ∞). The voltage gain is determined by substituting the assumed values into Equation 9.37.

$$A_V = \frac{(15.6 + 1)(1.67)}{(15.6 + 1)(0.117) + 5.57 + 1.67} = 3.02$$

For this example, μ is not large enough to permit the assumption of $r_d + R_d$ as negligible. It is more likely that $r_d + R_d$ can be neglected for the high values of μ that are obtained with a vacuum tube. The current gain for this amplifier is obtained by substituting values in Equation 9.41. If Z_L = 0 (a shorted output), then the input current is equal to the output current and $A_I \approx 1$.

The input impedance for the amplifier is:

$$Z_i = R_s + r_d + R_d \qquad\qquad \text{Eq. 9.44}$$

$$Z_i = 0.117 + 5.57 + 1.67 = 7.36 \text{ kilohms}$$

Note the 50-ohm source resistance is not included in Z_i because this component is not part of the amplifier. If the input terminals are open, the output

CIRCUIT CONFIGURATION

Amplifier Parameter	Source Follower		Cathode Follower	
Voltage Gain, A_V	Eq. 9.25	$\dfrac{\mu Z_D}{(1+\mu)Z_D + r_d}$	Eq. 9.45	$\dfrac{\mu Z_p}{(1+\mu)Z_p + r_p}$
Current Gain, A_I	Eq. 9.31	$\dfrac{\mu R_g R_s}{(1+\mu)R_s Z_L + r_d(R_s + Z_L)}$	Eq. 9.46	$\dfrac{\mu R_g R_k}{(1+\mu)R_k Z_L + r_p(R_k + Z_L)}$
Input Impedance, Z_i		R_g		R_g
Output Impedance, Z_o	Eq. 9.32	$\dfrac{R_s r_d}{r_d + R_s}$	Eq. 9.47	$\dfrac{R_k r_p}{r_p + R_k}$
	Common Source		**Common Cathode**	
Voltage Gain, A_V	Eq. 9.8	$\dfrac{-\mu Z_D}{r_d + Z_D}$	Eq. 9.48	$\dfrac{-\mu Z_p}{r_p + Z_p}$
Current Gain, A_I	Eq. 9.15	$\dfrac{\mu R_g}{Z_L + \dfrac{r_d(R_d + Z_L)}{R_d}}$	Eq. 9.49	$\dfrac{\mu R_g}{Z_L + \dfrac{r_p(R_p + Z_L)}{R_p}}$
Input Impedance, Z_i	Eq. 9.17	R_g	Eq. 9.50	R_g
Output Impedance, Z_o	Eq. 9.18	$\dfrac{r_d R_d}{r_d + R_d}$	Eq. 9.51	$\dfrac{r_p R_p}{r_p + R_p}$

Amplifier Parameter	CIRCUIT CONFIGURATION			
	Common Gate		Common Grid	
Voltage Gain, A_V	Eq. 9.37	$\dfrac{(\mu + 1)Z_D}{(\mu + 1)R_s + r_d + Z_D}$	Eq. 9.52	$\dfrac{(\mu + 1)Z_P}{(\mu + 1)R_k + r_p + Z_P}$
Current Gain, A_I	Eq. 9.41	$\dfrac{R_d}{Z_L + R_d}$	Eq. 9.53	$\dfrac{R_p}{Z_L + R_p}$
Input Impedance, Z_i	Eq. 9.42	$R_s + r_d + \dfrac{R_d Z_L}{R_d + Z_L}$	Eq. 9.54	$R_k + r_p + \dfrac{R_p Z_L}{R_p + Z_L}$
Output Impedance, Z_o	Eq. 9.43	$\dfrac{R_d(r_d + R_s)}{R_d + r_d + R_s}$	Eq. 9.55	$\dfrac{R_p(r_p + R_k)}{R_p + r_p + R_k}$

Table 9-1 (continued)

impedance is R_d = 1.67 kilohms. If the input terminals are shorted, the output impedance becomes:

$$Z_o = \frac{1.67(5.57 + 0.117)}{1.67 + 5.57 + 0.117} = 1.29 \text{ kilohms}$$

In the previous example for a no load condition, $\mu = 120$. Refer to figure 9-4. Recalculate the voltage gain (A_V) if the other constants remain unchanged.
(R9-12)

The equations derived for the three circuit configurations contain the symbols that are commonly used for FET amplifier circuit parameters. The equations for vacuum tubes are the same except that the symbols reflect vacuum tube amplifier parameters. The equations are summarized in Table 9-1, on pages 346 and 347. Note that the equations for the vacuum tube configurations are identical in form with those for the FET configurations, but have different symbols and thus new equation numbers.

Assume that a vacuum tube is to be used in the circuit shown in figure 8-8, instead of the FET. The device has the following parameters: $\mu = 100$ and $r_p = 50$ kilohms. The load impedance is assumed to be $Z_L = 120 + j0$ kilohms. $R_K = 60$ kilohms and $R_g = 300$ kilohms. The signal source has a low amplitude with negligible source impedance and is at a midband frequency. Calculate the voltage gain, current gain, and the input and output impedances for this amplifier.
(R9-13)

OPERATING THE HIGH-IMPEDANCE AMPLIFIER AT LOW FREQUENCIES

The operation of a high-impedance amplifier was discussed for signals in the intermediate frequency range. In this range, the amplifier performance is independent of the signal frequency. For ac coupled amplifiers, the ac voltage drop across a capacitor and the dc voltage drop across a transformer winding are considered to be zero. However, for signals at frequencies below the midband range, these assumptions cannot be made. Refer to figure 5-1. The high-impedance amplifier has a frequency response graph similar to the one shown. The discussion at the beginning of Chapter 5 is also applicable to ac coupled FET and vacuum tube amplifiers. Briefly review pages 152 to 154 before continuing with the following discussion.

This analysis considers the common source amplifier shown in figure 6-20. The input impedance for FETs and vacuum tubes is high, and the resistor R_g usually determines the resistance across the input terminals. The low-frequency equivalent circuit is shown in figure 9-6. The input impedance

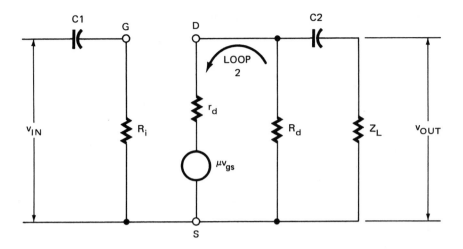

Fig. 9-6 Low-frequency equivalent circuit for the ac coupled amplifier

(Z_i) for this circuit is defined as it was previously, but now it contains reactance as well as resistance. It is given by:

$$Z_i = R_i + \frac{1}{j\omega C1}$$

Eq. 9.56

R_i is the resistive portion of the input impedance.

The output impedance is:

$$Z_o = \frac{r_d R_d}{r_d + R_d} + \frac{1}{j\omega C2}$$

Eq. 9.57

How do the coupling capacitors affect the voltage gain? The input voltage, in terms of v_{gs}, is:

$$v_{in} = \frac{v_{gs}\left(R_i + \frac{1}{j\omega C1}\right)}{R_i}$$

Eq. 9.58

The output voltage is not affected by capacitor C2 when the amplifier is at a no load condition. The output voltage is given by Equation 9.19:

$$v_{out} = v_{NL} = \frac{-\mu v_{gs} R_d}{r_d + R_d}$$

Thus, the voltage gain at a no load condition is expressed as follows:

$$A_V = \frac{v_{out}}{v_{in}} = \frac{-\mu R_d j \, \omega C1 R_i}{(r_d + R_d)(j \, \omega C1 R_i + 1)} \qquad \text{Eq. 9.59}$$

If the frequency is too low ($\omega \approx 0$), the voltage gain is 0. If the frequency is above the low-frequency range, then the $j \, \omega C1 R_i$ terms cancel and the voltage gain becomes the intermediate range gain. This gain is similar to the value determined from Equation 9.8.

$$A_V = \frac{-\mu R_d}{r_d + R_d} \qquad \text{Eq. 9.60}$$

At a frequency of $\omega = 1/C1R_i$, the voltage gain is 3 decibels down from the intermediate frequency range gain.

Current Gain at Low Frequency

For the intermediate frequency range, the current gain is described by Equation 9.15. At low frequencies, the equation is different and therefore must be developed for this condition. For a resistive load (R_L), the equivalent load impedance is:

$$Z_L = R_L + \frac{1}{j \, \omega C2} \qquad \text{Eq. 9.61}$$

The current gain at a low frequency can be determined by analyzing figure 9-6. The output current is expressed in terms of the circuit parameters and the input current. v_{gs} is expressed in terms of the input current as:

$$v_{gs} = i_{in} R_i \qquad \text{Eq. 9.62}$$

The current divider theorem can be used to express the drain current in terms of the output current.

$$i_d = \frac{[(R_d + R_L)j \, \omega C2 + 1]i_{out}}{j \, \omega C2 R_d} \qquad \text{Eq. 9.63}$$

Kirchhoff's Voltage Law is applied around Loop 2. When the terms of the resulting expression are rearranged, it becomes:

$$\mu v_{gs} - i_d \left(\frac{j \, \omega C2 R_d R_L + R_d}{j \, \omega C2 R_d + R_L + 1} \right) - i_d r_d = 0 \qquad \text{Eq. 9.64}$$

Substitute Equations 9.62 and 9.63 into Equation 9.64

$$\mu i_{in} R_i - \frac{i_{out}(j \omega C2 R_d R_L + R_d)}{j \omega C2 R_d} - \frac{i_{out}[(R_d + R_L)j \omega C2 + 1]r_d}{j \omega C2 R_d} = 0$$

Eq. 9.65

Solving Equation 9.65 for the output current and rearranging the terms yields:

$$i_{out} = \frac{\mu i_{in} R_i j \omega C2 R_d}{j \omega [C2 R_d R_L + (R_d + R_L)C2 r_d] + R_d + r_d}$$

Eq. 9.66

The current gain at low frequency is obtained by substituting Equation 9.66 into Equation 9.9 and rearranging the terms.

$$A_I = \frac{\mu R_i j \omega C2 R_d}{(R_d + r_d)\left\{ j \omega \left[\dfrac{C2 R_d R_L + (R_d + R_L)C2 r_d}{R_d + r_d} \right] + 1 \right\}}$$

Eq. 9.67

If $R_L = 0$, the terms containing R_L drop out of Equation 9.67 and A_I reduces to:

$$A_I = \frac{\mu R_i R_d j \omega C2}{(R_d + r_d)\left(\dfrac{j \omega R_d C2 r_d}{R_d + r_d} + 1 \right)}$$

Eq. 9.68

The corner frequency is controlled by the time constant, $R_d C2 r_d /(R_d + r_d)$. The current gain increases by 20 decibels per decade of frequency until it reaches the corner frequency. At the corner frequency, the current gain is down by 3 decibels. This frequency occurs at $\omega = (R_d + r_d)/R_d C2 r_d$ radians per second. When $\omega = \infty$, Equation 9.68 reduces to Equation 9.16:

$$A_I = \frac{\mu R_i}{r_d}$$

Amplifiers having a source or cathode resistor bypassed by a capacitor will have their performance adversely affected by low signal frequencies. For example, consider capacitor C3 in figures 8-23 and 6-21. The purpose of such a capacitor is to prevent degradation of the amplifier gain due to the bias resistors. These capacitors work well at midband frequencies, but lose their effect at low frequencies. As a result, the bias resistors add a voltage drop in the output circuit that reduces the amplifier voltage gain.

Voltage Gain for FET Amplifier at Low Frequency

An equation will be derived for the voltage gain at low frequency for the circuit shown in figure 6-21. Figure 9-7 is the low frequency ac equivalent circuit for the circuit in figure 6-21. The derivation is accomplished by: (1) writing an equation for the input voltage, (2) writing an equation for the output voltage in terms of the input voltage, and then (3) substituting these expressions in the basic voltage gain equation, Equation 9.3. According to the voltage divider theorem, the voltage across R_g is:

$$v_s = \frac{v_{in} R_g}{R_g + \dfrac{1}{j \omega C1}} = \frac{v_{in} j \omega C1 R_g}{j \omega C1 R_g + 1}$$

Eq. 9.69

Kirchhoff's Voltage Law can be applied around Loop 1 to yield:

$$v_{gs} = v_s - \frac{i_d R_s \left(\dfrac{1}{j \omega C3} \right)}{R_s + \dfrac{1}{j \omega C3}}$$

Eq. 9.70

Substitute Equation 9.69 into Equation 9.70.

$$v_{gs} = \frac{v_{in} j \omega C1 R_g}{j \omega C1 R_g + 1} - \frac{i_d R_s}{j \omega C3 R_s + 1}$$

Eq. 9.71

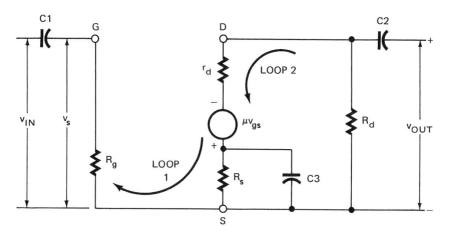

Fig. 9-7 Low-frequency ac equivalent circuit for FET amplifier using source resistor bias

To write an equation for the output voltage in terms of v_{gs}, apply Kirchhoff's Voltage Law around Loop 2.

$$\mu v_{gs} - \frac{i_d R_s}{j\omega C3R_s + 1} + v_{out} - i_d r_d = 0 \qquad \text{Eq. 9.72}$$

Substitute the following expression for i_d to remove the i_d terms in Equation 9.72.

$$i_d = \frac{-v_{out}}{R_d} \qquad \text{Eq. 9.73}$$

Substitute Equations 9.73 and 9.71 into Equation 9.72.

$$\frac{\mu v_{in} j\omega C1R_g}{j\omega C1R_g + 1} + \frac{\mu v_{out} R_s}{R_d(j\omega C3R_s + 1)} + \frac{v_{out} R_s}{R_d(j\omega C3R_s + 1)} +$$

$$v_{out} + \frac{v_{out} r_d}{R_d} = 0 \qquad \text{Eq. 9.74}$$

Collecting terms and solving for v_{out},

$$v_{out} = \frac{\dfrac{-\mu v_{in} j\omega C1R_g}{j\omega C1R_g + 1}}{\dfrac{\mu R_s}{R_d(j\omega C3R_s + 1)} + \dfrac{R_s}{R_d(j\omega C3R_s + 1)} + 1 + \dfrac{r_d}{R_d}} \qquad \text{Eq. 9.75}$$

Now substitute Equation 9.75 into Equation 9.3 and collect the terms to obtain the following expression for the voltage gain.

$$A_V = \frac{-\mu j\omega C1R_g R_d(j\omega C3R_s + 1)}{[j\omega C1R_g + 1][(\mu + 1)R_s + R_d(j\omega C3R_s + 1) + r_d(j\omega C3R_s + 1)]} \qquad \text{Eq. 9.76}$$

If ω approaches infinity, then R_s is shorted by C3. If a high value of ω is substituted into Equation 9.76, it is reduced to Equation 9.60. If R_s is zero, the circuit is reduced to an ordinary common source amplifier and Equation 9.76 reduces to Equation 9.59.

An FET is used in the circuit shown in figure 6-20. The following values apply to the circuit: $\mu = 10$, $r_d = 10$ *kilohms,* $R_i = R_g = 1$ *megohm,* $R_d = 1.5$ *kilohms,* $Z_L = \infty$, *C1* = 10 *microfarads, and C2 = 10 microfarads. For a signal frequency of 100 hertz, compute the input impedance, the output impedance, and the voltage gain of the amplifier.* *(R9-14)*

Assume the same circuit and data that were used for review problem (R9-14). Change the signal frequency to 10 kilohertz and repeat the calculations made for (R9-14). *(R9-15)*

Assume the same circuit and data that were used for review problem (R9-14), except that $Z_L = R_L = 0$. *What is the current gain at 100 Hz for this amplifier?* *(R9-16)*

What is the corner frequency of the voltage gain for the amplifier described in review problem (R9-14) at the low-frequency end of the frequency response graph? *(R9-17)*

Refer to figure 6-21. If the selected value of the bypass capacitor C3 is too small, what is the effect on the amplifier performance? *(R9-18)*

A common source amplifier circuit, such as the one shown in figure 6-20, is being used at a low frequency. Assume that $R_d = 10$ *kilohms,* $r_d = 12$ *kilohms,* $g_m = 1.1$ *millimhos,* $R_g = 300$ *kilohms, and C1 = C2 = C3 = 1 microfarad.*

a. Calculate the voltage gain in the intermediate frequency region. (The coupling capacitances ≈ 0.) *Assume* $Z_L = \infty$.
b. Calculate the current gain in the intermediate frequency region. Assume $Z_L = 0$.
c. Calculate the corner frequency for the current gain when the amplifier is operating in the low-frequency region.
d. Calculate the corner frequency for the voltage gain when the amplifier is operating in the low-frequency region. *(R9-19)*

A similar analysis can be made for the other amplifier connection configurations, by including the effect of all capacitances in the circuit. In each case, an impedance is formed where once only a resistance existed. The equations for each analysis are developed using impedances rather than resistances. The gain equations are then manipulated so as to more easily interpret the effect of low-frequency impedances on the frequency response graphs.

Transformer Coupled Amplifiers at Low Frequency

The other form of ac coupling is transformer coupling. A transformer coupled amplifier is shown in figure 9-8. In a transformer coupled circuit there is very little voltage input and output developed at low frequencies. This situation is due to the fact that the magnetic flux does not change rapidly enough between the primary and secondary windings at low frequencies. An ac equivalent circuit for the amplifier (figure 9-8) is shown in figure 9-9. The output voltage for a resistive load is:

$$v_{out} = i_{out}R_L \qquad \text{Eq. 9.77}$$

The voltage across L4 is the output voltage of the amplifier. The ac voltage from the drain to the source is the output voltage multiplied by the ratio of turns. This is expressed as:

$$v_d = \frac{N_3 v_{out}}{N_4} \qquad \text{Eq. 9.78}$$

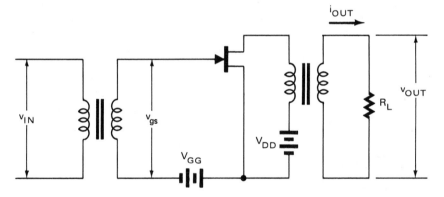

Fig. 9-8 Transformer coupled FET amplifier

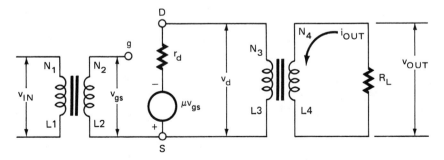

Fig. 9-9 Ac equivalent circuit of a transformer coupled amplifier

where N is the number of turns in the winding and the subscripts (1,2,3, or 4) identify the specific winding in figure 9-9.

The input voltage is the voltage developed across the primary winding of the input transformer. This can be defined by:

$$v_{in} = \frac{N_1 v_{gs}}{N_2}$$

Eq. 9.79

The impedance of the primary winding of the output transformer is Z_d.

$$Z_d = R_d + j \omega L3$$

Eq. 9.80

The voltage gain from v_{gs} to v_d is the same as the gain expressed by Equation 9.8.

$$\frac{v_d}{v_{gs}} = \frac{-\mu Z_d}{r_d + Z_d}$$

Eq. 9.81

The overall voltage gain from the input voltage to the output voltage is:

$$A_V = \frac{v_{out}}{v_{in}} = \frac{v_{gs}}{v_{in}} \cdot \frac{v_d}{v_{gs}} \cdot \frac{v_{out}}{v_d}$$

Eq. 9.82

Substitute Equations 9.81, 9.78, 9.80, and 9.79 into Equation 9.82 to obtain the voltage gain for the transformer coupled amplifier.

$$A_V = \frac{-\mu N_2 N_4 (R_d + j \omega L3)}{N_1 N_3 (r_d + R_d + j \omega L3)}$$

Eq. 9.83

R_d is usually very low for a transformer winding, so at low frequencies the voltage gain is low. For dc signals the gain is exactly zero. This value is due to the fact that dc voltages cannot induce a voltage across the transformer windings. The voltage gain equation does not include the fact that the input voltage (v_{in}) is low at low frequencies because the source is being shorted.

Chapter 12 investigates transformer coupled amplifiers in more detail.

Why does the frequency response of a transformer coupled amplifier drop at low frequencies?　　　　　　　　　　　　　　　　　　　　　*(R9-20)*

What is the slope of the dc and ac load lines for a transformer coupled FET amplifier, if the only impedance in the drain circuit is the transformer primary winding impedance of 10 + j6000 ohms?　　　*(R9-21)*

EQUIVALENT CIRCUITS FOR HIGH-FREQUENCY OPERATION

High-frequency amplifier performance is affected by the shunt capacitances in the amplifying devices. These capacitances can be neglected in the midband range of frequencies. At high frequencies, the capacitances tend to short out the signal, reduce the input impedance, and reduce the gain.

The shunt capacitances in the FET are due to different mechanisms than those in vacuum tubes. However, the effect of these capacitances is the same in both types of devices. In schematic form, the capacitances are shown as in figure 9-10. The subscripts identify the locations of the capacitances, such as the capacitance between the grid and the plate, C_{gp}, and the capacitance between the drain and the source, C_{ds}.

A capacitor is formed wherever there are two conducting surfaces separated by a dielectric. In the case of the vacuum tube, the grid, the plate, and the cathode are separated from each other by a vacuum which represents a dielectric medium. Capacitances then exist between these components and are known as the tube capacitances C_{gp}, C_{gk}, and C_{pk}.

The FET also has capacitances between the terminals. The capacitance between the gate and the drain (C_{gd}) and between the gate and the source (C_{gs}) is the capacitance previously considered in Chapter 5 where it was identified as the capacitance between the collector and base of a bipolar transistor. These capacitances are due to the dielectric medium formed by a lack of mobile charges in the region of the junction between the gate and the channel of the FET. These capacitances are referred to as C_{gs} and C_{gd}. The subscripts represent the terminals between which the capacitance appears.

A capacitance also exists along the channel and represents the capacitance between the drain and the source. The value of this capacitance usually is very small because of the low dielectric constant that exists in a doped semiconductor

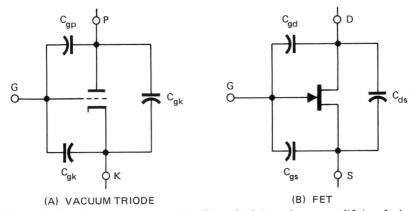

(A) VACUUM TRIODE   (B) FET

Fig. 9-10 High-frequency schematic of two high-impedance amplifying devices showing shunt capacitances

crystal. (Capacitance is directly proportional to the dielectric constant.) The channel capacitance includes (1) the capacitance between the insulated drain and source terminals, and (2) the capacitance between the channel and the substrate region upon which the semiconductor crystal is built.

The high-frequency equivalent circuit for an FET is shown in figure 9-11. The output circuit represented by r_d and $g_m v_{gs}$ is the Norton equivalent circuit of the FET. By shorting terminals A and B together in figure 9-1, the Norton equivalent current source is found to be $g_m v_{gs}$. The Norton resistance is the same as the Thévenin equivalent resistance, r_d. The input resistance r_{is} is related to the input admittance y-parameter, y_{is}.

Miller's theorem, as discussed in Chapter 5, was developed to be used with high-frequency vacuum tube circuits. By utilizing figures 5-4 and 5-5 and Miller's theorem, the equivalent circuit of figure 9-11 can be converted to the circuit shown in figure 9-12. Substitute $1/j\,\omega C_{gd}$ for Z in Equation 5.19 to obtain:

$$Z_1 = \frac{Z}{1 - A_V} = \frac{1}{(1 - A_V)j\,\omega C_{gd}} \qquad \text{Eq. 9.84}$$

In this expression, A_V is the midband frequency voltage gain and Z_1 is an impedance indicated in figure 5-5.

A capacitance C_{eqi} is defined so that Equation 5.23 results:

$$Z_1 = \frac{1}{j\,\omega C_{eqi}}$$

Substitute this expression in Equation 9.84. Then define an equivalent capacitance as follows:

$$C_{it} = C_{gs} + C_{eqi} \qquad \text{Eq. 9.85}$$

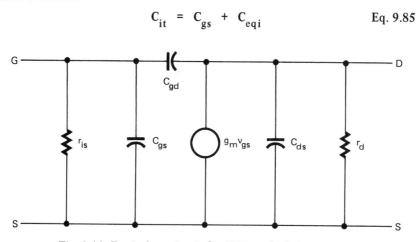

Fig. 9-11 Equivalent circuit for FETs at high frequencies

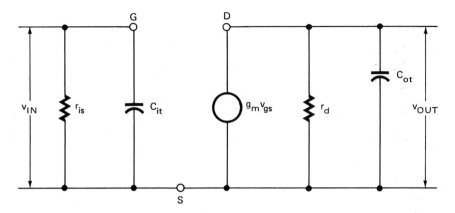

Fig. 9-12 Approximate equivalent circuit for a high-impedance amplifying device operated at high frequencies

Combine Equations 9.84, 5.23, and 9.85 to obtain the following expression:

$$C_{it} = C_{gs} + C_{gd}(1 - A_V) \qquad \text{Eq. 9.86}$$

Similarly, substitute the term $1/j\,\omega C_{gd}$ for Z in Equation 5.20 to obtain:

$$Z_2 = \frac{ZA_V}{A_V - 1} = \frac{A_V}{j\,\omega C_{gd}(A_V - 1)} \qquad \text{Eq. 9.87}$$

where Z_2 is the Miller impedance shown in figure 5-5. Define a capacitance C_{eqo} such that Equation 5.25 is again presented:

$$Z_2 = \frac{1}{j\,\omega C_{eqo}}$$

Substitute this expression in Equation 9.87. Then obtain an equivalent output capacitance C_{ot}, such that

$$C_{ot} = C_{ds} + C_{eqo} \qquad \text{Eq. 9.88}$$

Thus, by combining Equations 9.87, 5.25, and 9.88, this expression is obtained:

$$C_{ot} = C_{ds} + \frac{C_{gd}(A_V - 1)}{A_V} \qquad \text{Eq. 9.89}$$

If A_V has a large value, then

$$C_{ot} \approx C_{ds} + C_{gd} \qquad \text{Eq. 9.90}$$

Data published by FET and MOSFET manufacturers frequently contain terminology for FET capacitances that differs from the previously defined capacitances C_{gd}, C_{ds}, and C_{gs}. The capacitances given are referred to as the *small signal short circuit capacitances,* obtained from tests by the manufacturer. These capacitances are related to C_{gd}, C_{ds}, and C_{gs} by the following equations.

$$C_{rss} \approx C_{gd} \qquad \text{Eq. 9.91}$$

$$C_{osp} \text{ or } C_{oss} \approx C_{ds} \qquad \text{Eq. 9.92}$$

$$C_{iss} \text{ or } C_{gss} \approx C_{gs} \qquad \text{Eq. 9.93}$$

Frequently, C_{rss} is described in the literature as the small signal, short circuit reverse or feedback capacitance. The reason for the name is that, in a common source amplifier, the capacitance feeds the output signal back into the input.

The voltage gain equation for midband frequencies, Equation 9.8, can be combined with Equations 8.19 and 9.1 to yield:

$$A_V = \frac{-g_m r_d R_d}{r_d + R_d} \qquad \text{Eq. 9.94}$$

In this case, no load is connected to the amplifier. Substitute Equation 9.18 into Equation 9.94 to obtain:

$$A_V = -g_m Z_o \qquad \text{Eq. 9.95}$$

Equation 9.95 can be written as follows:

$$A_V = -g_m r_L \qquad \text{Eq. 9.96}$$

This expression applies when a load is connected and r_L is the parallel combination of R_L and Z_o. Then, Equation 9.86 can be changed to

$$C_{it} = C_{gs} + C_{gd}(1 + g_m r_L) \qquad \text{Eq. 9.97}$$

FET small signal behavior can be expressed by the use of y-parameters. These parameters represent the reciprocal of the impedances for the high-frequency FET. At a low signal frequency, the y-parameter admittances are conductances only and have no imaginary parts. At high frequencies, the y-parameters consist of a real part (a conductance) and an imaginary part (a susceptance). For example, the input admittance y_{is} can be derived from figure 9-12. The parameter y_{is} is the reciprocal of the input impedance for the high-frequency FET connected as a common source amplifier. The input imped-ance for the equivalent circuit of figure 9-12 is:

$$Z_{is} = \frac{r_{is}}{j\omega C_{it} r_{is} + 1}$$

Eq. 9.98

The input admittance, originally defined by Equation 8.26 is determined from Equation 9.98.

$$y_{is} = \frac{1}{r_{is}} + j\omega C_{it}$$

Eq. 9.99

The term $1/r_{is}$ is the input conductance (g_{is}) and ωC_{it} is the input susceptance (b_{is}). The resistance r_{is} (or r_{iss}) is the apparent small signal gate resistance of the FET. This parameter can be compared to h_{ie} for the bipolar transistor.

The parameter y_{os} can be defined in a similar manner for high-frequency operation. The output impedance from figure 9-12 is:

$$Z_{os} = \frac{r_d}{j\omega C_{ot} r_d + 1}$$

Eq. 9.100

The output admittance was defined first in Equation 8-27. This parameter is the reciprocal of Equation 9.100. It is

$$y_{os} = \frac{1}{r_d} + j\omega C_{ot}$$

Eq. 9.101

$$y_{os} = g_{os} + jb_{os}$$

Eq. 9.102

where g_{os} and b_{os} are the output conductance and the output susceptance, respectively.

The dynamic drain resistance (r_d) is sometimes given in the literature as r_{os}, r_{oss}, or $1/g_{os}$.

The forward transfer admittance (y_{fs}) generally is considered to have only a real part, g_{fs} or g_m. However, test results on some MOSFETs indicate that y_{fs} may have an imaginary part, b_{fs}. In this case, the equivalent circuit of figure 9-12 has g_m replaced by y_{fs}. In addition, the phase angle is included in the performance calculations.

Based on figure 9-12, an equivalent circuit can be drawn for a complete common source amplifier. Such a circuit is shown in figure 9-13, page 362. The equivalent circuit of figure 9-12 can be used for any of the three amplifier circuit configurations, just as in figure 9-13.

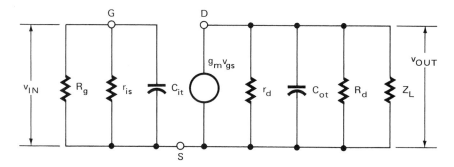

Fig. 9-13   Approximate ac equivalent circuit for a common source amplifier operated at a high frequency

The equivalent circuits of figures 9-11 and 9-12 are also applicable to high-frequency applications of vacuum tube amplifiers. Table 9-2 lists the appropriate symbols and subscripts for vacuum tubes along with the analogous symbols and subscripts for FETs.

It is not necessary to use the y-parameter concept for vacuum tubes because the dynamic grid and plate resistances are so high and the electrode capacitances are so low. However, the equivalent circuit of figure 9-12 and the preceding equation development can be applied to vacuum tubes by using the subscripts listed in Table 9-2 and the necessary tube test data.

Refer to FET specifications given in a handbook published by any of the large semiconductor manufacturers (such as General Electric Company, Motorola, or RCA). Find and list the high-frequency parameters for three types of MOSFETs and JFETs. Define each term used by the manufacturer, using the information suppled in the text. *(R9-22)*

FET	Vacuum Tube
r_d	r_p
C_{gd}	C_{gp}
C_{gs}	C_{gk}
C_{ds}	C_{pk}
r_{is}	r_g

Table 9-2

The specifications for a 2N4222 n-channel JFET list the following values: C_{rss} = 1.2 picofarads, C_{osp} = 1.5 picofarads, C_{iss} = 4.5 picofarads, y_{fs} = 2.4 millimhos, and r_d = 6.33 kilohms. The voltage gain of a common source amplifier using this device is to be computed at a high frequency. The circuit to be considered is shown in figure 6-20, with R_g = 300 kilohms, R_d = 1670 ohms, and Z_L = 5 + j0 kilohms. The coupling capacitors are too large to have any effect on the calculations. Circuit components other than the FET, such as R_d, Z_L, and R_g, are added to figure 9-12 to complete an equivalent circuit for the entire amplifier and its load. This equivalent circuit is shown in figure 9-13. The parameter y_{fs} is approximately equal to g_m. The output impedance (Z_o) is the combination of r_d and R_d in parallel when no load is connected. R_d, r_d, and Z_L can be combined as follows:

$$\frac{1}{r_L} = \frac{Z_o + Z_L}{Z_o Z_L} = \frac{1}{r_d} + \frac{1}{R_d} + \frac{1}{Z_L} \qquad \text{Eq. 9.103}$$

Substitute the given values to obtain:

$$\frac{1}{r_L} = \frac{1}{6.33} + \frac{1}{1.67} + \frac{1}{5} = 0.958$$

Thus, r_L = 1.044 kilohms. To find the voltage gain, substitute the appropriate values into Equation 9.95, using $y_{fs} = g_m$ = 2.4 millimhos.

$$\left| A_V \right| = g_m r_L = (2.4)(1.045) = 2.51$$

The sign of A_V is negative. The equivalent input capacitance (C_{it}) is determined as follows:

$$C_{it} = C_{gs} + C_{gd}(1 + g_m r_L)$$
$$= C_{iss} + C_{rss}(1 - A_V)$$
$$= 4.5 + (1.2)(1 + 2.51)$$
$$= 8.71 \text{ picofarads}$$

The equivalent output capacitance (C_{ot}) can be calculated as well. Because the value of A_V calculated in this example is so low, the exact equation for C_{ot} must be used.

$$C_{ot} = C_{ds} + \frac{C_{gd} A_V}{A_V - 1} = C_{osp} + \frac{C_{rss} A_V}{A_V - 1}$$
$$= 1.5 + \frac{(1.2)(-2.51)}{-2.51 - 1} = 2.36 \text{ picofarads}$$

Figure 9-12 can be simplified even more as shown by the circuit in figure 9-14. The value of r_{is} drops drastically at high frequencies for the majority of FETs. Therefore, R_g is too high by comparison to be considered in the simplified circuit.

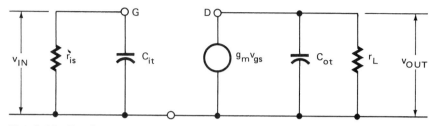

Fig. 9-14 Further approximation for a common source amplifier operated at high frequency

An FET has the following values: C_{rss} = 0.01 picofarad, C_{oss} = 1 pico-farad, and C_{iss} = 5 picofarads. For a value of A_V = -5, compute the input and output capacitances for the FET when it is used in a common source circuit.
(R9-23)

The output voltage for the common source amplifier is the drain current times the parallel combination of r_L and the C_{ot} reactance:

$$v_{out} = \frac{-i_d r_L}{j \omega C_{ot} r_L + 1} \qquad \text{Eq. 9.104}$$

The input voltage (v_{in}) is also v_{gs}. The drain current is:

$$i_d = g_m v_{gs} = g_m v_{in} \qquad \text{Eq. 9.105}$$

Combine Equations 9.104 and 9.105 and substitute the result into the standard voltage gain equation (Equation 9.3) to obtain:

$$A_V = \frac{-g_m r_L}{j \omega C_{ot} r_L + 1} \qquad \text{Eq. 9.106}$$

The equation for voltage gain indicates a single time constant that causes a drop in gain of 20 decibels per decade above the frequency at which $\omega = 1/C_{ot}r_L$. At this frequency, the drop is exactly 3 decibels. Substitute the given values and calculate the voltage gain for the example given on page 363.

$$A_V = \frac{-(2.4)(1.045)}{j(2.0\pi)(2.36)(10^{-12})(1.044)(10^3)f + 1}$$

$$= \frac{-2.51}{j(1.55)(10^{-8})f + 1}$$

The reduction in gain at high frequency begins at a frequency equivalent to:

$$f = \frac{1}{(1.55)(10^{-8})} = 64.5 \text{ megahertz}$$

This amplifier has a flat frequency response until about 64.5 megahertz. At this frequency the gain is down by 3 decibels. Above 64.5 megahertz, the voltage gain decreases by 20 decibels per decade. Because this is a low-gain amplifier (a gain of only 2.5 in the midfrequency range), a drop in gain is unacceptable. Thus, the signal frequencies must remain well below 64.5 megahertz.

When calculating the high-frequency current gain for a high-input impedance amplifier, complex terms are introduced because of the parallel circuits shown in figure 9-13. The input impedances R_g, r_{is}, and $1/j\,\omega C_{it}$ are combined into a parallel impedance which becomes:

$$\frac{R_g r_{is}\left(\dfrac{1}{j\,\omega C_{it}}\right)}{R_g r_{is} + R_g\left(\dfrac{1}{j\,\omega C_{it}}\right) + r_{is}\left(\dfrac{1}{j\,\omega C_{it}}\right)}$$

The input voltage is also v_{gs} and is the input current times this parallel impedance term:

$$v_{gs} = \frac{i_{in} R_g r_{is}}{R_g r_{is} j\,\omega C_{it} + R_g + r_{is}} \qquad \text{Eq. 9.107}$$

The terms r_d, R_d, and C_{ot} can be described as a complex impedance equal to:

$$\frac{r_d R_d\left(\dfrac{1}{j\,\omega C_{ot}}\right)}{r_d R_d + \dfrac{r_d}{j\,\omega C_{ot}} + \dfrac{R_d}{j\,\omega C_{ot}}} = \frac{r_d R_d}{r_d R_d j\,\omega C_{ot} + r_d + R_d}$$

The output current is the current through the load (Z_L). The output current can be calculated by combining the current source ($g_m v_{gs}$) with the current divider formed by Z_L and the complex impedance described previously that contains r_d, R_d, and C_{ot}. Thus,

$$i_{out} = g_m v_{gs}\left(\frac{\dfrac{r_d R_d}{r_d R_d j\,\omega C_{ot} + r_d + R_d}}{\dfrac{r_d R_d}{r_d R_d j\,\omega C_{ot} + r_d + R_d} + Z_L}\right)$$

$$i_{out} = g_m v_{gs} \left[\frac{r_d R_d}{r_d R_d + Z_L (r_d R_d j \omega C_{ot} + r_d + R_d)} \right]$$
Eq. 9.108

Solve Equation 9.107 for the input current and substitute this expression into Equation 9.108. This yields

$$i_{out} = g_m i_{in} \left(\frac{R_g r_{is}}{R_g r_{is} j \omega C_{it} + R_g + r_{is}} \right) \left[\frac{r_d R_d}{r_d R_d + Z_L (r_d R_d j \omega C_{ot} + r_d + R_d)} \right]$$
Eq. 9.109

Substituting Equation 9.109 into Equation 9.9 results in an expression for the current gain:

$$A_I = \frac{g_m R_g r_{is} r_d R_d \left(\dfrac{1}{R_g + r_{is}} \right) \left(\dfrac{1}{r_d R_d + r_d Z_L + R_d Z_L} \right)}{\left(\dfrac{R_g r_{is} j \omega C_{it}}{R_g + r_{is}} + 1 \right) \left(\dfrac{Z_L r_d R_d j \omega C_{ot}}{r_d R_d + r_d Z_L + R_d Z_L} + 1 \right)}$$
Eq. 9.110

At low frequencies, the $j \omega$ terms drop out and

$$A_I = \frac{g_m R_g r_{is} r_d R_d}{(R_g + r_{is})(r_d R_d + r_d Z_L + R_d Z_L)}$$
Eq. 9.111

Since $r_{is} \gg R_g$ at low frequencies, Equation 9.111 reduces to:

$$A_I = \frac{g_m R_g r_d R_d}{r_d R_d + r_d Z_L + R_d Z_L}$$
Eq. 9.112

When the output is shorted, $Z_L = 0$ and

$$A_I = g_m R_g$$
Eq. 9.113

If Equation 9.113 is multiplied and divided by r_d and is then combined with Equation 8.18, the result is Equation 9.16 again:

$$A_I = g_m R_g = (g_m r_d) \frac{R_g}{r_d} = \frac{\mu R_g}{r_d}$$

However, r_{is} can be very much smaller than R_g at very high frequencies. When this occurs, and the output is shorted, the high-frequency current gain is:

$$A_I = \frac{g_m r_{is}}{r_{is} j \omega C_{it} + 1}$$

Eq. 9.114

At very high frequencies, the current gain is inversely proportional to frequency, because

$$A_I = \frac{g_m r_{is}}{r_{is} j \omega C_{it}} = \frac{g_m}{j \omega C_{it}}$$

$$= \frac{g_m}{2 \pi f C_{it}}$$

Eq. 9.115

One precaution that should be observed when calculating amplifier performance at high frequencies is that the data used should be suitable for high frequencies. The values of the y-parameters and other information for an FET (or a MOSFET or a vacuum tube) should be values obtained at frequencies approximately near those of interest. For example, the value of r_{iss} for one popular MOSFET changes from 6 megohms at 1 megahertz to 1 kilohm at 200 megahertz.

With the use of a high-gain amplifying device, such as some types of pentodes, which has an amplification factor (μ) of 50 or 70, the input capacitance is increased by a larger amount due to the feedback effect from the plate to the grid through the grid-to-plate capacitance C_{gp}. The vacuum tube form of the input and output capacitances are given by the following relationships.

$$C_{it} = C_{gk} + C_{gp}(1 - A_V)$$

Eq. 9.116

$$C_{ot} = C_{pk} + \frac{C_{gp} A_V}{A_V - 1}$$

Eq. 9.117

For a high-gain amplifying device, Equation 9.117 is approximately:

$$C_{ot} = C_{pk} + C_{gp}$$

Eq. 9.118

A vacuum tube amplifier, such as the one shown in figure 7-21, uses a tube with the following characteristics: $\mu = 96$, $r_p = 53.5$ kilohms, $C_{gk} = 1.6$ picofarads, $C_{gp} = 1.7$ picofarads, $C_{pk} = 0.46$ picofarad, $R_p = 121$ kilohms, and $R_g = 500$ kilohms. Compute the no load voltage gain at a signal frequency of 1 megahertz. (R9-24)

A 3N200 dual insulated gate FET is used in a high-frequency amplifier. The amplifier is tuned so that it is fed from a resistive source and feeds a resistive load of 100 kilohms. The circuit is given in figure 6-26(a). Gate 1 is used for a high-frequency signal. There is no signal at Gate 2. Use $R_g = 1$ megohm and $R_d = 2$ kilohms. R_s is bypassed by a capacitor and has no effect at the signal frequency. The specifications for the 3N200 DIGFET include these values: $C_{rss} = 0.02$ picofarads, $C_{iss} = 6$ picofarads, $C_{oss} = 2$ picofarads, $g_{fs} = g_m = 15$ millimhos, $r_d = r_{os} = 1/g_{os} = 6.7$ kilohms, and $r_{is} = 1/g_{is} = 4$ kilohms. Compute the voltage gain for this amplifier when a small 100-megahertz signal is applied to Gate 1. (R9-25)

Compute the current gain for the amplifier described in review problem (R9-25) at the signal frequency. (R9-26)

Compute the power gain for the amplifier, using the results of problems (R9-25) and (R9-26). (R9-27)

The equivalent circuit of figure 9-12 is that of a bare amplifying device. Space does not permit the discussion of all of the circuits in which this device can be connected. However, for any amplifier being considered, an equivalent circuit like the one in figure 9-13 can be drawn based on figure 9-12 as a model. The complete amplifier is then further simplified as much as possible. An example of this is figure 9-14, which was based on the equivalent circuit shown in figure 9-13. This circuit is typical of the simplification that can be achieved. However, the amount of simplification depends on the relative values of the parameters involved

The common source amplifier was studied for high-frequency use because it is most commonly used and the y-parameters given by the manufacturer are expressed in terms of the common source circuit. If it is necessary to study a common gate or common drain amplifier at high frequencies, then it is a simple matter to use the parameters defined in terms of the common source circuit as building blocks for the construction of equivalent circuits of common

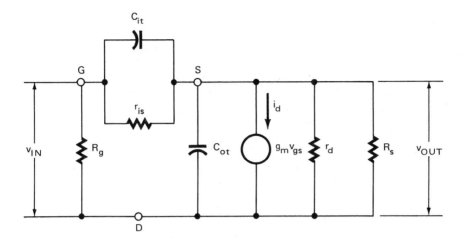

Fig. 9-15 Approximate ac equivalent circuit for a common drain amplifier operated at high frequency

drain and common gate amplifiers. An ac equivalent circuit for the source follower amplifier, utilizing these parameters, is shown in figure 9-15. To analyze the circuit, the relative values of the parameters for a specific problem are studied and a decision is made on how the circuit can be simplified. The method of analysis for the circuit is the same as for the previous circuits studied. The quantity $g_m v_{gs}$ is a perfect current source having infinite internal resistance. Keeping this fact in mind, another look at the circuit shows that it is merely a series-parallel arrangement of resistances and reactances. Such a circuit can be solved by ordinary ac network analysis. An equation for the voltage gain is obtained by using Kirchhoff's Voltage Laws and algebraic manipulation.

The equation for the voltage gain for the high-frequency common drain amplifier has the same form as Equation 9.26 for the voltage gain at an intermediate frequency.

$$A_{VH} = \frac{g_m z_L}{1 + g_m z_L} \qquad \text{Eq. 9.119}$$

Equation 9.119 can be compared to Equation 9.26 after Equation 8.19 ($\mu = g_m r_d$) is inserted into it. The term z_L is similar to r_L at intermediate frequencies. At high frequencies, the load may include a reactance.

The voltage gain at high frequency is related to the voltage gain at intermediate frequency by the expression:

$$A_{VH} = \frac{A_V}{j\omega\tau + 1} \qquad \text{Eq. 9.120}$$

where

$$\tau = \frac{C_{ot}R_s r_d}{R_s + r_d + g_m R_s r_d}$$ Eq. 9.121

τ is derived from an analysis of figure 9-15.

An ac equivalent circuit is shown in figure 9-16 for the common gate amplifier operated at high frequencies. The analysis of the circuit of figure 9-16 is similar to that of the circuit shown in figure 9-15. If R_s is a finite value, the expression for the voltage gain is so complex that it is difficult to describe in this text. However, if $R_s \approx 0$ and no load is connected to the output terminals, then the voltage gain is:

$$A_V = \frac{\dfrac{g_m R_d r_d}{R_d + r_d}}{\dfrac{j\omega C_{ot} r_d R_d}{r_d + R_d} + 1}$$ Eq. 9.122

The gain falls by 3 decibels at a frequency of $\omega = (r_d + R_d)/C_{ot} r_d R_d$. Therefore, the operation of the amplifier must be well below this frequency. Equation 9.122 reduces to Equation 9.37 at low frequencies, when $R_s = 0$.

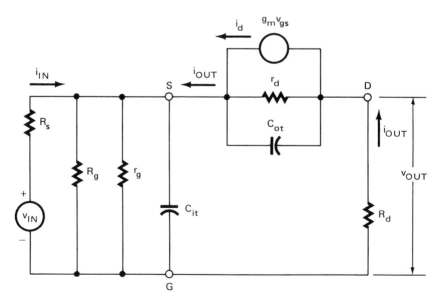

Fig. 9-16 Approximate ac equivalent circuit for a common gate amplifier operated at high-frequency

An FET is used in a common drain amplifier, such as the circuit shown in figure 8-8. The circuit values are as follows: $R_{source} = 0$, $R_g = 200$ kilohms, $R_s = 3$ kilohms, $g_m = 6000$ micromhos, $C_{ot} = 10$ picofarads, $R_d = 0$, and $r_d = 5$ kilohms.

   a. Compute the time constant (τ) for the voltage gain.

   b. If $A_V = 0.7$ for this amplifier, compute the frequency at which the voltage gain drops by 3 decibels.   (R9-28)

LABORATORY EXPERIMENT 9-1

The following laboratory experiment was tested and found useful for illustrating the theory in this Chapter.

OBJECTIVE

To measure the frequency response of a capacitively coupled FET amplifier and to find the corner frequencies for this amplifier.

EQUIPMENT AND MATERIALS REQUIRED

1   Field effect transistor, 2N4224, n-channel

1   Oscilloscope, time based, dual channel, frequency response up to 35 megahertz

1   Signal generator, 0 to 5 volt peak-to-peak adjustable, capable of frequency control of a sinusoidal signal from 3 hertz to 3 megahertz

3   Resistance substitution boxes, Heathkit IU-28A or equivalent

3   Capacitors, 0.1 microfarad, 50 volts

2   Resistors, 1 kilohm, 1%, 1/2 watt

1   Power supply, dc, regulated, 0-30 volts dc, 1/2 ampere

1   Volt-ohm-milliammeter, ac and dc, with a lowest full scale reading of 10 millivolts, and 2% accuracy. (A decibel scale on the meter will help to determine the frequency response, but is not required for the experiment.)

PROCEDURE

1.   Obtain a 2N4224 transistor that is either new or tested to insure that it is useable.

2.   Construct a circuit like the one shown in figure 9-17, page 372.

3.   Apply a sinusoidal signal to the input terminals of the circuit. Use a frequency of approximately 1 kilohertz at an amplitude that is well within the linearity limits of the FET. Select an input amplitude that can be kept

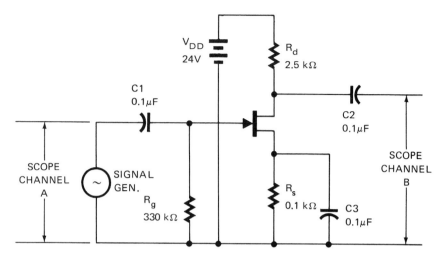

Fig. 9-17 Circuit for voltage gain frequency response

constant using the signal generator controls. In addition, the amplitude value should allow the voltage decibels to be computed easily. (For example, a value of 1 volt is appropriate.) For high-frequency input signals, there is a high noise level. Therefore, the signal generator should be first adjusted to its highest frequency level to determine if the amplitude is sufficient to overcome the noise level.

4. Reduce the frequency to a level at which the output is almost zero. (This situation occurs at a frequency for which the reactances of the circuit are large with respect to the resistances in the circuit.)

5. Increase the frequency in approximately 3 steps per decade. Measure the voltage output at each step. Take more readings at those frequencies where the output seems to be more sensitive to frequency. These points will occur at the highest and lowest levels of the signal frequency. The signal generator output may change with frequency at the lowest and highest ends of the frequency spectrum. The signal generator output must therefore be checked and readjusted for each reading at these frequency levels.

6. Calculate the gain in voltage decibels.

$$dB = 20 \log \left(\frac{v_o}{v_i} \right)$$

Eq. 9.123

7. Plot the gain in decibels versus frequency in hertz on 5- or 6-cycle semi-logarithmic graph paper.

8. The gain drops to a low value at the highest and lowest frequency levels. Mark the places on the graph drawn in step 7 where the voltage gain at high- and low-frequency levels is 3 decibels below the gain at intermediate frequency levels. There will be two places at which these values occur: once at a high frequency and once at a low frequency. These places mark the corner frequencies for the frequency response of the circuit. The high-frequency cutoff point depends upon the shunt capacities for components other than the FET. The high-frequency cutoff point for the circuit will be much lower than the potential cutoff frequency for the FET itself, as a result of the construction details of the student's circuit. However, the corner frequency for the high-frequency response of the circuit can still be determined from the frequency response curve drawn in step 7.

The low-frequency corner generally occurs at a more predictable value than does the high-frequency cutoff. The voltage gain at low frequency behaves according to Equation 9.76.

In experiments conducted by the author, a resonant condition was frequently exhibited near the high-frequency cutoff point. As a result, the corner frequency measurements were difficult or impossible to make at the high-frequency end. The time constants due to the shunt capacitances of the circuit may be concentrated at the same point. This point may be well below the corner frequency caused by the FET itself. This situation is not covered by the voltage gain equation derived in this chapter (Equation 9.106) and it is not applicable to the experiment.

9. Calculate the upper and lower corner frequencies in radians per second and hertz. Calculate the time constants corresponding to these frequencies. These time constants are given by:

$$\tau = \frac{1}{2\pi f} \text{ seconds} \qquad \text{Eq. 9.124}$$

where f is the upper or lower corner frequency in hertz.

10. Change the circuit to that of figure 9-18, page 374.

11. This portion of the experiment is performed to demonstrate and analyze the frequency response for the current gain. Set the input signal at about 1 kilohertz at an amplitude which is well within the limits of linearity for the JFET. An input signal of 1 volt applied across the 1-kilohm series resistor (as measured by an ac VOM), provides a 1-milliampere input into the circuit. If the output current through the 1-kilohm load resistance (R_L) is also measured in milliamperes, then the current gain ratio can be read directly and the decibel calculation is easier. The student should recall that the meter reads rms values and the oscilloscope reads peak-to-peak values. Thus, the student must convert from rms values to peak values (or vice versa) before the gain can be calculated.

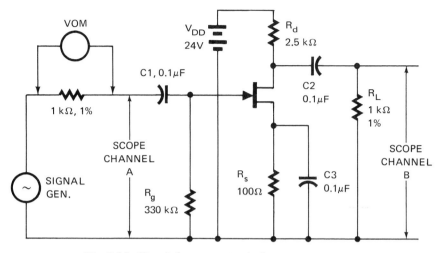

Fig. 9-18 Circuit for current gain frequency response

12.   Repeat steps 4 and 5.

13.   Calculate the gain in current decibels using the following equation:

$$dB = 20 \log \left(\frac{i_o}{i_i} \right)$$

Eq. 9.125

where i_o is the current through the 1-kilohm load resistor (R_L) and i_i is the current through the 1-kilohm series input resistor. The currents i_o and i_i are best determined by measuring the voltage across their respective resistors and dividing by the value of these resistors. This method is preferred to the use of the ammeter setting and a series connection for the VOM. Changing the scales on an ammeter causes the series resistance to change as well, and the performance of the circuit is altered.

14.   Repeat steps 7, 8, and 9, using the results of the current gain calculations.

EXTENDED STUDY TOPICS

1.   Compare the common gate, common source, and source follower amplifiers for (1) voltage gain, (2) current gain, (3) input impedance, and (4) output impedance.

2.   Discuss the criteria used to select coupling capacitors and capacitors used to bypass the source or cathode bias resistors.

3.   What advantage is there in the use of Miller's theorem to construct the ac equivalent circuit for an amplifier operated at high frequency?

4. Explain the source of the interterminal capacitances of the FET.

5. Refer to the amplifier circuit shown in figure 6-31. (a) Draw the ac equivalent circuits for low-frequency, intermediate frequency, and high-frequency operation. (b) Determine the voltage gain at various frequencies in the low-, intermediate, and high-frequency regions. (c) Express the gain values in decibels and plot the complete frequency response curve for the amplifier from a frequency close to zero to frequencies well beyond the high-frequency cutoff. Use multicycle semilogarithmic graph paper and a broken scale for the wide intermediate frequency region. Use the following parameters for the FET: C_{iss} = 5.5 picofarads, R_g = 500 kilohms, C_{rss} = 0.25 picofarads, C_{oss} = 2 picofarads, g_{fs} = 7.5 millimhos, g_{os} = 40 micromhos, and g_{is} = 0.14 millimhos. In actual practice, these parameters can vary with frequency. The values used will be considered constant for the frequency ranges being considered.

6. Compare the four y-parameters for the FET (y_{is}, y_{os}, y_{fs}, and y_{rs}) to the corresponding h-parameters for the bipolar transistor. Make a table listing the analogous h-parameters corresponding to the four y-parameters.

10

Modern
Amplifying Devices

After studying this chapter, the student will be able to:

- explain the processes used in the manufacture of integrated circuit amplifiers
- use integrated circuit amplifiers in more complex amplifying systems
- explain how resistors, capacitors, and inductors are included in integrated circuits
- explain the terms used on integrated circuit amplifier specifications
- describe special nonlinear semiconductor amplifying devices
- describe how maser and laser systems function as amplifiers
- describe other types of amplifying systems, including magnetic amplifiers, rotating amplifiers, and ultra high-frequency amplifying systems

CONSTRUCTION OF SEMICONDUCTOR DEVICES

Six basic processes used in the construction of semiconductor devices are:

1. growing a single multielement crystal from a molten mass of semiconductor material;

2. melting a small amount of semiconductor material placed on a solid crystal having majority carriers of the opposite type (n or p);

3. etching a depression into the semiconductor crystal and electroplating impurity atoms into the depressed surface;

4. subjecting a semiconductor crystal to a gaseous atmosphere containing atoms of the correct impurity, the impurity atoms then diffuse into the crystal in a process known as *gaseous diffusion;*

5. subjecting two or more semiconductor crystals having opposite majority carriers to extreme pressures so that the crystals are forced tightly together to form a cold weld connection;

6. growing a semiconductor material on the surface of another semiconductor crystal; this is accomplished by allowing a vaporized or liquid semiconductor material to crystallize on the base material; this process is called *epitaxial growth.*

The epitaxial growth process differs from gaseous diffusion in that the vapor is a gaseous form of a second complete semiconductor material which is then changed to a solid state. The diffusion process only adds gas atoms which supply the proper impurity to make the semiconductor crystal an *n*- or *p*-type material.

These manufacturing processes can be used separately or in combination to build a semiconductor diode or a bipolar transistor. The processes of epitaxial growth and gaseous diffusion are combined to construct single field effect transistors and large scale integrated circuits. Integrated circuits, which can contain a large number of semiconductor diodes and transistors, are constructed entirely of one semiconductor crystal containing various combinations of *n*- and *p*-type materials. Several processes are combined to manufacture these devices. The crystal growing method is used to form the basic crystal which is called the *substrate.* Transistors and complete integrated circuits are made from the substrate. The substrate usually is a silicon crystal doped with some *n*- or *p*-type impurity. Silicon is used almost exclusively to make field effect transistors and integrated circuits. Germanium is sometimes used when diodes and bipolar transistors are to be made. Semiconductor materials other than germanium and silicon are used for specialized devices, such as thermistors (temperature sensitive resistors) and electro-optic semiconductor devices.

The first steps in the crystal growing process are to melt a large block of pure semiconductor silicon and add impurities. A small, very pure *seed* crystal is inserted in the melt. This crystal is gradually pulled out as it is rotated, as shown in figure 10-1, page 378. As the seed crystal is drawn out, molten silicon sticks to it and then solidifies. A cylindrical crystal is formed which is between one and four inches in diameter and eight to ten inches long. The crystal is cut into thin wafers. The wafers are then smoothed and polished until they have a fine mirrorlike finish.

A process is then selected for the next step in the construction of a semiconductor device. Semiconductor diodes and bipolar transistors can be constructed entirely by using any one of the six processes listed, including the growth process just described. The processes of gaseous diffusion and epitaxial

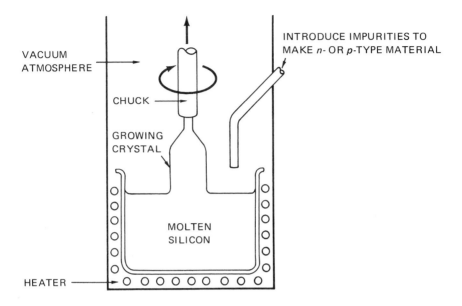

Fig. 10-1 Diagram of how a crystal is grown

growth will be described in detail, since they are the most commonly used methods of manufacturing semiconductor devices.

The following description of the manufacture of a semiconductor device concentrates on an integrated circuit. Since integrated circuits are complete electronic circuits, such a description will help in understanding them.

The polished wafer that is produced by the first manufacturing process is subjected to a pure oxygen atmosphere while at a temperature of about $1000°C$. (This temperature is close to the melting point of the material.) As a result of this treatment, the surface of the wafer is oxidized; that is, an oxide layer is built up on the wafer, as shown in figure 10-2. The oxide surface is painted with a material known as *photoresist*. The chemical composition of this material changes when exposed to light. A mask or template is placed over the surface of the photoresist-covered wafer. Part of the mask admits light and part does not. A typical pattern for a mask is shown in figure 10-3. The transparent portions of the mask admit light to the light-sensitive painted surface of the wafer. Wherever light is admitted, the photoresist becomes impervious to a liquid cleansing solution. When such a solution is applied, the remaining areas of photoresist are washed off, reexposing some of the oxide surface. This oxide surface area is then etched with a corrosive solution that does not affect the light-hardened photoresist area. The semiconductor material of the wafer is now exposed in specific areas, as shown in figure 10-4, page 380. The process of gaseous diffusion is used to add *n*- or *p*-type impurities to the exposed portion

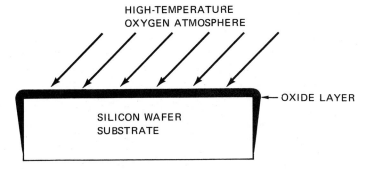

Fig. 10-2 Semiconductor wafer covered by an oxide layer

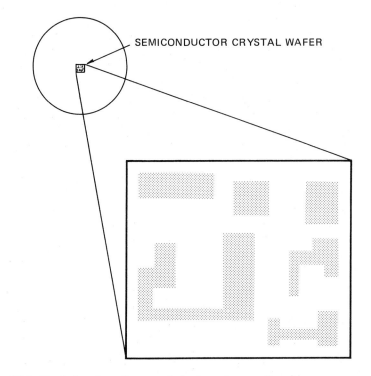

Fig. 10-3 Typical pattern for a mask (enlarged many times)

of the semiconductor material. In figure 10-4, page 380, n-type impurities are added to a p-type substrate. The depth to which the impurity atoms penetrate the substrate is controlled by the temperature, gas pressure, and the time of exposure. The thickness of the n- or p-type region can be controlled with precision by this process.

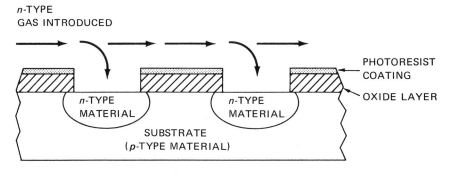

Fig. 10-4 Semiconductor wafer is oxide covered, treated with photoresist and then has oxide removed at designated points; wafer is then treated with a gas to produce an *n*-type region by gaseous diffusion

For the next step in the process, a new layer of oxide is formed over the entire surface of the wafer, the photoresist material is painted over the oxide surface, and a mask, consisting of transparent and opaque regions, is placed over the photoresist material. Again, light admitted to the masked portion sensitizes the photoresist. The photoresist and oxide layers can then be removed at selected locations, exposing the semiconductor crystal at those locations. The gaseous impurities are reintroduced and portions of the crystal nearest the surface are converted to *p*-type impurity, as shown in figure 10-5.

The process is repeated as many times as necessary to build up a given configuration. In figure 10-5, the *n*-type layer is exposed again by the masking and etching process. When the terminals are attached, the semiconductor crystal can be used as a pnp transistor. The *p*-type substrate is used as a collector, followed by the *n*-type base and the *p*-type emitter. More transistors can be built onto the same substrate by (1) using a more complex mask or (2) repeating the entire process as many times as necessary to achieve the desired number of transistors.

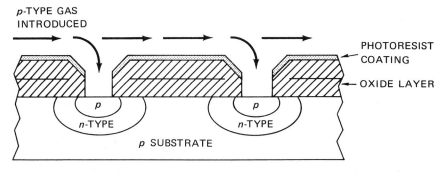

Fig. 10-5 Wafer of figure 10-4 being treated to produce a second *p*-type layer

The process of vapor deposition or epitaxial growth can be used alternately with gaseous diffusion. Generally, the epitaxial layer is the first layer of material deposited on the substrate. For example, the substrate may be a *p*-type material, with an *n*-type layer formed on it by depositing a vaporized form of an *n*-type material on the substrate. This process provides the most precise way of controlling the thickness of a particular *n*- or *p*-type layer. After the last oxide film layer is etched away, the exposed semiconductor material is the surface to which contacts can be attached. Aluminum is introduced as a gas and is deposited on the semiconductor surface. The resulting pattern of aluminum provides (1) a surface to which wires can be attached and (2) interconnections between the transistors and diodes built from the basic wafer.

BUILDING LINEAR DEVICES USING SEMICONDUCTOR TECHNIQUES

Transistors and diodes are not the only circuit elements that can be made using the techniques that have been described. Resistors can be constructed by controlling (1) the concentration of the impurities diffused or condensed on the semiconductor surface and (2) the length pattern that is made on the surface. A maze can be constructed as shown in figure 10-6. The ends of the maze form the points to which external connections are made.

Larger resistance values are obtained by thinning the pattern using techniques that involve abrasion, oxidation, or laser beam techniques. To obtain higher currents, thick film techniques are used. In thick film technology, a *silk screen* process is used to make resistors. In addition, low-to-moderate quality semiconductors can be made in limited quantities by this method. The basic silk screen process is similar to the one used in the printing industry for making colored pictures. In this process, the basic sequence of steps is the same as that of the epitaxial growth process, except that the semiconductor materials are applied to a ceramic substrate.

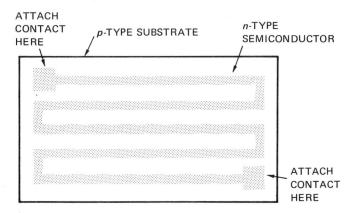

ATTACH CONTACT HERE

p-TYPE SUBSTRATE

n-TYPE SEMICONDUCTOR

ATTACH CONTACT HERE

Fig. 10-6 Top view of an integrated circuit resistor, magnified many times

A capacitor exists wherever a pair of conducting plates is separated by a dielectric material. The region of immobile free charges at the junction of a reverse biased diode forms an effective insulating dielectric material. Therefore, capacitors can be made by providing reverse biased *p-n* junctions in the integrated circuit. The capacitors exist as long as the circuit biasing and internal layout of the integrated circuit keep the junctions reverse biased. The values of the capacitors formed in this manner are very small because the capacitance is proportional to area.

Insulating regions in the circuit are also produced by reverse biasing *p-n* junctions. For example, the *p*-type substrate in figure 10-6 is insulated from the rest of the circuit by the *n*-type region above it.

The only circuit element that is difficult or impossible to fabricate using integrated circuit techniques is an inductor. A circuit requiring an inductance, a large capacitance, or large resistors must use standard discrete components connected externally to the integrated circuits.

A complete integrated circuit amplifier can be constructed using the methods just described. Figure 10-7 shows an integrated circuit crystal or *chip*.

Fig. 10-7

List six basic processes used in the construction of semiconductor devices.
(R10-1)

Which of the processes listed in question (R10-1) are used in the construction of integrated circuit amplifiers? *(R10-2)*

What is the purpose of the photoresist material applied to the semiconductor crystal during the construction of an integrated circuit? *(R10-3)*

Briefly state what principles are involved in making resistors and capacitors in an integrated circuit. *(R10-4)*

INTEGRATED CIRCUIT AMPLIFIERS

One obvious advantage of integrated circuits over discrete components is that a small housing can be used for a complex amplifying system. If a large number of the same design of integrated circuit can be built, then the resulting cost to the consumer is low and the reliability of the device is excellent. The reliability of integrated circuits is far better than that of either tubes or circuits using discrete components. Additional advantages of integrated circuits include the following:

- close matching of devices over a wide temperature range can be accomplished;

- the relative cost of transistors and diodes in the circuit is very low when compared to that of resistors, capacitors, and inductors; a large number of transistors and diodes can be used in a single circuit with very little additional cost; and

- excellent thermal coupling is achieved since the entire circuit can be fabricated on a very tiny chip of crystal material.

In integrated circuit technology, it is often simpler to add another transistor on the substrate than it is to provide for other components in the circuit or to design for close tolerances in the components. The most frequently used components in integrated circuits are diodes, resistors, and transistors.

An integrated circuit amplifier typically contains a differential amplifier section. By including this section, use is made of the thermal and performance matchups that are possible with integrated circuits. The differential amplifier, covered in Chapter 3, eliminates some of the inherent defects of transistor behavior. This amplifier is inexpensive and easy to include in the integrated circuit amplifier.

Figure 10-8, page 384, shows a differential amplifier consisting of transistors Q1 and Q2. A controlled constant current source for the emitters of Q1 and Q2 is provided by transistor Q3. Q3 helps to reject the *common mode* signal at the collectors of transistors Q1 and Q2 of the differential amplifier. This signal exists at these points (terminals 6 and 8) when the input signals (at terminals 5 and 1) are zero.

Some integrated circuits do not include the resistors shown in figure 10-8. Such circuits consist of the transistors only and all of the separate terminals are brought out.

Figure 10-9, page 384, shows the Darlington connection of a pair of transistors (Q3 and Q4) which are to be used in a circuit similar to that of figure 3-20. The same integrated circuit also includes transistors Q1 and Q2 with all of their leads brought out. A feature of this arrangement is that transistors Q1 and Q2 are matched from the standpoint of performance and thermal effects with each other and with the Darlington transistors. As a result, bias stabilization circuits can be built more easily. Note that the substrate is on a separate terminal (terminal 10). This terminal is connected to the most negative point in the external circuit to maintain isolation between the transistors. The diode symbols are used because the substrate behaves like a reverse biased diode when it is connected as recommended.

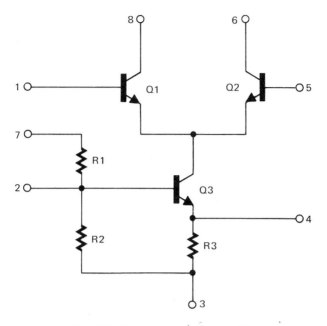

Fig. 10-8 Integrated circuit amplifier

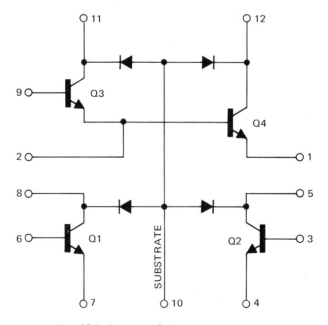

Fig. 10-9 Integrated circuit transistor array

Why is the substrate in the previous discussion connected at all? (To answer this question, refer to pages 376 to 383.)   *(R10-5)*

Examples of Integrated Circuit Amplifiers

Figure 10-7 is the RCA CA3020 integrated circuit wideband power amplifier. This device is a multistage difference amplifier whose schematic is shown in figure 10-10. This integrated circuit is sold either as a bare chip for customers to do the packaging or as a package similar to that of a single transistor. For a comparison of size, figure 10-11 shows a single transistor and the integrated circuit of figure 10-7, packaged for sale and use in a practical amplifying system. This amplifier can be used at frequencies down to dc because of the differential amplifier stage and the dc coupling between stages. The differential amplifier (transistors Q2 and Q3) assures that any thermally generated

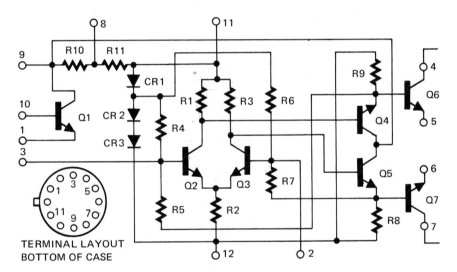

Fig. 10-10 Schematic of an integrated circuit amplifier

Fig. 10-11

offset signal in Q2 is duplicated and canceled in Q3, and vice versa. The second stage of amplification (Q4 and Q5) can also be treated as a differential amplifier. This statement is true if the output is taken between terminals 4 and 7 of transistors Q6 and Q7 and the appropriate connections are made to the collectors and emitters of Q4 and Q5. Q4 and Q5 are emitter follower stages and Q6 and Q7 are connected in the common emitter mode in the usual application. The various bias levels are supplied through terminal 9. The bias voltages are dropped across resistors R10 and R11 and forward biased diodes CR1, CR2, and CR3. Because there is relatively little change in the voltage drop across a forward biased semiconductor diode, the diodes help to provide a regulated bias voltage. The base of transistor Q1 can be used as the input to provide a greater overall gain.

Diodes CR1, CR2, and CR3 in figure 10-10 are forward biased by an external bias voltage connected to terminal 9. Each diode has a voltage drop of 0.7 volt. What is the value of bias voltage V_{CC} applied to the collector circuit of the first differential amplifier stage (the voltage between terminals 11 and 12)? (R10-6)

One fairly complex integrated circuit amplifier is the operational amplifier for high output current applications, figure 10-12. This circuit contains many stabilization features. For example, consider what happens if there is an increase in temperature. The collector current for Q3 and Q4 will increase. At the same time, V_{BE} for Q7, Q9, Q8, and Q10 decreases. This decrease is due to the increasing temperature and the resulting lower base signal to Q7 and Q8. As a result, transistor Q5 moves toward shutoff, reducing the signal to transistor Q6 through resistor R9. Thus, Q6 is closer to cutoff, reducing the collector current through Q6. (This current is also the emitter current for Q3 and Q4.) This action counteracts the effect of the temperature increase. Q7 and Q9 in cascade and Q8 and Q10 in cascade are the two halves of another difference amplifier configuration. Q12 acts as the constant current source which is used to reduce the common mode signal of this section.

In an integrated circuit multistage amplifier, a dc voltage buildup occurs as the signal proceeds from the input to the output stages. This buildup is the result of direct coupling between stages. Thus, for a zero input signal, the output signal is not zero. The circuit containing transistors Q11 and Q13 in figure 10-12 is a dc level shifter that compensates for this voltage buildup. Q14 is an emitter follower stage that follows the dc level shifter stage. This signal, plus the current flowing through R15, make up a balanced input to a two-stage push-pull amplifier. Transistors Q16 and Q17 provide the "push" and transistors Q15 and Q18 provide the "pull" for this amplifier. (This circuit is not the same type of push-pull amplifier shown in figure 3-21 and described in Chapter 3. However, there is some similarity. In figure 3-21, the center tap of the transformer windings

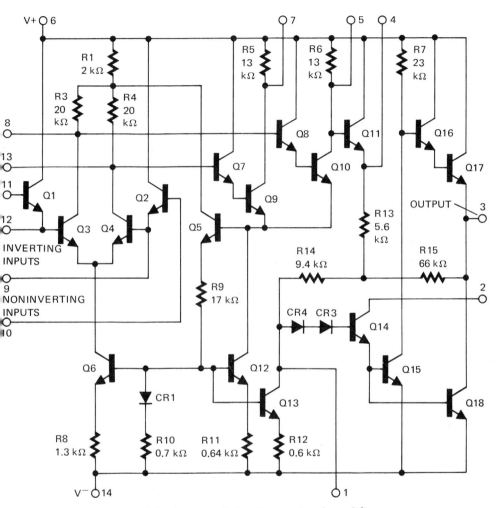

Fig. 10-12 Integrated circuit operational amplifier

provide for a 180 degree difference in polarity between transistors Q1 and Q2. This means that one transistor is on while the other is off, alternately. The same type of operation is possible without transformers by reversing the connections of one transistor with respect to another. This latter configuration is represented in the final stage of amplification in figure 10-12.) CR3 and CR4 provide additional temperature compensation. The output of the amplifier is taken between terminals 2 and 3. Additional terminals (4, 5, 7, and 1) are available for any desired feedback signals through external components.

Another family of integrated circuit amplifiers is also available. Such amplifiers are called *hybrids,* and include integrated circuit chips packaged with capacitors, transistors, or diodes that are made as separate discrete devices. Since these components are packaged together as integrated circuits, externally they appear to be single monolithic circuit units. Therefore, they are connected with external discrete components as if they were monolithic units. Examples of hybrid amplifiers include the LH201 operational amplifier by National Semiconductor. This device is an LM201 operational amplifier with a 30-picofarad capacitor incorporated within it to stabilize the circuit against internal oscillations. Another example is the LH0005 operational amplifier that contains a number of discrete silicon semiconductors within its case.

Assume the integrated circuit amplifier of figure 10-12. If the forward biased p-n junctions have a 0.7-volt drop across them, what is the approximate voltage between terminals 14 and 1 when this circuit is in use? (R10-7)

APPLYING INTEGRATED CIRCUIT AMPLIFIERS

If integrated circuits are used, the construction of a complete amplifier is reduced to the addition of the voltage dividing resistors, ac coupling capacitors, and other components necessary to make the amplifier function correctly. A typical integrated circuit amplifier does not have the coupling capacitors or transformers that the complete circuit may require. Additional resistors may be added to bias auxiliary transistors that are built into the integrated circuit.

A typical complete amplifier built around the RCA integrated circuit amplifier model CA3020 is shown in figure 10-13. The output is fed to a transformer to make a push-pull amplifier connection. The resistors R1 and R2 are used to provide a bias for transistor Q1. (Refer to figure 10-10 as well as figure 10-13.) Transistor Q1 can be connected as an emitter follower preamplifier to provide a higher amplifier input impedance. External bias resistors are added to establish the Q-point for this preamplifier. Since all three terminals of transistor Q1 are brought out, this amplifier can be treated as a standard transistor problem. Capacitors in the circuit restrict the low-frequency range of the amplifier and insure that dc currents do not flow through the signal input leads.

A circuit using the CA3020 amplifier without Q1 is shown in figure 10-14. The typical input resistance of this connection is 700 ohms as compared to 50,000 ohms for the circuit in figure 10-13.

There is a voltage drop of 4 volts across resistor R2 in figure 10-13. What is the approximate value of the dc base current in transistor Q1? (R10-8)

Figure 10-15, page 390, shows the CA3033 integrated circuit amplifier connected in an instrument preamplifier circuit. The zener diodes provide voltage regulation for the bias voltages supplied to the amplifier. Transistors Q19

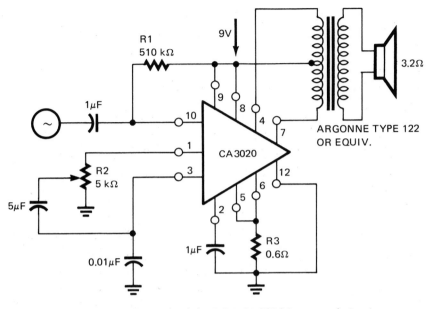

Fig. 10-13 Application of the RCA CA3020 integrated circuit

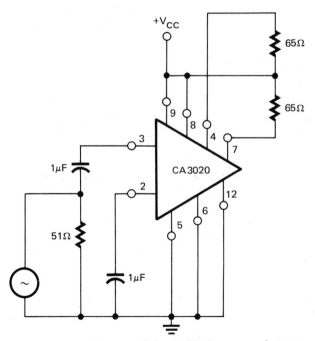

Fig. 10-14 Application of the CA3020 integrated circuit

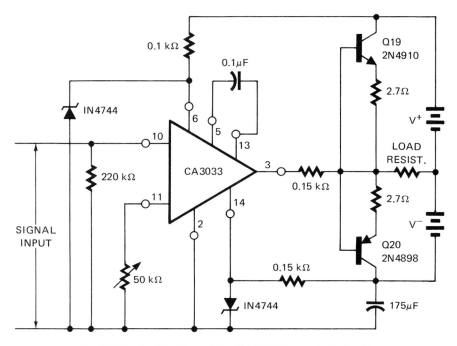

Fig. 10-15   Application of the CA3033 integrated circuit

and Q20 are matched complementary transistors (that is, they have the same characteristics although one is npn and the other pnp). These transistors are connected in a push-pull configuration. This feature eliminates the need for a center tapped transformer for the alternately conducting transistors. A circuit of this type can be used for frequencies down to dc.

Although an integrated circuit amplifier may not require a heat sink to limit the temperature rise, such a device often is added. The heat sink limits the temperature rise due to the internal power dissipation and the dc offset voltage drift in the amplifier output.

Some linear integrated circuit amplifiers have the advantage of nonsaturation. This feature does not mean that the amplifier cannot saturate. Rather, it means that although the amplifier may be overdriven, there is no storage time. If the signal suddenly comes within the linear range of the amplifier after having momentarily overdriven the amplifier, no time is lost in regaining linear operation.

When using integrated circuit amplifiers to make weak signal instrumentation measurements with a high accuracy, there may be a low signal-to-noise ratio. It is difficult in this situation to achieve the desired accuracy. In this case, sophisticated noise minimization techniques must be used. Some integrated circuit amplifiers are equipped with extra terminals for this purpose.

These terminals often are not connected to anything, but are located in the integrated circuit so that the input terminals can be surrounded by special connections to minimize the noise. The extra terminals are used to make special grounding connections or to intercept stray leakage currents that circulate through the insulation between terminals or between the terminals and ground. When using integrated circuit amplifiers for sensitive instrumentation signals, it is recommended that ungrounded sensors be used or that the integrated circuit itself be ungrounded.

When experimenting with integrated circuits, the dc bias voltages must *not* be reversed accidentally. A very small dc bias voltage can cause the internal connections to fuse and destroy the unit. This situation may occur if the integrated circuit is installed backwards in a test socket. Remember that a specification sheet presents a diagram of the terminal connections for an integrated circuit as viewed from the bottom of the package. The test socket terminal numbers are reversed from the numbers on the integrated circuit itself, when viewed from above by the experimenter. Many integrated circuits have met untimely ends because of a careless interpretation of the terminal connection arrangements.

Capacitive loads applied directly across the output terminals of an integrated circuit amplifier may cause the amplifier to oscillate when in use. Even if oscillations do not occur, these capacitive loads may reduce the margin of stability of the amplifier. Some feedback signal adjustment or resistive load adjustment may be required in these situations.

Many integrated circuit amplifiers have short circuit protection for the output terminals. However, even these amplifiers can fail if the short is allowed to remain for a long period of time. The resulting power dissipation eventually changes the material so that excessive junction temperatures are reached. However, if low bias voltages are used, there is less chance of failure due to a shorted output. For this reason, minimum bias voltages for a circuit should be used wherever possible. A failure due to a short can occur more readily when there is more than one complete amplifier in an integrated circuit package and two or more of the amplifiers are being used simultaneously. Such a failure is due to the greater amount of heat that must be dissipated by the entire package.

When soldering to integrated circuit terminals, care must be taken to insure that the high temperatures reached during soldering do not destroy or damage the integrated circuit. A safe rule to use is to not subject the terminals of the integrated circuit to the temperature of molten solder for more than ten seconds.

SETTING BIAS LEVELS FOR INTEGRATED CIRCUITS

Bias levels for the integrated circuits are set according to the manufacturers' recommendations. Sufficient information is supplied in most cases to permit the proper biasing of the integrated circuit amplifier. Additional information is required when the integrated circuit is used in circuits not recommended by the manufacturer. In this situation, the technician must rely on his or her knowledge of semiconductor electronics. By combining theory and experiment, the technician must be able to produce the required information.

To prevent the introduction of high-frequency noise into integrated circuits, it is sometimes helpful to connect a small capacitance (in the order of 0.01 microfarad) across the bias supply terminals of the integrated circuit amplifier. This capacitor acts as a "final filter" to eliminate any stray high-frequency signals that are picked up in the power supply leads.

Modern integrated circuits will operate within their specifications for a wide range of dc bias voltages. Many integrated circuit amplifiers have internal regulating circuits so that it is possible to use dc voltages varying between 2 and 20 volts and the amplifier will still perform to specifications. There is an obvious advantage to the use of unregulated and unfiltered power sources. Some integrated circuit amplifiers have special terminals which provide regulated dc voltage levels. These voltages can be used as the bias voltages for other semiconductor circuits.

In many cases, integrated circuit amplifiers may be connected like single three-terminal devices. Such a connection uses a dc bias voltage, an input, an output, and a common connection between the input and output. The integrated circuit amplifier can be treated as a generalized amplifying device, as in figure 10-16. (Refer to figures 1-14 and 1-15 for the generalized amplifying device.) In addition, a dc bias voltage connection can be made to the input lead to improve the amplifier performance. Recall that this type of connection was also used for the gate of an FET and the base of a bipolar transistor.

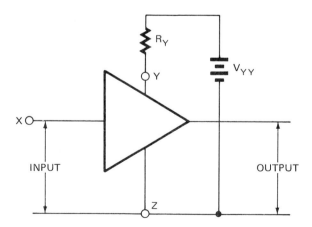

Fig. 10-16   Integrated circuit amplifier, connected as a generalized amplifying device

In the circuit of figure 10-13, resistor R3 provides added stability against temperature changes. This resistor operates much as the resistor R_e does in an emitter self-bias circuit for common emitter transistor amplifiers. (Refer to figure 2-22.) If terminals 8 and 9 are connected together, resistor R10 (figure 10-10) is shorted out. To achieve both a lower idling current and power dissipation, the bias supply voltage may be applied to terminal 9, leaving terminal 8 open. However, experiments show that less signal distortion occurs when the same voltage is applied to terminal 8. Terminal 2, which is connected through a capacitor to ground in figure 10-13, provides a zero input signal voltage to the base of Q3. The effect of temperature and power supply voltage variations is then cancelled because the extraneous signal inputs caused by variations in the supply voltage and temperature are applied to both Q2 and Q3. By taking the output signal between the collectors of Q2 and Q3, these effects are cancelled. This operation is similar to the performance of the dc differential amplifier described in Chapter 3.

The differential amplifier section of an integrated circuit amplifier frequently has a standard form. This section often incorporates temperature stabilization and eliminates the *common mode* signal that exists at the output of the differential amplifier. In the discussion of figure 3-16, Chapter 3, it was shown that if the resistor R_e is made large, the amplifier drift is reduced. The effect of the drift is often the common mode signal. To reduce this problem even further, a transistor can be added in the emitter circuit, as shown in figure 10-17, page 394. The resistance is high between the collector of Q3 and ground. As a result, a low current is maintained through the emitters of Q1 and Q2 to ground. Note that I_{E1} and I_{E2} are opposite in direction and tend to cancel each other. Transistor Q3 acts like a constant current source for the emitters of Q1 and Q2. Resistor R3 provides the temperature and bias stabilization for Q3. The temperature dependent characteristics of transistor Q3 in figure 10-17 can be compensated further by connecting an additional transistor or diode into the base-to-emitter circuit of Q3, as shown in figure 10-18, page 394. The collector-to-base circuit of this transistor (Q4) is shorted out. Kirchhoff's Voltage Law can be applied to sum the voltage drops around the circuit. (Follow the arrow in figure 10-18.)

$$V_{CC} - (I_{C4} + I_{B4} + I_{B3})R4 - V_{BE4} = 0$$

Solving for the collector current of Q4,

$$I_{C4} = \frac{V_{CC} - V_{BE4}}{R4} - I_{B4} - I_{B3} \qquad \text{Eq. 10.1}$$

Since I_{B4}, I_{B3}, and V_{BE4} are very small, I_{C4} reduces to:

$$I_{C4} \approx \frac{V_{CC}}{R4} \qquad \text{Eq. 10.2}$$

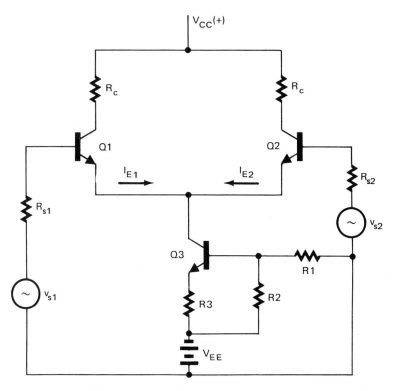

Fig. 10-17 Differential amplifier with constant current source added

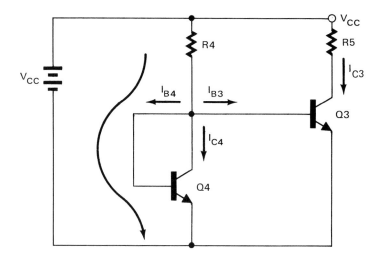

Fig. 10-18 Temperature stabilization for integrated circuit amplifier

Transistors Q3 and Q4 in figure 10-18 are made identical. If R4 and R5 are also identical, then $I_{C4} = I_{C3}$, $V_{BE4} = V_{BE3}$, and $I_{B4} = I_{B3}$. V_{CC} and R4 can be made insensitive to temperature, which means that collector currents also will be insensitive to temperature. Since the collector and base terminals of transistor Q4 are common, Q4 can be replaced by a diode. Refer to the integrated circuit amplifier shown in figure 10-12 where diode CR1 is used in a similar manner.

In figure 10-18, $I_{B4} = I_{B3} = 10$ microamperes, $V_{CC} = 10$ volts, R4 = 100 kilohms, and $V_{BE4} = 0.7$ volt. Compute the error in calculating I_{C4}, if I_{B4}, I_{B3}, and V_{BE4} are assumed to be negligible. (R10-9)

Refer to figure 10-19. (a) With V_{REF} to common = 5.1 volts, compute V_{MIN} to common. (b) With $V_{REF} = 5.1$ volts and $V_2 = 1.0$ volt, calculate the current through R_{os}. (R10-10)

Because of direct coupling in a multistage amplifier, a dc voltage buildup occurs at the output terminals, causing an output voltage to exist even though the input voltage is zero. What type of circuit can be added after the amplifier to eliminate this problem? (R10-11)

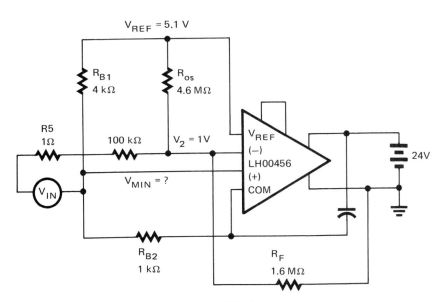

Fig. 10-19

ANALYSIS OF INTEGRATED CIRCUIT AMPLIFIERS

A performance index for differential amplifiers is the *common mode rejection ratio*. This index is the ratio of the voltage gain of the difference signal (A_D) to the voltage gain of the common mode signal (A_C). The voltage gain of the difference signal is:

$$A_D = \frac{v_{out}}{v_{in1} - v_{in2}}$$

Eq. 10.3

The symbols in Equation 10.3 refer to the quantities given in figure 3-16. A voltage gain is defined as:

$$A_1 = \frac{v_{out}}{v_{in1}}$$

Eq. 10.4

with v_{in1} applied and v_{in2} terminals shorted to ground. Similarly, a voltage gain A_2 is defined as:

$$A_2 = \frac{v_{out}}{v_{in2}}$$

Eq. 10.5

with v_{in2} applied and v_{in1} terminals shorted to ground. These voltage gains appear in the output voltage as a common mode gain (A_C).

$$A_C = A_1 + A_2$$

Eq. 10.6

Experimentally, the output voltage can be expressed as a function of A_C and A_D.

$$v_{out} = A_D(v_{in1} - v_{in2}) + A_C \frac{(v_{in1} + v_{in2})}{2}$$

Eq. 10.7

A useful experiment with a differential amplifier is to take readings of these voltages and verify the above relationships. The ideal value of the common mode rejection ratio, A_D/A_C, is infinity.

In testing a differential amplifier, the voltage gain of the difference signal is 100 and the common mode voltage gain is 0.1. Calculate the common mode rejection ratio as a numeric and in decibels. *(R10-12)*

For the amplifier described in review problem (R10-12), assume that $v_{in1} = 1.05$ millivolts and $v_{in2} = 0.95$ millivolt. Calculate the output voltage. *(R10-13)*

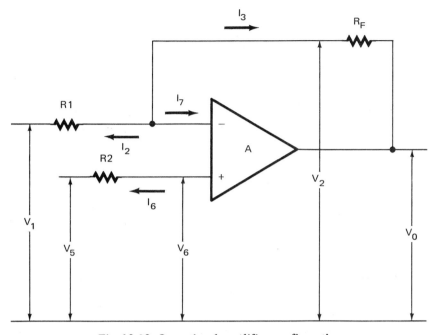

Fig. 10-20 Operational amplifier configuration

Many integrated circuit amplifiers include FETs in their first stage. These amplifiers can have an extremely high input impedance. Some types of these amplifiers are known as operational amplifiers. These devices have a high input impedance, a low output impedance, and an extremely high open loop gain. (*Open loop gain* is gain without any feedback of the output signal into the input. Feedback is covered in detail in Chapter 11.) High-impedance circuits of this sort are of particular interest because y-parameters can be used in the analysis of these circuits.

If an integrated circuit amplifier is a high-gain, high-input impedance differential amplifier, as in figure 10-20, a group of equations can be written, using Kirchhoff's Current Laws. To see how these equations can be written, the resistive branches in figure 10-20 are shown as separate circuits in figure 10-21, page 398. The equations are:

$$i_2 = y_{21}v_1 + y_{22}v_2 \qquad\qquad \text{Eq. 10.8}$$

$$i_3 = y_{33}v_2 + y_{34}v_o \qquad\qquad \text{Eq. 10.9}$$

$$i_6 = y_{65}v_5 + y_{66}v_6 \qquad\qquad \text{Eq. 10.10}$$

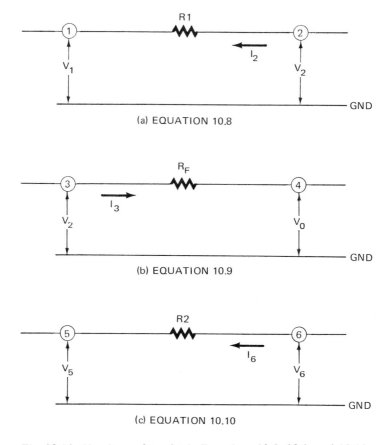

(a) EQUATION 10.8

(b) EQUATION 10.9

(c) EQUATION 10.10

Fig. 10-21   Circuits used to obtain Equations 10.8, 10.9, and 10.10

Equations 10.8, 10.9, and 10.10 can be obtained by referring to figure 10-21. For example, Equation 10.8 is a general expression that describes the current through the circuit of figure 10-21(a), and is appropriate when there are two voltage sources in the network.   The admittance symbols used in Equation 10.8 to 10.10 are defined as follows:

y_{21} = forward transfer admittance of the network between ① and ② with the output terminals shorted.

y_{22} = output admittance of the network between terminals ① and ② with its input terminals shorted.

y_{33} = input admittance of the feedback network between terminals ③ and ④ with its output terminals shorted together.

y_{34} = reverse transfer admittance of the network between terminals ③ and ④ with the input terminals shorted.

y_{65} = forward transfer admittance of the network between ⑤ and ⑥ with the output terminals shorted.

y_{66} = output admittance of the network between ⑤ and ⑥ with the input terminals shorted.

The input admittance of the amplifier is so high that currents i_6 and i_7 can be neglected. Then, $i_3 + i_2 = 0$. Equations 10.8 and 10.9 can be combined to yield:

$$y_{21}v_1 + y_{22}v_2 = y_{33}v_2 + y_{34}v_o \qquad \text{Eq. 10.11}$$

If $i_6 = 0$, then v_6 can be determined using Equation 10.10.

$$v_6 = -v_5\left(\frac{y_{65}}{y_{66}}\right) \qquad \text{Eq. 10.12}$$

Since the voltage gain of the operational amplifier is very high, the input voltages are very small when compared to the output voltages. As a result, the differential input,

$$v_6 - v_2 \approx 0$$

Therefore,

$$v_6 \approx v_2 \qquad \text{Eq. 10.13}$$

Combine Equations 10.13, 10.12, and 10.11 to obtain:

$$y_{21}v_1 - y_{22}\left(\frac{y_{65}v_5}{y_{66}}\right) = y_{33}v_2 + y_{34}v_o \qquad \text{Eq. 10.14}$$

This expression is solved for v_o:

$$v_o = -\left(\frac{y_{21}v_1}{y_{34}}\right) + \frac{y_{65}(y_{22} + y_{33})v_5}{y_{66}y_{34}} \qquad \text{Eq. 10.15}$$

The y-parameters just defined can be expressed in terms of the resistances of the respective networks of figure 10-21.

$$y_{33} = \frac{1}{R_F} \qquad \text{Eq. 10.16}$$

$$y_{66} = \frac{1}{R2} \qquad \text{Eq. 10.17}$$

$$y_{21} = \frac{1}{R1}$$

Eq. 10.18

$$y_{22} = \frac{1}{R1}$$

Eq. 10.19

$$y_{34} = \frac{1}{R_F}$$

Eq. 10.20

$$y_{65} = \frac{1}{R2}$$

Eq. 10.21

Substitute Equations 10.16 through 10.21 into Equation 10.15 to obtain an expression for v_o.

$$v_o = -\frac{R_F v_1}{R1} v_1 + \frac{(R_F + R1)v_5}{R1}$$

Eq. 10.22

Added to v_o is the common mode signal developed within the integrated circuit (if it is considered significant).

THE LH0036 INTEGRATED CIRCUIT AMPLIFIER

Some integrated circuit amplifiers are capable of operating as complete amplifying systems. One example is the National Semiconductor LH0036 integrated circuit amplifier. A simplified schematic of this device is shown in figure 10-22. This amplifier will be examined as an example of a high-input impedance operational amplifier that uses both inverting and noninverting inputs. (In the simple form considered previously, only the inverting input of the amplifier was studied.) The equations developed in the foregoing analysis can be applied to the analysis of this amplifier.

The LH0036 is a two-stage amplifier with a high input impedance, high-gain stage that consists of A1 and A2, and a differential-to-single ended, high-gain stage, A3. A1 and A2 are both operational amplifier stages. Each of these stages can be treated as the equivalent of the amplifier shown in figure 10-20. In addition, A1 and A2 will each process a common mode signal, e_{CM}. According to Equation 10.22, the signal out of A1 is:

$$v_1 = \frac{(R1 + R_G)e_1}{R_G} - \frac{R1e_2}{R_G} + e_{CM}$$

Eq. 10.23

Similarly, the signal out of A2 is:

$$v_2 = \frac{(R2 + R_G)e_2}{R_G} - \frac{R2e_1}{R_G} + e_{CM}$$

Eq. 10.24

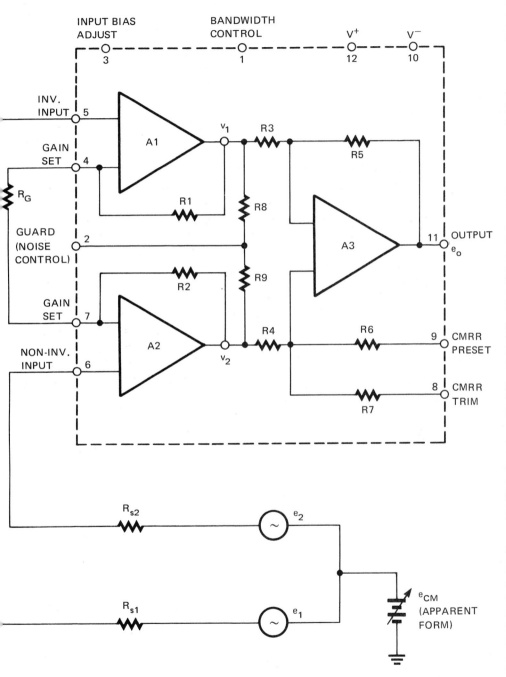

Fig. 10-22 Simplified schematic of LH0036 amplifier, with external connections

In the LH0036 amplifier, R1 = R2, and R3 = R4 = R5 = R6. When Equations 10.23 and 10.24 are combined, the voltage appled to the input of A3 is defined as:

$$v_2 - v_1 = \left(\frac{2R1}{R_G} + 1\right)(e_2 - e_1) \qquad \text{Eq. 10.25}$$

Note that the common mode voltage (e_{CM}) is canceled. Since all of the input and feedback resistors of A3 are equal, the gain of A3 is one. Therefore, the output voltage of the LH0036 is:

$$e_o = (1)(v_2 - v_1) = (e_2 - e_1)\left(1 + \frac{2R1}{R_G}\right) \qquad \text{Eq. 10.26}$$

The manufacturer of the LH0036 amplifier says that the voltage gain becomes:

$$A_V = \frac{e_o}{e_2 - e_1} = 1 + \frac{50K}{R_G} \qquad \text{Eq. 10.27}$$

What is the voltage gain of the LH0036 when it is operated in the circuit shown in figure 10-22 and R_G = 5 kilohms?   *(R10-14)*

DEFINITION OF TERMS USED IN INTEGRATED CIRCUIT AMPLIFIER SPECIFICATIONS

The following definitions are for those terms frequently used in integrated circuit specifications.

- **Total quiescent device dissipation.** The total power drain of the device when no signal is applied and there is no external load current.

- **Quiescent operating voltage.** The dc voltage at the output terminal, with respect to ground, with no signal applied.

- **Quiescent operating current.** The average dc value of the current in an output terminal, with no signal applied.

- **Output offset voltage.** The dc voltage between the output terminals with no signal applied.

- **Total harmonic distortion.** The ratio of the total rms (root mean square) voltage of all harmonics to the rms voltage of the fundamental, expressed in percent. The rms voltage is the voltage deflection normally seen on an ac voltmeter. The harmonic voltages are measured at the output terminals of the device with respect to ground.

- **Noise figure (NF).** There are two definitions that are in use. (1) Noise is the term used to describe an unwanted signal that is not part of the signal to be amplified. The term "noise figure" or NF specifies how noisy the circuit is. NF is the ratio of the actual noise power output of the circuit to the noise power output obtained over the same frequency range if the only source of noise is the thermal noise in the internal resistance of the signal source. Basically, the noise figure compares the noise in an actual amplifier to that in an ideal (noiseless) amplifier.

   (2) The noise figure is the ratio of the total noise power of the device and a resistive signal source to the noise power of the signal source alone. The signal source represents a generator of zero impedance in series with the source resistance.

- **AGC (automatic gain control).** The gain of the amplifier is held constant by internal circuit features. Some of the more sophisticated amplifier designs have this feature.

- **AGC range.** The total change in voltage gain from the maximum output to complete cutoff that can be achieved by applying the specified range of dc voltage to the AGC input terminal.

- **Power gain.** The ratio of the signal power developed at the output of the device to the signal power applied to the input. The power gain frequently is expressed in decibels. (It must be noted that the definition of power decibels is *not* the same as that of voltage or current decibels; see the Appendix for a discussion of decibels.)

- **"Single-ended"** or **"double-ended" output** or **input.** When applied to an integrated circuit amplifier, a "single-ended" or "double-ended" circuit configuration refers to whether a signal is being taken with reference to a common ground (single-ended) or with reference to another signal which is not grounded. An example of a double-ended output is the difference signal output from the amplifier described by figure 3-16,

$$v_{out} = v_{c1} - v_{c2} \qquad \text{Eq. 10.28}$$

   The single-ended signal in this case is v_{c1} alone with respect to common (the + side of V_{EE}).

- **Bandwidth.** The usable frequency range of an amplifier. The upper and lower limits of the bandwidth occur at the frequency at which the gain drops by 3 decibels.

- **Input offset voltage.** The dc voltage that must be applied across the input terminals to obtain a zero voltage at the output terminals.

- **Input offset current.** The difference in the currents at the two input terminals when the quiescent operating voltages at the two output terminals are equal.

- **Input bias current**. The average value of the currents at the two input terminals of a device, when the quiescent operating voltages at the two output terminals are equal.

- **Quiescent operating current ratio**. The ratio of the quiescent operating currents in the two output terminals.

- **Slew rate**. The slew rate of an amplifier is the rate at which the output voltage of the amplifier rises toward its maximum possible level when one or more stages of the amplifier reach cutoff or saturation. The slew rate is usually expressed in volts per microsecond.

The slew rate of an amplifier is important when the amplifier is expected to receive and amplify pulses that rise quickly. The slew rate limits the speed with which these fast-rising pulses can be amplified. Consequently, the amplifier will not operate in its linear mode for these signals. A compensating capacitor may be placed across the inputs of amplifiers whose slew rates are too slow. This feature permits linear amplifier operation over a wider range of input signals.

- **Power supply rejection ratio**. The ratio of the change in the input offset voltage to the change in the power supply voltages that produce it. This ratio is a measure of how well the circuit behaves during a change in the power supply voltage. This term is frequently abbreviated to PSRR in integrated circuit specifications.

- **Input impedance**. The ratio of the input voltage to the input current under some stated conditions for source and load resistance levels.

- **Input voltage range**. The range of voltages at the input terminals for which the amplifier operates within specifications.

- **Output impedance**. The ratio of the output voltage to the output current under some stated conditions of source resistance and load resistance.

- **Settling time**. The time from the initiation of an input step function to the time at which the output voltage settles to within specified limits.

- **Supply current**. The current from the power supply required to operate the amplifier with no load and the input at zero.

Different manufacturers may describe their products by using symbols that are similar. However, the circuit parameters that are indicated by the symbols may have different definitions, or they may be obtained in a different manner. The standardization of the symbols lags behind the technology in many instances. This situation may lead to confusion. Therefore, a symbol should be viewed with suspicion until its meaning is made clear.

The output offset voltage is reduced in many integrated circuit amplifiers by making special external connections to some of the connecting terminals supplied by the manufacturer. The manufacturer's specifications should be consulted when there is a circuit problem with the output offset voltage in the event that a circuit improvement can be made. The possibility of adding a heat sink was suggested previously. Other circuit changes include the addition of potentiometers across some of the integrated circuit terminals or changing the + and – bias voltage levels slightly, depending on the device used.

Many integrated circuit amplifiers contain a provision to adjust the common mode rejection ratio (CMRR). However, the output offset voltage and CMRR adjustments often are interactive. In other words, one value must be sacrificed for the other.

To obtain a better CMRR in a double-ended operational amplifier, a dc voltage is applied across the + and – input terminals. Any adjustment capability that is provided in the device can then be made. After adjusting to obtain the best CMRR for dc or low-frequency signals, an ac common mode signal is applied at higher frequencies and adjustments made to obtain the best CMRR for these frequencies. To reduce input bias currents to an integrated circuit amplifier with a floating input, one input terminal is tied to common. (Common is the low side of the power supply; or, if there are both + and – voltage supplies, common means the common connection between them.)

The signal power output of an amplifier is 100 milliwatts. The signal input power is 10 milliwatts. What is the power gain in decibels? *(R10-15)*

In figure 10-12, the output is taken between terminals 3 and 2. If terminal 2 is tied to ground (as in figure 10-15), is this a double-ended output or a single-ended output? *(R10-16)*

What is the value of the noise figure (NF) for an ideal (noiseless) amplifier? *(R10-17)*

Define the quiescent operating voltage for an integrated circuit amplifier. *(R10-18)*

What is the ideal value of the common mode rejection ratio for an integrated circuit difference amplifier? *(R10-19)*

The output voltage of an amplifier rises by 3 volts in 1 microsecond, toward its maximum level, when one stage is driven to saturation. What is the slew rate? *(R10-20)*

The power supply voltage for an amplifier changes by 5 volts. There is a resulting change in the input offset voltage of 100 millivolts. What is the power supply rejection ratio for this amplifier?   *(R10-21)*

NONLINEAR AMPLIFIERS AND AMPLIFYING DEVICES

Many types of amplifying devices are used in electronics in addition to the ones described so far in this text. However, most of these other amplifiers are for special applications and nonlinear amplifier circuits. Many devices that function like switching devices or relays can also be considered to be amplifiers. These devices use a weak, low-power signal to produce a large change in the power output. Many of these devices are semiconductor switching units. These semiconductor nonlinear amplifying devices may be discrete components or they may be tiny portions of an integrated circuit, programmed to perform a complex series of operations.

The category of *switching amplifiers* includes many circuits, each designed for a specific use. The modern digital computer is an example of thousands of complex switching circuits, arranged to provide computation and data storage.

Silicon Controlled Rectifier and Other PNPN Devices

One popular semiconductor switching device is the *silicon controlled rectifier* (SCR) or the *silicon controlled switch* (SCS). This device belongs in a class known as thyristors. Thyristors are pnpn devices. That is, they consist of a single crystal with three *p-n* junctions in series, as shown in figure 10-23. This device may be viewed as two transistors having a common collector-to-base junction. Figure 10-24 shows the connection of the two transistors. Two output terminals and one or two control terminals are brought out of a pnpn device. The emitter current in the pnp transistor is the current I_1. The collector current in the pnp transistor is given by $a_1 I_1$, where a_1 is the common base

p	*n*	*p*	*n*
ANODE END	ANODE GATE	CATHODE GATE	CATHODE END

Fig. 10-23 PNPN device

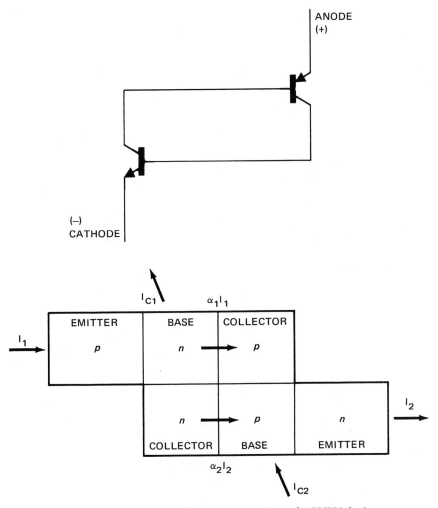

Fig. 10-24 A two-transistor representation of a PNPN device

gain. I_2 is the emitter current of the npn transistor and $a_2 I_2$ is the collector current of the npn transistor. The term a_2 is the common base gain for the npn transistor. Two components of current cross the collector-to-base junction. Currents I_{C1} and I_{C2} are control currents. That is, either one of these currents can turn the device *on*. If only I_{C2} is applied, I_2 splits into two components crossing the collector-to-base junction.

$$I_2 = I_{C2} + I_1 \hspace{2cm} \text{Eq. 10.29}$$

Consider the total current input,

$$I_1 = a_1 I_1 + a_2 I_2 \qquad \text{Eq. 10.30}$$

Substitute Equation 10.29 into 10.30, and then rearrange the terms and solve for I_1.

$$I_1 = \frac{I_{C2} a_2}{1 - (a_1 + a_2)} \qquad \text{Eq. 10.31}$$

If $a_1 + a_2 \approx 1$ by design, I_1 is a very large current, even though I_{C2} is very small. A pnpn device made in this manner is a very efficient nonlinear amplifier. Normally, there is no current flowing between the anode and the cathode unless a signal is applied to the gate. Only a momentary control current signal is needed to cause the device to act like a forward biased semiconductor diode. Once the junction currents begin to flow, they replace the control current through the gate and force conduction to continue. Unfortunately, although a small control current is needed to turn the device *on,* a current comparable in magnitude to the anode current is required to switch it *off.* This feature usually means that the primary load current must be switched off by another, separate switching device. While this arrangement may seem to be a serious disadvantage, the device is useful in alternating-current control applications. For applications involving small currents and low level logic and switching schemes, this type of pnpn device is called the silicon controlled switch (SCS). In this form, both of the control signal terminals of the device may be brought out. A larger form of the device is the silicon controlled rectifier (SCR). Only the control terminal at the cathode end is brought out on this device. The SCR is used in motor controls and light dimmer controls and may operate at currents greater than 100 amperes and at voltages up to 1000 volts.

A typical circuit for a light dimming control using 115 volts ac and a pnpn device is shown in figure 10-25. This circuit uses a General Electric number C106 SCR.

Other names for pnpn devices include binistor, trigistor, and transwitch. These devices are all forms of the SCR or SCS, and their operation is the same. The characteristic curves of this family of devices are like those in figure 10-26. When the forward anode voltage reaches a value V_{BR}, the device conducts. V_{BR} can be reached by applying (1) a sufficient anode-to-cathode voltage or (2) a triggering voltage to one of the gate terminals. When the anode current drops below the holding current (I_H), the device switches off again.

Referring to figure 10-24, what polarity of voltage is required at the cathode gate to cause the SCR to switch on? (R10-22)

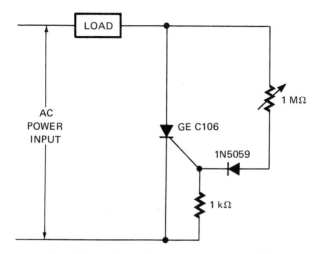

Fig. 10-25 Light dimming circuit using SCR

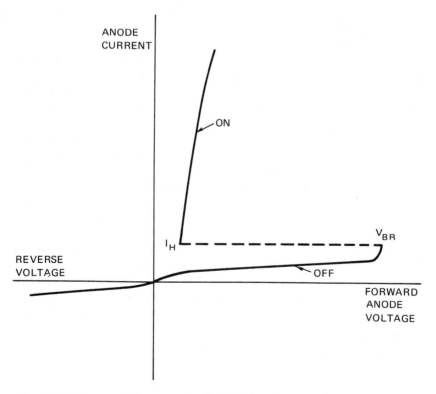

Fig. 10-26 Characteristic curves for the SCR (anode current vs. anode voltage)

Diacs and Triacs

The *triac* is another pnpn device. Unlike the SCR, a voltage of either polarity may be connected from the anode to the cathode of the triac. As a result, conduction can take place in either direction. A triggering voltage having a corresponding polarity at the gate causes the triac to switch on. Current then flows between the anode and the cathode. This device can be viewed as two SCRs in parallel, with one SCR pointed one way and the other SCR pointed the other way. Figure 10-27 shows a triac used in a light dimming circuit. As seen in this figure, there is a special symbol for the triac. The characteristic curves of the triac are the same as those for the SCR. However, the triac has a negative polarity curve that is a mirror image of the positive polarity curve for the SCR. The characteristic curves are shown in figure 10-28.

The *diac* is another pnpn device that is similar to the triac. The diac does not have a trigger or gate terminal. The characteristic curves for this device are the same as those of the triac. When the forward or reverse voltage reaches the value V_{BR}, the diac automatically switches on. Note the symbol for the diac used in the circuit shown in figure 10-27.

The silicon unilateral switch is a pnpn device that does have a gate terminal. The operation of this device is similar to that of the triac, but it is restricted to low-voltage, low-current applications.

The diac in figure 10-27 is used as a trigger for the triac. When the diac turns on at a voltage selected by the potentiometer, it allows voltage to be applied to the triac gate terminal. This turns the triac on and completes the light circuit. The switching operation occurs twice during each cycle of the ac voltage. The potentiometer determines the point in the waveform at which the switching will occur. This is known as *phase control.* The brightness of the light is determined by the point in the cycle at which the triac is switched

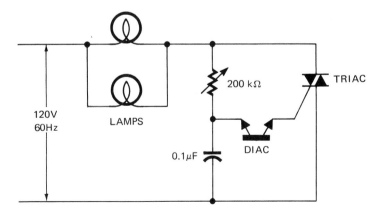

Fig. 10-27 Light dimmer circuit using a diac and a triac

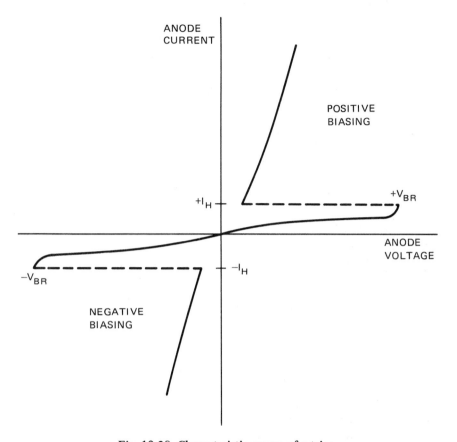

Fig. 10-28 Characteristic curves of a triac

on. The triac is off during a portion of each cycle. If the triac is switched on earlier in the cycle, the current flows for a longer period of time and the light is brighter.

The Tunnel Diode

Another interesting switching device is the tunnel diode. This device conducts heavily in the forward direction as the anode-to-cathode voltage is increased. When a peak current is reached, the tunnel diode switches off. The device conducts heavily in the reverse direction without limit. The tunnel diode operates on an entirely different principle when compared to other devices. It operates at very low voltages and currents. This device is described here only because of the nature of its behavior. It acts like a normally closed relay that opens when a sufficient voltage is applied. A typical tunnel diode characteristic curve is shown in figure 10-29, page 412.

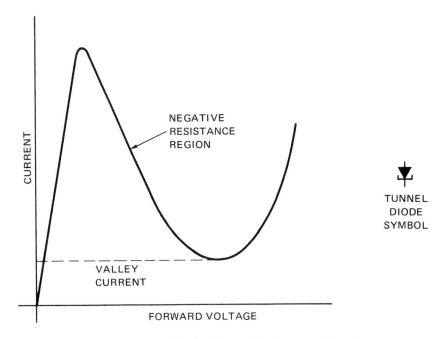

Fig. 10-29 Tunnel diode characteristic curve and symbol

Triggerless devices like the tunnel diode and the diac act like amplifiers that derive their input signal from the changing level of the power supply voltage. This action is an advantage in that it provides simpler switching for oscillator circuits. However, it is also a major disadvantage because the input signal is too intimately related to the output signal.

Other available devices include light operated transistors and pnpn devices that can be used in light activated amplifiers. Light operated semiconductor devices are based on the principle that light causes the generation of free carriers within a material. When light energy is directed at the semiconductor crystal, the energy level of outer orbit electrons is changed. As a result, these electrons escape the interelectronic bonds existing between the electrons and the nucleus of the atom. A similar action occurs when heat is applied to the semiconductor crystal. Light is energy that has a higher frequency in the electromagnetic spectrum of frequencies, whereas heat is energy that is at a low frequency in the spectrum. If the semiconductor has a transparent covering, light can affect the operation, just as heat does. A photosensitive device can be thought of as a *transducer*, rather than an amplifier, since the input signal is light and the output signal is electric current or voltage. However, photosensitive devices can be used to build an amplifier based on the definition given in Chapter 1. For example, assume that a light source is driven by an electric signal. The light from this source can be beamed on a photosensitive device to produce an electric current. This type of arrangement provides complete electrical isolation between the input and output signals. In fact, the input signal and the light source can be many feet away from the light sensitive device that serves as the output circuit. This concept has many uses in industry and can be applied to the construction of both linear and nonlinear amplifiers.

The light source may be either an ordinary light bulb or a semiconductor device called a *light emitting diode* (LED). This type of diode is constructed from a semiconductor material in which the applied voltage raises the electrons to a momentarily higher energy level. The energy is released as light. This light is visible if the wall of the casing covering the semiconductor crystal is transparent.

The light sensitive device in the output circuit of this type of amplifier can be a photoconductor, a photodiode, a phototransistor, or a light activated silicon controlled rectifier (LASCR). An amplifier can be constructed using a photoconductor or a photodiode, but additional semiconductor amplifying circuits are required. The LASCR acts like an ordinary SCR, but the gate signal consists of a beam of light.

In some ways the phototransistor looks and acts like an ordinary transistor and a common emitter amplifier can be built using this device. However, the phototransistor is packaged in a transparent plastic case so that light can serve as the input signal rather than a base current. A relationship exists between the base current and the collector current, and a base terminal is furnished, but it usually is not used. The LASCR and the phototransistor both have high-gain characteristics and thus can be used to drive a load directly. The set of characteristic curves shown in figure 10-30, page 414, is typical for the phototransistor. The phototransistor can be made into a linear amplifier using the light beam for isolation. An amplifier using a light emitting diode and a phototransistor is shown in figure 10-31, page 414.

Low light intensities can be amplified directly by means of a photomultiplier tube. This tube contains a material that easily emits secondary electrons when struck by light. The secondary electrons thus produced are directed to another emissive surface by means of a high dc voltage potential. The second surface emits more secondary electrons which are then directed to a third emissive surface. This process continues until the required amplification is achieved. In a commercial photomultiplier, a change of 0.001 lumens of light can produce a change of 100 milliamperes in the output current.

A silicon controlled rectifier, as described in figure 10-24, is designed to have a common base current gain (α) of 0.5 for the pnp section and for the npn section. What is the overall current gain from the device? Theoretically, what value of gate current is needed to switch the device on? (R10-23)

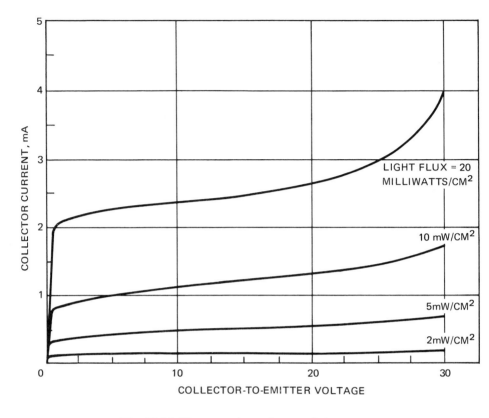

Fig. 10-30 Phototransistor characteristic curves

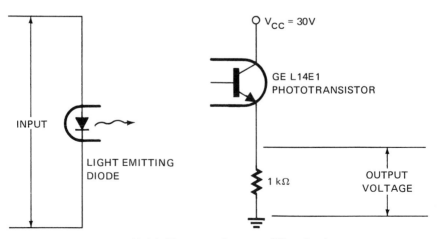

Fig. 10-31 Phototransistor amplifier circuit

An SCS needs a gate current of 1 milliampere to switch it on. The load current is 50 milliamperes. Without knowing anything else about the unit, what can be done to insure that it will be switched off? (R10-24)

A load line can be drawn on a set of curves for a phototransistor just as it is drawn on the curves for any other transistor. The phototransistor represented by figure 10-30 is to be used in the circuit of figure 10-31. For a constant light flux of 10 milliwatts per square centimeter, what is the collector current and the collector-to-emitter voltage at the quiescent point? (R10-25)

The output voltage in the amplifier of figure 10-31 is taken across the 1-kilohm resistor. What name is applied to this amplifier configuration? (Refer to Chapter 4.) (R10-26)

The amplifier shown in figure 10-31 is given an ac light input of ± 5 mW/cm^2. Using the curve and Q-point determined in problem (R10-25), what is the change in V_{CE} for this input? (R10-27)

Refer to figure 10-25. The 1-kilohm resistor prevents the full ac voltage from being impressed across the gate of the SCR. Assume that 1 volt is required across the 1-kilohm resistor to fire (switch on) the SCR. The drop across the diode is zero when it is conducting. To what resistance must the 1-megohm potentiometer be adjusted to cause the SCR to fire when the power input voltage reaches +80 volts? (R10-28)

LASERS AND AMPLIFIERS FOR MICROWAVE FREQUENCIES

Recent developments in electronics include masers and lasers. *Laser* is the abbreviation of *l*ight *a*mplification by *s*timulated *e*mission of *r*adiation. This device produces a beam of light at one very well defined frequency. *Maser* is the abbreviation for *m*icrowave *a*mplification by *s*timulated *e*mission of *r*adiation. The maser produces microwaves at one very well defined frequency.

When energy (light or heat) is applied to a material, three actions can occur on an atomic scale:

1. the energy is absorbed so that the energy levels of an atom are increased, thereby raising the electrons of the atom to an excited state and a more unstable orbit;

2. the atom is momentarily raised to an excited state, but then returns to its ground or unexcited state; in this case, the applied energy tends to pass through the material with only a slight loss;

3. some of the applied energy affects a previously excited atom; in this case, the energy is not absorbed, and not only passes through the material, but it may be amplified. This phenomena is known as *stimulated emission.*

Amplification can be made to occur if either of these conditions exist.

1. If more atoms are in the excited state than are in the ground state when the external energy is applied, then amplification occurs. Otherwise, more energy is absorbed by the atoms than is radiated. Normally, an intense light source is used to achieve this condition.

2. Some of the energy radiated is reflected back into the materials which have the appropriate atomic structure and are of a suitable shape or geometry. Amplification occurs because the radiated light is reflected back into the material by providing a polished mirrorlike surface that provides further stimulated emission. A typical laser system using a ruby rod is shown in figure 10-32.

There are three types of lasers. A ruby or glass laser is represented by figure 10-32. The second type is the gas laser which uses carbon dioxide or a mixture of neon and helium. The electrically excited atoms of the gas emit

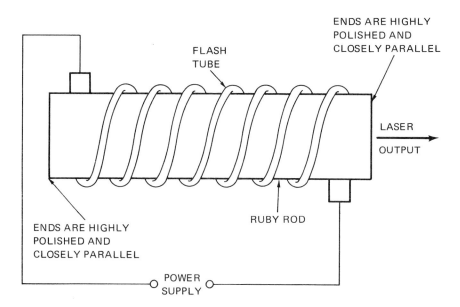

Fig. 10-32 Ruby rod laser system

radiation and mirrors are used as the reflecting surface. A third type of laser is a semiconductor *p-n* junction device, as shown in figure 10-33. A *p-n* junction laser is similar to an ordinary semiconductor diode. In this type of laser, the atoms are raised to the excitation levels by a large forward bias current through the diode. Light is emitted and reflected back from the polished surface at both ends. The reflecting surfaces stimulate further emission. The output radiation is emitted from the junction area as shown in figure 10-33. The forward biasing current of the *p-n* junction laser is analogous to the collector bias current in a transistor amplifier. A light from an external signal or a modulation of the bias supply can serve as a signal source. The signal is amplified in the laser system and emerges from the output surfaces.

The light emitted from lasers is at just one energy level and therefore has only one frequency. This is called monochromatic light. The frequency of light is detected by the human eye as light of a specific color. Light normally contains a broad spectrum of colors, resulting in a wide range of frequencies. A characteristic of monochromatic light is that it can be directed easily over great distances. Monochromatic light can be focused at one spot in a concentrated beam that can be made very narrow. Light produced by lasers is commonly used for surgery and welding. Because this light can be directed over long distances, it is also used for communications systems.

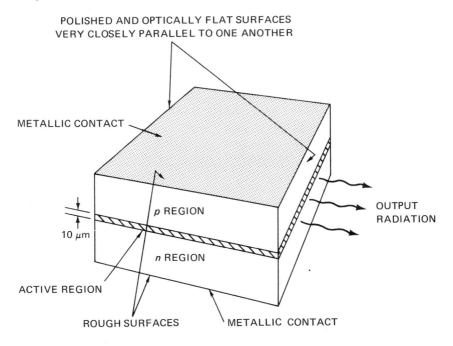

Fig. 10-33 Construction of a *p-n* junction laser

The Maser Amplifier

The theory of the maser (meaning *m*icrowave *a*mplification by *s*timulated *e*mission of *r*adiation) is very similar to that of the laser. In fact, the maser was the first of this class of devices to be developed. The most outstanding difference between the two devices is the frequency of the radiation. In the electromagnetic frequency spectrum of energies, the microwave region is at frequency levels below that of light. Visible light is at frequencies of approximately 10^{15} hertz and microwave frequencies are of the order of 10^{10} hertz. (The term microwave is used because the energy at this frequency has a very short wavelength. Wavelength is directly proportional to the reciprocal of the frequency.)

Microwave amplification also results from the following conditions:

1. more atoms are in an excited state than are in a ground state, and

2. some of the radiated energy is reflected back into the material causing further emission.

The applied and radiated energy in a laser is light; in a maser, the applied and radiated energy is an electromagnetic field. The student should be familiar with the electromagnetic fields produced and used at low frequencies to drive motors, transformers, relays, and similar devices. At microwave frequencies, the electromagnetic fields can transmit energy over much longer distances.

Types of Masers

A typical ruby maser is shown in figure 10-34. The atoms are set at the required magnetic energy levels by applying a strong dc magnetic field. The ruby is exposed to the microwave signal to be amplified. The structure is made so that it resonates at the frequency of the microwave signal. The microwave signal affects the excited atoms by interacting with the resonating structure. As a result, the signal is greatly amplified. Like the laser, the amplified output signal has a single frequency. The signal amplitude can be modulated to carry information or perform a control function.

An ammonia maser, sketched in figure 10-35, uses the excited atoms of ammonia gas to achieve microwave amplification. The gas atoms are excited not by a magnetic field, but rather by a high-voltage electrostatic field that separates low-energy atoms from those having higher energy. While the high energy atoms are reverting to their ground state, a microwave signal is applied to stimulate the further emission of electrons. This emission is amplified further by resonance in the specially designed structure. The gas is recirculated so that it can supply high-energy atoms continuously to the resonator.

The Parametric Amplifier

The parametric amplifier can also be used with microwave signals. This amplifier uses a varactor diode which acts like a voltage variable capacitor. The

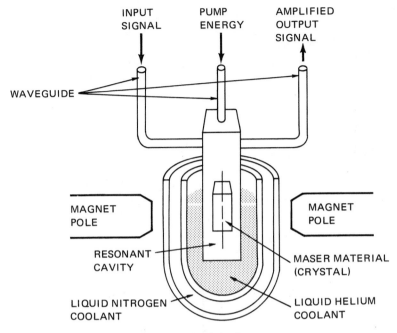

Fig. 10-34 Ruby maser

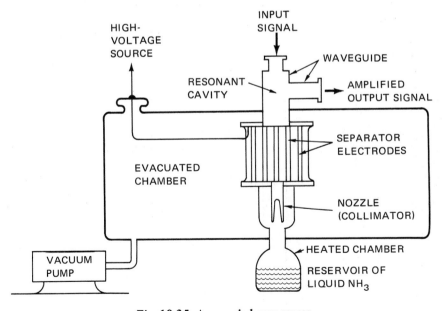

Fig. 10-35 Ammonia beam maser

parametric amplifier also uses a resonating parallel LC circuit to assist the amplification. In its simplest form, the capacitor has variable plate spacing, as shown in figure 10-36. Using a sinusoidally varying signal, the capacitance is changed at points a, b, c, and so on. Capacitance varies inversely according to the distance between two conducting plates. The voltage across the capacitor is:

$$V = \frac{q}{C}$$

Eq. 10.32

where q is the charge on the plates. An increase in the plate spacing causes an increase in the voltage across the plates (points a and c in figure 10-36). The resulting voltage wave is then amplified. The source of energy is derived from the means by which the capacitance is changed. If the plate distances are physically changed, the energy source is the power used to drive the plates. If the capacitor is a semiconductor diode varactor, the energy source is a radio-frequency oscillator which changes the reverse bias voltage of the diode. Figure 10-37 is a block diagram of the parametric amplifier.

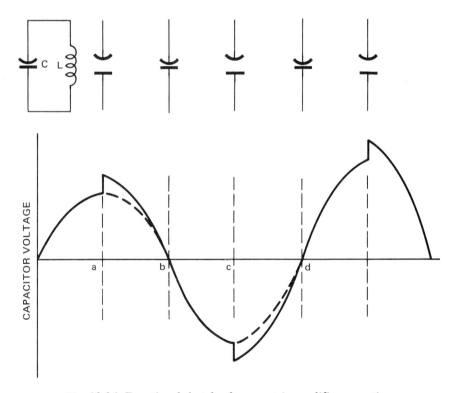

Fig. 10-36 Functional sketch of parametric amplifier operation

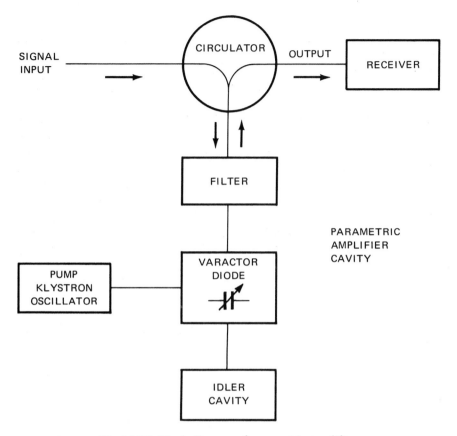

Fig. 10-37 Block diagram of parametric amplifier

The Traveling Wave Tube Amplifier

Amplification at microwave frequencies over a wide frequency range is achieved by a *traveling wave tube amplifier.* The basic form of this device consists of an electron gun which projects a focused electron beam through a helically wound coil to a collector electrode, as shown in figure 10-38, page 422. The electrons are focused into a tiny beam through the center of the helix by the field of a magnet that runs the full length of the tube. (The magnet is not shown in figure 10-38.) A continuous wave radio-frequency signal is introduced through the input directional coupler coil at the cathode end of the tube. The signal then travels around the turns of the helical coil toward the collector at the other end of the tube. The electric field from the continuous wave signal on the helical coil interacts with the electron beam and draws energy from it. This causes an increase in the magnitude of the signal that is traveling around the

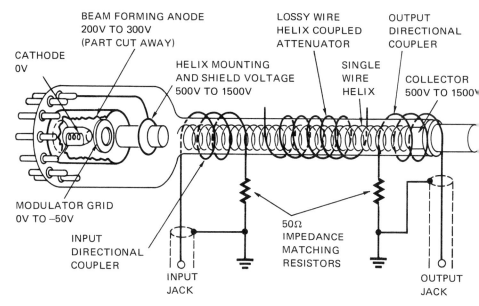

Fig. 10-38 The traveling wave tube amplifier (the cylindrical magnet, concentric to the tube, is not shown)

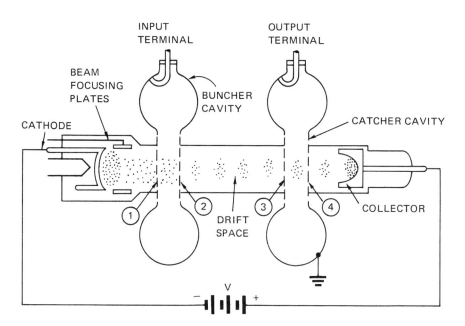

Fig. 10-39 Two-cavity Klystron amplifier

helical coil. The output signal is transferred by the output directional coupler coil to the output jack at the collector end of the coil.

The Klystron Amplifier

A *klystron amplifier* can also be used to amplify signals in the microwave region. In operation, the density of a stream of electrons is modulated in an evacuated tube. A diagram of a two-cavity klystron amplifier is shown in figure 10-39. A beam of electrons is emitted from the cathode and accelerated toward the collector at the other end of the tube. The dashed lines (①, ②, ③, and ④) between the cavities represent grids. The cavities are designed to resonate at the signal frequencies to provide for the maximum amplitude of the input signal and the minimum amplitude of other frequencies. The high-frequency input signal is introduced between grids ① and ② of the buncher cavity. The field produced by this signal modulates the beam of electrons so that the electrons are transported in bunches. In other words, the electron density is modulated. The electron bunches are slowed at grids ③ and ④ and then resonate in the catcher cavity. This cavity is also the output cavity. The power output is equal to the difference in the kinetic energy of the electrons averaged before and after passing through the output cavity. The energy source for amplification is taken from the accelerating potential, V. Little or no power is taken from the input signal. The amplifier can be made as a source of radio-frequency power by feeding a portion of the output power back from the output cavity to the input cavity. This power is fed back at the proper phase and with sufficient amplitude to overcome the losses in the system.

What source of energy provides power amplification in the (a) parametric amplifier, (b) klystron amplifier, (c) traveling wave tube amplifier, and (d) ruby maser?   *(R10-29)*

*Where is the dc bias level connected in the **p-n** junction laser?*   *(R10-30)*

What are the advantages of the light from a laser?   *(R10-31)*

Name three types of lasers. What material provides the excited atoms in each?   *(R10-32)*

What three actions can occur in the atoms of a material that is subjected to light energy?   *(R10-33)*

Laser action occurs when more atoms are in the _____ state than are in the _____ state.   *(R10-34)*

ELECTROMECHANICAL AND ELECTROMAGNETIC AMPLIFYING DEVICES

According to the general definition of an amplifier in Chapter 1, any device or circuit that takes a small, weak, low-power signal and converts it to a large power signal is an amplifier. A number of electromechanical and electromagnetic amplifying devices are to be described briefly in this section, because the electronic technician frequently uses these devices or supplies preamplifiers or control circuits for them.

The relay is a prime example of an electromechanical or electromagnetic amplifying device. Chapter 1 explained that the relay is a nonlinear amplifier. However, there are linear amplifiers based on electromagnetic theory. One example is the saturable reactor. A magnetic amplifier can be constructed from such a device. Before semiconductors, this type of amplifier was considered to be the most reliable of the amplifiers then available. In addition, it had the longest life when compared to other amplifiers.

The saturable reactor is actually a current amplifier. It consists of a closely wound pair of coils on a steel core. One coil is the control winding and the other coil is the load winding. The simplest arrangement is shown in figure 10-40. The steel core has a B-H or magnetization curve as shown in figure 10-41. The load winding carries alternating current of a particular frequency, usually sinusoidal. When there is no current in the control winding, the core is not magnetized. Under these conditions, the inductance of the load coil is a maximum. This result is due to the slope of the B-H curve, which actually is the permeability of the steel core material. The inductance is expressed by

$$L = \frac{\mu N^2 A}{1} = \frac{\Delta B N^2 A}{\Delta H 1}$$ Eq. 10.33

where μ is the permeability, B is the magnetic flux density, A is the cross-sectional area, 1 is the length of the flux path, and N is the number of turns. H is the magnetic field intensity, as defined by:

$$H = \frac{NI}{1}$$ Eq. 10.34

where I is the dc current through the control winding.

When the control current is at the lowest level, the inductive reactance of the load coil is high because the operation of the reactor is along the steep

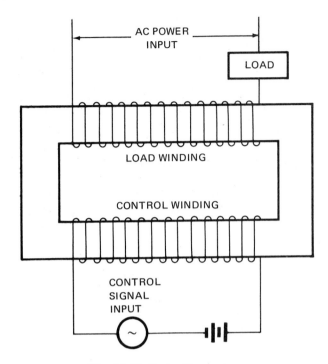

Fig. 10-40 Saturable reactor

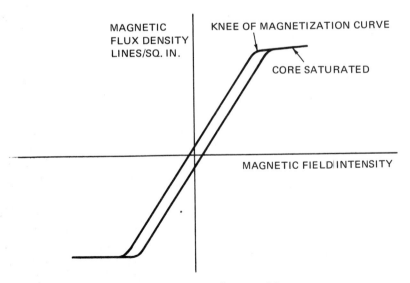

Fig. 10-41 Magnetization curve for saturable reactor core

portion of the magnetization curve. (In other words, $\Delta B/\Delta H$ is high.) When dc control current is increased so that the core is saturated, the curve has the least amount of slope ($\Delta B/\Delta H$ is low) and the load current is at a maximum. However, with this arrangement, an ac voltage is induced in the control winding. Such a voltage may be objectionable. The saturable reactor shown in figure 10-42 is a more practical device than the one in figure 10-40 because the ac

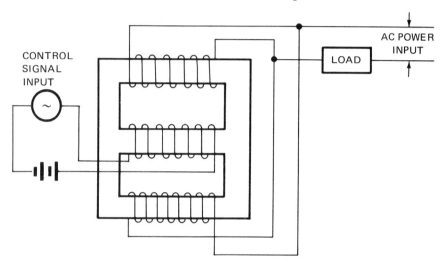

Fig. 10-42 Practical saturable reactor

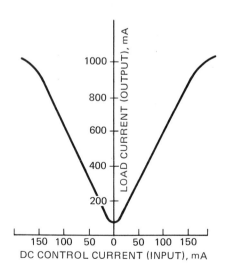

Fig. 10-43 Input-output curves for a saturable reactor

voltage induced in the dc coil is cancelled through the action of two oppositely connected ac coils. The input-output curve of a saturable reactor is shown in figure 10-43. The current gain is equal to the change in the load current divided by the change in the control current.

Compute the current gain of the saturable reactor having the curve of figure 10-43, if the control current changes from 50 milliamperes to 100 milliamperes. *(R10-35)*

An increase of control current causes the permeability to (increase) (decrease) (stay the same) when the core is driven to saturation. *(R10-36)*

A higher current gain can be obtained by adding diodes in series with the ac windings, figure 10-44, page 428. A saturable reactor connected in this way is called a magnetic amplifier. The diodes rectify the current through the reactor output windings while the ac waveform is maintained through the load. The ac induced voltage in the control winding is still cancelled. The steel core is saturated by the rectified load current, instead of the dc control current. With this arrangement the Q-point or bias level of the amplifier is established without drawing current from the input signal source. The control signal is only required to take the steel core to just below and just above the knee of the magnetization curve near the point of saturation, figure 10-41. As a result, there is a considerable improvement of the current gain as compared to the simple saturable reactor. The use of a magnetic amplifier means that the input-output curve of the saturable reactor is changed to the solid line curve shown in figure 10-45, page 428.

For the magnetic amplifier of figure 10-45, the control signal current changes from 1 milliampere to 3 milliamperes. Compute the current gain and compare this answer with the result of review problem (R10-35). *(R10-37)*

The control coil polarity is selected so that the core is brought out of saturation when the control current is increased. A requirement for load current is that it must be high enough to saturate the core when the control signal is at a minimum. A much higher gain is achieved, although the amplifier gain has a negative polarity. That is, an increase in the control current results in a load current decrease. Additional windings can be added to the core to (1) take the difference or sum of two or more signals, (2) achieve positive or negative feedback, or (3) provide a fixed bias setting. (Feedback is examined in detail in

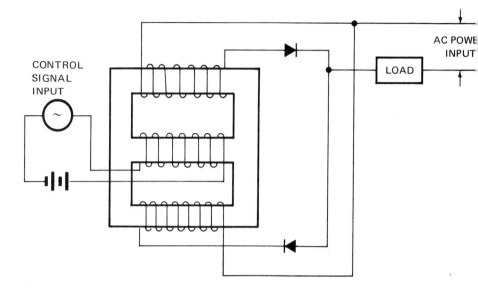

Fig. 10-44 Simple magnetic amplifier

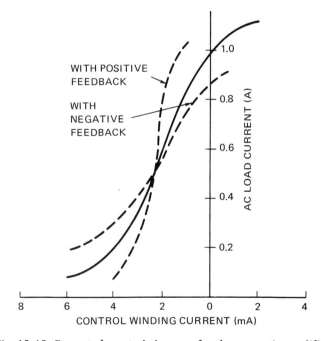

Fig. 10-45 Current characteristic curve for the magnetic amplifier

the next chapter.) Figure 10-46 shows a magnetic amplifier arrangement with additional windings. The figure shows how a dc load is connected into the magnetic amplifier using two additional diodes. These diodes form a bridge which rectifies the load current and saturates the steel core at the same time.

A separate winding can be used with a fixed bias current so that the load current decreases with a decrease in the control winding current. The bias winding current desaturates the core. Then a control current with a polarity opposite to that of the bias current, saturates the core and allows more load current to flow. The same device can use both of its control windings and act as a differential amplifier if a varying control signal is inserted in the bias winding.

The feedback winding makes the magnetic amplifier more versatile. The dashed lines on the curve in figure 10-45 show the effects of positively or negatively connecting the feedback winding into the load circuit. Negative feedback

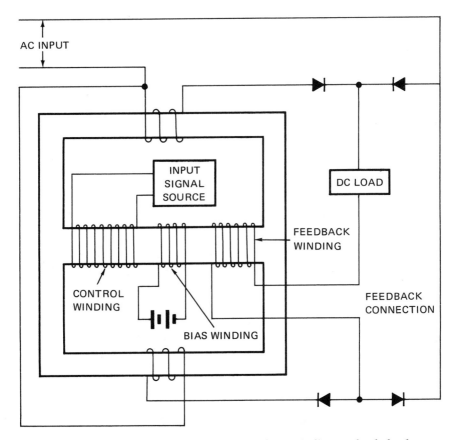

Fig. 10-46 **Magnetic amplifier with auxiliary windings and a dc load**

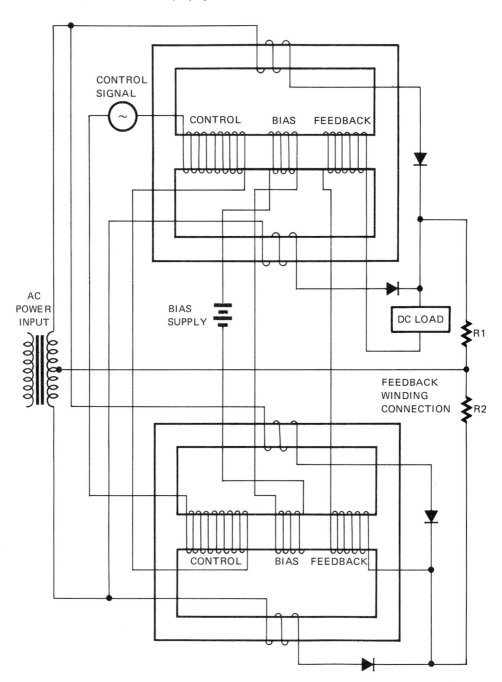

Fig. 10-47 Push-pull magnetic amplifier

provides the amplifier with greater stability, faster response, and more linear operation. Amplifier operation with positive feedback is less linear. However, the gain is greater with positive feedback and the magnetic amplifier can be operated as a switching device when the feedback winding has a large number of ampere-turns.

Two magnetic amplifiers can be combined to form a push-pull amplifier. This circuit makes it possible to bring the load current down to zero at a zero control current. At negative values of the control current, the polarity of the load current can be made to reverse. The schematic of a push-pull amplifier feeding a dc load is shown in figure 10-47.

How many windings are used with a magnetic amplifier having bias and feedback windings? What is the purpose of each winding? (R10-38)

Figure 10-48 shows the schematic of another form of amplifier, called the *amplidyne*. This amplifier is basically a dc generator that is driven at a constant speed by a motor that serves as the source of energy for the device. The output voltage is controlled by the field current through the control winding. The amplidyne has two pairs of brushes riding on a commutator as in any dc machine. By supplying current to the field winding, dc generator action causes a voltage to develop across each pair of brushes. One pair of brushes is shorted so that a large current occurs in the windings connected to these brushes. The other pair of brushes is connected to the load through a compensating winding. The windings are arranged so that the strong magnetic field developed by the shorted pair of brushes reinforces the control field excitation. In this way, a greater voltage output is produced than could be obtained without the shorted brushes. The power delivered to the load is an amplified function of a signal

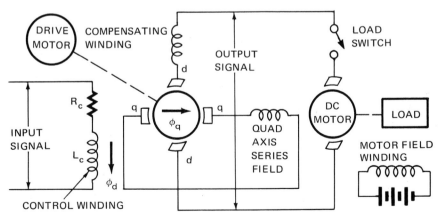

Fig. 10-48 Amplidyne system with connected load

applied to the control winding. The resulting amplification is almost linear. The amplidyne can be used to construct very large power amplifiers. Operation of the amplidyne is restricted to low frequencies because of the long time constants associated with an electromechanical type of amplifying system.

The large alternating-current generators in electric power plants can also be viewed as amplifiers even though they are not operated as amplifiers. They require a dc control field current which is much smaller than the ac output current. In general, the control fields for these generators are varied only by the small amount necessary to regulate the output voltage. The voltage produced by generators for power plants generally is required to be constant regardless of load fluctuations. Thus, field circuit controls are used as voltage regulator circuits.

Servomechanism systems can be considered as amplifiers, according to the general definition of an amplifier given in Chapter 1. For example, the steering mechanism of a large ship or even a large luxury automobile is a servomechanism system which functions as an amplifier. That is, the use of a small amount of physical effort to move the steering mechanism can control the rudder of a large ship or the front wheels of a land vehicle. The gain or amplification achieved is the output force (or travel or position) divided by the input force (or travel or position). A servomechanism "amplifier" can have phase shift and frequency response limitations just as electronic amplifiers do. Even the design techniques for servomechanisms are copied from those developed for electronic amplifiers. (See Chapter 11.)

In the design of servomechanisms, a broader definition of gain is used. Gain in a servomechanism can be expressed in a number of ways, including the output position divided by the input voltage or the output current divided by the input voltage. In other words, gain is usually, *but not necessarily,* a dimensionless amplifier characteristic. The gain can be given in volts per ampere, amperes per volt, or radians per second per inch.

Although this text concentrates on electronic amplifiers, the discussion of amplifiers is not complete without briefly considering these other types of amplifiers. The electronic technician may be called upon to supply electronic hardware for these amplifiers, as well as analyze, build, or test circuits that are to serve as preamplifiers or intermediate stages of amplification for other forms of amplifying systems.

A giant telescope in an astronomical observatory is moved by 10 minutes of arc when a hand crank is turned 10 revolutions. What is the gain of this amplifying system?   (R10-39)

An amplidyne can have a very large power gain. What is the source of the majority of the power that appears at the output terminals?   (R10-40)

LABORATORY EXPERIMENTS

The following laboratory experiments were tested and found useful for illustrating the theory in this chapter.

LABORATORY EXPERIMENT 10-1

OBJECTIVE

To connect and operate a CA3028A integrated circuit amplifier.

EQUIPMENT AND MATERIALS REQUIRED

1	Integrated circuit amplifier, RCA CA3028A
1	Resistor, 6.8 kilohms, 10%, 1/2 watt
1	Resistor, 2.2 kilohms, 10%, 1/2 watt
1	Resistor, 47 ohms, 10%, 1/2 watt
1	Resistor, 4.7 kilohms, 10%, 1/2 watt
1	Audio signal generator, 0-1 volt, 0.3 hertz to 3 megahertz
1	Oscilloscope, dual channel, time based, with provision for measuring the difference between the signals on both channels
4	Power supplies, dc, 1/2 ampere, filtered, regulated, and adjustable from 0-20 volts dc
3	Capacitors, 10 microfarads, 50 volts
1	Integrated circuit socket, eight contacts, constructed so that a connection can be made to each of the contacts with clip leads
1	Volt-ohm-milliammeter, battery operated FET type, capable of measuring voltages down to 1 millivolt, dc or ac

PROCEDURE

1. Obtain a CA3028A integrated circuit amplifier. The schematic and terminal identification for the device are shown in figure 10-49, page 434. The following maximum values should be observed:
   Pins 8 to 3 or 6 to 3: -0 volt, +30 volts
   Current through pin 8 or pin 6: less than 3 milliamperes

The CA3028A can be used as a differential amplifier or cascode amplifier. This device can be used at radio frequencies with little attenuation. When it is used as a differential amplifier, Q3 acts as a constant current source. This arrangement allows both amplifier inputs (① and ⑤) to have a high input impedance, and keeps the common mode gain at a low value. A cascode amplifier consists of a common emitter amplifier and a common base amplifier connected in series. This type of amplifier behaves like a common base amplifier with a high current gain and is frequently used as a tuned amplifier in a practical circuit.

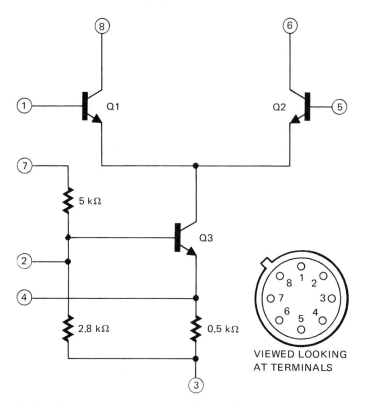

Fig. 10-49   Schematic and terminal layout for CA3028A integrated circuit amplifier

Transistor Q3 is biased by 5-kilohm and 2.8-kilohm internal resistors. The 500-ohm resistor produces stabilization against changes due to temperature. Q1 and Q2 must be biased by external circuits.

2. Connect the CA3028A into the circuit shown in figure 10-50.

3. Adjust the external bias resistors and voltages for a maximum voltage gain from ⑥ to ⑦ or ⑧ to ⑦, and yet still maintain an undistorted output. Use any convenient frequency that can be pictured easily on the scope. Calculate the voltage gain for each stage.

4. The 47-ohm resistor acts as a dummy signal source. The only function of the Q1 section is to cancel out the effects due to temperature and power supply voltage variation. The output from ⑥ to ⑦ is put on channel A of the oscilloscope and the output from ⑧ to ⑦ is put on channel B of the oscilloscope. Use the A-B function on the scope and measure the difference between the two signals. Calculate the voltage gain of the amplifier as a differential amplifier according to:

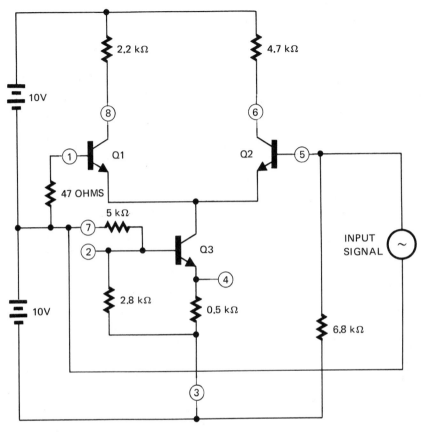

Fig. 10-50 CA3028A integrated circuit amplifier connected for experiment 10-1

$$A_V = \frac{A - B \text{ output}}{\text{⑤ to ⑦ input}} \qquad \text{Eq. 10.35}$$

5. Change the circuit to the one shown in figure 10-51, page 436.

6. Apply a low-voltage dc input signal from the base to common for both sides of the differential amplifier. Adjust V_{in2} to zero. Set V_{in1} at 0.5 volt.

7. Measure V_{in1} and the voltage from ⑧ to ⑥. Use a VOM or the A - B function on the scope.

8. Compute A_1 where

$$A_1 = \frac{V_{8-6}}{V_{in1}} \qquad \text{Eq. 10.36}$$

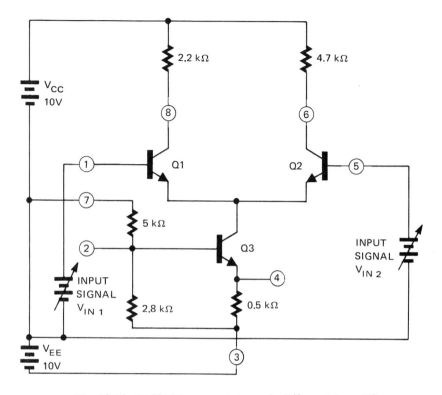

Fig. 10-51   CA3028A connected as a dc differential amplifier

9.   Adjust V_{in1} to zero, and V_{in2} to 0.5 volt.

10.   Measure V_{in2} and the voltage from ⑧ to ⑥.

11.   Calculate A_2 where

$$A_2 = \frac{V_{8-6}}{V_{in2}}$$ Eq. 10.37

Use the new value of V_{8-6} measured in step 10.

12.   Calculate the common mode gain

$$A_c = \frac{A_1 + A_2}{2}$$ Eq. 10.38

13.   Adjust V_{in1} to 1 volt and again measure the voltage from ⑧ to ⑥.

14.   Using this value of V_{8-6}, perform the following calculation for the voltage gain of the difference signal.

$$A_D = \frac{V_{out} - \frac{(A_1 + A_2)(V_{in1} + V_{in2})}{2}}{V_{in1} - V_{in2}}$$

Eq. 10.39

15. Calculate the common mode rejection ratio, A_D/A_C. This ratio is a measure of the performance of the CA3028A as a difference amplifier. Ideally, A_D/A_C is infinite. The lower the value of this ratio, the worse is the performance of the circuit as a differential amplifier.

16. Connect the CA3028A into the cascode amplifier circuit shown in figure 10-52. The resistors in the circuit should provide for the proper biasing of all of the transistors. Transistor Q1 is not used in this circuit and is shorted out to reduce potential noise problems.

17. Apply a sinusoidal input signal of 2 kilohertz or higher. Keep the amplitude below the point at which the output becomes distorted; that is, maintain linear operation of the amplifier. Adjust the external resistors if necessary to achieve distortion-free operation.

18. Measure the voltage input and the voltage output on the oscilloscope. Calculate the voltage gain.

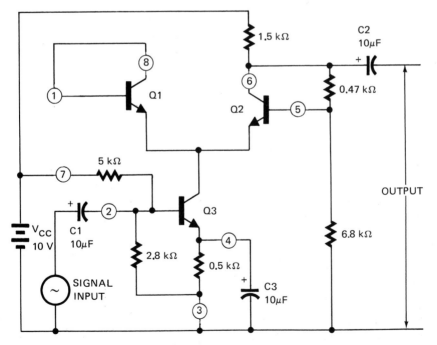

Fig. 10-52 CA3028A amplifier connected as a cascode amplifier

19. Remove the bypass capacitor between terminals ③ and ④ and note the drop in the voltage gain with this negative feedback in the circuit. Measure and record the voltage gain for this condition.

LABORATORY EXPERIMENT 10-2

OBJECTIVE

To connect and operate a CA3020 as an ac wide-band power amplifier measure its gain, and determine its frequency response.

EQUIPMENT AND MATERIALS REQUIRED

1	Multipurpose wide-band power amplifier, RCA No. CA3020
1	Power supply, dc, 1/2 ampere filtered and regulated, adjustable from 0-20 volts dc
1	Oscilloscope, dual channel, time based
2	Resistors, 65 ohms, 7 watts, 10%
2	Capacitors, 1 microfarad, 50 volts
1	Resistor, 51 ohms, 1/2 watt, 10%
1	Volt-ohm-milliammeter, battery operated FET type, capable of measuring voltages down to 1 millivolt, dc or ac
1	Audio signal generator, 0-1 volt, 3 hertz to 3 megahertz
1	Integrated circuit socket, 12 contacts, with terminals of sufficient size and spacing to connect to each of the contacts using clip leads

PROCEDURE

1. Obtain a CA3020 integrated circuit amplifier. Consult figure 10-10 for the schematic and terminal layout. Observe the following maximum values:
   Input signal voltages: ±3 volts
   Terminals ② and ③: 0 volt
   Terminals ⑧ or ⑨ to common: +9 volts
   Do not apply external voltages to any other terminals.

2. Connect the device into the circuit of figure 10-53. All voltage sources are to be off and turned to zero until the circuit is connected and checked by the instructor or laboratory partner.

3. Turn on the dc power supply and adjust V_{CC} to 9 volts, as indicated by the VOM.

4. Measure the bias voltages for the transistors in the integrated circuit by taking dc voltage measurements from ③ to ⑫, ④ to ⑫, ⑦ to ⑫, ⑪ to ⑫, ② to ⑫, and ⑪ to ⑫. Try to justify the values obtained by studying the complete schematic for the CA3020 device.

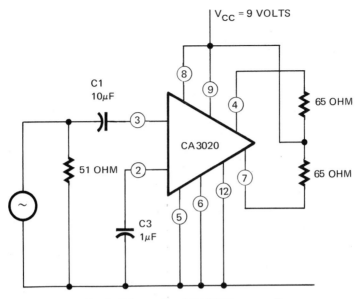

Fig. 10-53 Schematic of CA3020 connections

5. Turn on the signal generator. Adjust the frequency to about 1000 hertz using a sinusoidal waveform. Adjust the amplitude to a level at which there is a distortion-free output from terminals ④ to ⑦.

6. Measure the input and output voltages on an oscilloscope.

7. Calculate the voltage gain as a ratio and as voltage decibels. Compare this value with the manufacturer's specified level of 44 decibels for this circuit. (Refer to RCA publication ICAN-5320, page 5.)

8. Change e_{in} to a value of less than 10 millivolts rms.

9. Record the resulting value of e_{out}.

10. Calculate the voltage gain in decibels.

11. Vary the input signal frequency while holding e_{in} constant. Measure e_{out} at each level of frequency. Vary the frequency above and below 1 kilohertz until the gain drops off significantly.

12. Calculate the voltage gain in decibels at each level measured. Plot the frequency response using semilogarithmic paper. Use semilogarithmic paper that permits five decades of frequency to be plotted.

13. Determine the bandwidth (in kilohertz) of the amplifier by measuring the frequency between the points at which the gain drops by 3 decibels.

LABORATORY EXPERIMENT 10-3

OBJECTIVE

To use an integrated circuit amplifier as one stage of an audio power amplifier, and to construct and evaluate the performance of the overall amplifier circuit.

EQUIPMENT AND MATERIALS REQUIRED

1   Audio amplifier, RS 377 dual, 2 watts
1   Dual in-line socket, 14 pin, with terminals large enough for connecting clip leads
1   Power supply, dc, 0-40 volts, regulated and filtered, 0.5 ampere
1   Capacitor, 1000 microfarads, 50 volts
1   Capacitor, 100 microfarads, 50 volts
1   Transistor, npn, S5004
1   Transistor, pnp, S5005
1   Resistor, 4 ohms, 2 watts, 10%
1   Resistor, 5 ohms, 2 watts, 10%
1   Resistor, 27 kilohms, 1/4 watt, 10%
1   Capacitor, 82 picofarads, 50 volts
1   Resistor, 100 kilohms, 10%, 1/4 watt
1   Resistor, 2 kilohms, 10%, 1/4 watt
1   Capacitor, 250 microfarads, 50 volts
1   Resistor, 1 megohm, 1/4 watt, 10%
1   Capacitor, 0.1 microfarad, 15 volts
1   Capacitor, 5 microfarads, 50 volts
1   Oscilloscope, dual channel, time based
1   Resistance substitution box, Heathkit EU-28A or equivalent
1   Resistance decade box, Heathkit DR-1 or equivalent

PROCEDURE

1. Obtain an RS-377, two-watt audio amplifier. This device is a dual unit used for stereo phonographs and receivers. This experiment uses only one unit of the RS-377 integrated circuit. The connections to the unit are shown in figure 10-54. The maximum voltage ratings from terminals $\overline{14}$, $\overline{6}$ or $\overline{9}$ to terminals $\overline{3}$, $\overline{4}$, $\overline{5}$, $\overline{10}$, $\overline{11}$, and $\overline{12}$ are 26 volts dc.

2. Connect the amplifier into the circuit shown in figure 10-55. All power should be disconnected and the voltage adjusting knobs should be turned to the minimum output level.

3. Operate the circuit as a linear amplifier, using a low-level sinusoidal input signal of about 5 kilohertz, and a V^+ or V_{CC} of 24 volts to common. Operate the circuit with R_L set at infinity and R_i set at 0 ohms.

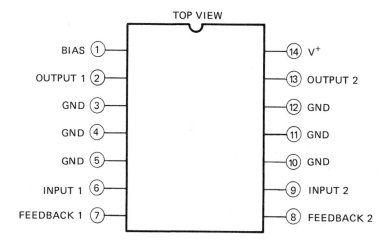

Fig. 10-54 Pin connections for RS-377

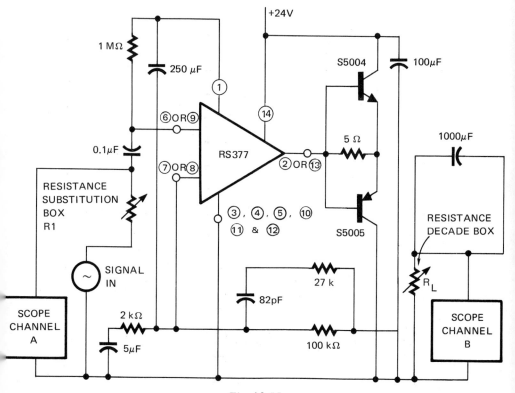

Fig. 10-55

4. Measure the input and output voltages using a calibrated oscilloscope.

5. Calculate the voltage gain.

6. Measure and calculate the input resistance, using the technique described in Laboratory Experiment 3-2, step 4.

7. Measure and calculate the output resistance, using the technique described in Laboratory Experiment 3-2, step 5.

EXTENDED STUDY TOPICS

1. Explain how the common mode signal can be minimized in a difference amplifier.

2. Describe some methods of increasing the common mode rejection ratio and reducing the effect of temperature variations for integrated circuit amplifiers.

3. Consult the specification sheet for an integrated circuit amplifier and write a summary of its performance characteristics. Do *not* use any of the manufacturer's symbols or subscripts.

4. Explain how a difference amplifier circuit can be used to minimize the effect of temperature on the performance of a transistor amplifier.

5. How are lasers related to masers? How are they different?

6. Light and heat are forms of _____ at different frequencies, where light has the (higher) (lower) frequency.

7. Describe the steps in the manufacture of an integrated circuit. The steps are to be taken in the correct sequence.

8. Describe two methods by which light may be amplified and converted to a usable electrical signal.

9. What is the difference between a saturable reactor and a magnetic amplifier?

10. What are some variations that can be made to the basic magnetic amplifier circuit of figure 10-44?

11. Sketch the transistor array of figure 10-9 as the side view of an integrated circuit crystal. Use figure 10-5 as a model. Show the terminal connections formed by the vapor deposition process and label the various areas.

12. What is meant by the term Noise Figure (NF)? Record the values of NF in some integrated circuit amplifier specifications.

13. What is the bandwidth of the integrated circuit amplifier that has the frequency response curve shown in figure 5-1?

14. Explain the difference between a nonlinear amplifier and a linear amplifier.

15. What are two methods for switching on an SCR? What is the easiest method of switching an SCR off?

16. Describe some ways in which light can be amplified.

17. List some advantages of integrated circuits over circuits made from discrete components.

18. Some integrated circuit amplifiers can operate within a wide range of dc bias voltages. What precautions should be observed when operating with a high level of dc bias voltage as opposed to a low level of dc bias voltage? What are the advantages of using integrated circuit amplifiers that can operate with a wide range of dc bias voltages?

11

Feedback
Theory

OBJECTIVES

After studying this chapter, the student will be able to:

- define feedback
- select and identify feedback amplifier circuits
- name the types of feedback amplifier circuits and their characteristics
- identify feedback element parameters for amplifiers having specific functions
- analyze feedback circuits using Bode and Nyquist graphs
- compare the Bode and Nyquist methods of analysis
- list the advantages and disadvantages of positive and negative feedback

BASIC FEEDBACK THEORY

In general terms, *feedback* is used to indicate that information or a command signal is directed back to its source in such a way that future information or commands are affected. To use the concept of feedback in electronics technology, an exact definition of feedback and its effects is required. Before defining feedback any further, two related concepts must be considered: transfer functions and steady state sinusoidal frequency response.

Amplifiers that are subjected to a time varying input frequently are analyzed by subjecting them to a steady state sinusoidal input, either in the laboratory or on paper. A sinusoidal input signal is supplied for a period of time long enough to achieve an equilibrium condition. The output response of the circuit is then compared to this steady state input. By applying sinusoidal signals for a number of different frequencies, the transfer function of the amplifier can be determined.

The transfer function is the ratio of the magnitude and phase of the output to the input magnitude and phase as a function of frequency. This ratio is the same as the gain, with the frequency as an independent variable. The frequency response chart of figure 5-1 actually is the magnitude of the transfer function plotted against frequency. The transfer function concept can be used for inputs other than sinusoidal, including a step input and a ramp input. However, these inputs are used only for transient response studies that require integral calculus and LaPlace transformations. The discussion of transfer functions is simplified by assuming a steady state sinusoidal response. This assumption does not restrict the discussion of feedback theory or analysis methods using feedback, because the most common analysis techniques are based on steady state sinusoidal inputs.

The following discussion generally is restricted to signals that are within the linear range of a device or circuit. In other words, there is no change in the gain due to the signal amplitude. That is, the amplification is linear. To insure linearity, relatively low-level signals are assumed.

Figure 11-1 shows an electronic amplifying device having an output signal (O) and an input signal (I). The transfer function or gain, as a function of frequency, is:

$$A = \frac{O}{I} \qquad \text{Eq. 11.1}$$

A portion of the output can be fed back into the input and subtracted from the input, figure 11-2, page 446. The forward gain of the device without feedback is A. The feedback gain is the feedback signal, X, divided by the output signal, O:

$$F = \frac{X}{O} \qquad \text{Eq. 11.2}$$

For *negative* feedback, the feedback signal is *subtracted* from the input signal. The signal applied to the device is now I-X. The gain A is redefined such that the output is:

$$O = A(I\text{-}X) \qquad \text{Eq. 11.3}$$

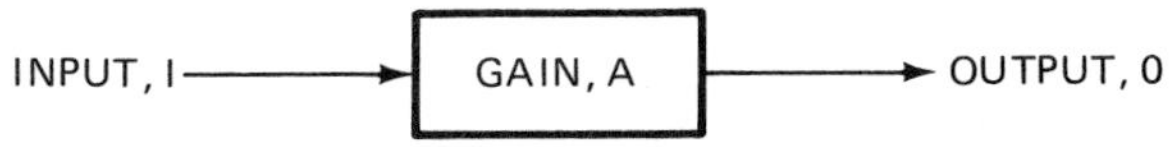

INPUT, I ———————→ | GAIN, A | ———————→ OUTPUT, 0

Fig. 11-1 Block diagram of an electronic device

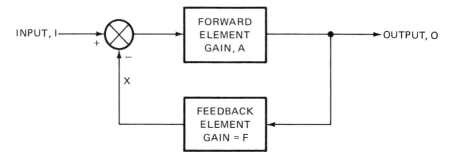

Fig. 11-2 Block diagram of electronic device with negative feedback

But, since X = F • O, Equation 11.2 can be substituted into Equation 11.3, and the resulting expression can be solved for the transfer function with feedback.

$$\frac{O}{I} = A_f = \frac{A}{1 + AF}$$

Eq. 11.4

A_f represents the gain of the amplifier with feedback. When A is large, the characteristics of the amplifying device do not affect the overall amplifier gain, since

$$A_f \approx \frac{1}{F}$$

Eq. 11.5

This situation frequently occurs for a well-designed amplifier using negative feedback. If there are any nonlinearities or parameter variations in the amplifying device, they are minimized by the use of negative feedback. The emitter self-bias circuit in Chapter 2 and figure 2-17 is an example of the use of negative feedback. Any changes in A are multiplied by the factor 1/(1 + AF), before they appear in the output.

The gain of an amplifier without feedback (A) drops by 20%. Let A = 100 initially. Calculate the gain from the output to the input when A = 100 and negative feedback is used. Assume F = 1. Recalculate the gain, using F = 1 and with A reduced by 20%. What is the percent change in the overall gain when feedback is used? *(R11-1)*

In a bipolar transistor, what are the nonlinearities and parameter variations that can be reduced by the use of negative feedback? *(R11-2)*

EXAMPLES OF NEGATIVE FEEDBACK

As an example of negative feedback, consider the very simple circuit in figure 11-3. This low-pass RC circuit is equivalent to a feedback amplifier having unity gain. The circuit is fed by a sinusoidal signal from a zero-impedance source and is applied to an infinite impedance load. A portion of the output of the circuit (the current through the capacitor) is fed back into the input. The output voltage is:

$$V_o = -IjX_c = \frac{I}{j\omega C} \qquad\qquad \text{Eq. 11.6}$$

where j is a complex operator whose value is $\sqrt{-1}$. This operator represents a $90°$ phase angle. The rms values of voltage and current are V_o and I, respectively. The input voltage is obtained by applying Kirchhoff's Voltage Law for steady state sinusoidal circuit analysis:

$$V_i = RI + \frac{I}{j\omega C} \qquad\qquad \text{Eq. 11.7}$$

The transfer function is obtained by dividing Equation 11.6 by Equation 11.7 and rearranging the terms. The transfer function is:

$$\frac{V_o}{V_i} = \frac{1}{1 + j\omega RC} \qquad\qquad \text{Eq. 11.8}$$

Equation 11.4 can be regarded as a generalized form of Equation 11.8. Figure 11-3 can be drawn as a specialized form of figure 11-2. This is shown in figure 11-4, page 448. The feedforward element in this circuit is the resistor shown by "A = 1" in figure 11-4. The feedback element is the term $j\omega RC$, formed by the capacitive reactance, X_c. If the frequency (ω) is zero, there is no feedback and the output is the same as the input.

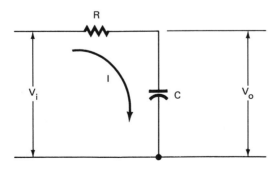

Fig. 11-3 Equivalent circuit of an amplifier having a gain of unity

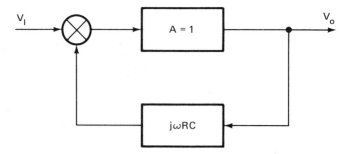

Fig. 11-4   Block diagram of the feedback system formed by the unity gain amplifier (Fig. 11-3)

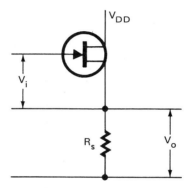

Fig. 11-5 JFET amplifier without feedback

What is the output of the circuit in figure 11-3 when the signal frequency is infinitely great? Explain this condition in terms of the feedback expression, Equation 11.4. (R11-3)

What is the phase shift between the input and the output for the circuit in figure 11-3, when the frequency is infinitely high? (R11-4)

An amplifier circuit with and without feedback is examined in the following paragraphs. It is assumed that the gain of this circuit is not sensitive to frequency. Because of this assumption, it is possible to relate the following analysis more easily to the theory developed in earlier sections. Refer to figure 11-5. V_o and V_i are defined as ac rms signals. The input voltage V_i is the ac gate-to-source voltage V_{gs}. According to Equation 8.13,

$$V_{ds} = \mu V_{gs} \qquad \text{Eq. 11.9}$$

In figure 11-5, $V_{ds} = V_o$ and $V_{gs} = V_i$. Therefore,

$$V_o = \mu V_i \qquad \text{Eq. 11.10}$$

Using the input and output voltages as defined, the gain of the amplifier is:

$$A = \frac{V_o}{V_i} = \mu \qquad \text{Eq. 11.11}$$

This expression is the amplifier gain without feedback. Negative feedback is added by applying an input voltage in the manner shown in figure 11-6. The gate-to-source voltage is:

$$V_{gs} = V_i - V_f \qquad \text{Eq. 11.12}$$

Figure 11-7 is a diagram of the circuit as a feedback amplifier. $V_o = V_f$ because the drop across R_s is the feedback voltage V_f as well as the output voltage V_o.

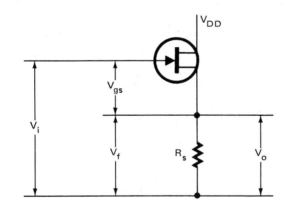

Fig. 11-6 JFET feedback amplifier

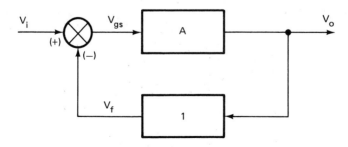

Fig. 11-7 Block diagram of amplifier with unity negative feedback

Thus, the feedback gain, as defined by Equation 11.2, is "1." By analogy with Equation 11.4, the overall gain of the amplifier with feedback is expressed as:

$$\frac{V_o}{V_i} = \frac{A}{1 + (A)(1)} = \frac{A}{1 + A}$$ Eq. 11.13

Substitute Equation 11.11 in Equation 11.13 to obtain:

$$\frac{V_o}{V_i} = \frac{\mu}{1 + \mu}$$ Eq. 11.14

Note that Equation 11.14 is the same as Equation 9.26. Refer back to Chapter 9 and note that Equation 9.26 was developed for the same circuit without the use of the feedback concept.

The Effects of Negative Feedback

The following list gives the characteristics of an amplifier that uses negative feedback.

- The amplifier has a lower gain than is possible without feedback. (This can be seen by examining Equation 11.14.)

- In general, the frequency response can be widened if the feedback element is not frequency sensitive. Changes in A with frequency are multiplied by the factor $1/(1 + AF)$. As a result, the amplifier bandwidth is increased.

- Nonlinear distortion as well as frequency distortion is reduced. Any distortion constitutes a change in A and the change is multiplied by the factor $1/(1 + AF)$.

- Temperature stabilization and bias voltage regulation are improved because any changes in A are multiplied by $1/(1 + AF)$. The emitter self-bias circuit for a transistor is an example of an amplifier using negative feedback for stabilization and regulation.

- Normally, the use of negative feedback means that the input resistance is made higher and the output resistance is made lower.

- Unstable oscillations can occur in the output as a result of an improper application of negative feedback. Such a situation can be detected using methods described later in this chapter.

An amplifier has an open loop gain A = 1000 ± 100. It is necessary to have an amplifier whose voltage gain varies by no more than ± 0.1%.
a. Find the feedback factor F of the feedback network used.
b. Find the average gain with feedback.   *(R11-5)*

Assume that $\mu = 50$ *for the FET used in figures 11-5 and 11-6. What is the voltage gain of the circuit with and without negative feedback?*   (R11-6)

POSITIVE FEEDBACK

Feedback can be positive or negative, depending upon whether the portion of the output fed back is added to or subtracted from the input. Positive feedback means that the signal applied to A in figure 11-2 is I + X. Thus, Equation 11.3 becomes:

$$O = A(I + X)$$ Eq. 11.15

Substituting Equation 11.2 into Equation 11.15, the gain with positive feedback is:

$$A_f = \frac{O}{I} = \frac{A}{I - AF}$$ Eq. 11.16

If the output is fed back into the input and added to it, it should be evident that the output is greater than if there is no feedback. Equation 11.16 expresses this fact. The larger the feedback gain, F, the smaller is the denominator and the greater is the output. In the usual application of the feedback principle, the elements that make up the feedback gain (F) do not include an amplifier. Generally, AF cannot be greater than one because the feedback signal is only a *portion* of the output. The feedback gain derives all of its power capability from the output signal itself. The elements comprising F are usually a network of resistances, capacitances, and inductances. There is no externally applied bias power. However, AF can *approach* unity or even equal unity under some conditions. This is an unstable situation and is usually undesirable.

Some positive feedback can be an advantage in a linear amplifier. This is true if the fraction fed back is not proportional to the frequency anywhere within the spectrum of possible signal frequencies. An example of such a circuit is a magnetic amplifier with the feedback winding connected so that the feedback signal adds to the control signal (see Chapter 10 and figure 10-46).

High positive feedback causes an amplifier to go to saturation, even at small input signals. Saturation results because the input signal is amplified according to the physical characteristics of the amplifying devices and then fed back to amplify itself. The effect is cumulative and the amplifying device operates in either its cutoff region or saturation region. The forward gain of the amplifier is reduced in the saturation and cutoff regions because of the shape of the characteristic curves in these regions. This results in severe distortion of the output signal.

For the amplifier having positive feedback, the gain goes to infinity at AF = 1, according to Equation 11.16. Practically, however, this situation is not possible because the equation does not account for saturation of the amplifier. A different technique is needed to analyze nonlinear amplification with feedback. An amplifier driven to saturation has an output which resembles figure 3-6(c).

What waveform is generated by driving an amplifier to saturation with a large sinusoidal input signal?   *(R11-7)*

Explain how a high-gain amplifier using positive feedback can be driven to saturation.   *(R11-8)*

An amplifier has an open loop gain of 100. One percent of the output is fed back to the input and added to the input. What is the theoretical gain of the amplifier with feedback? Is it still operating as a linear amplifier?   *(R11-9)*

Some circuits are designed with sufficient positive feedback to saturate or cut off the active amplifying devices and hold them at that condition. Such an approach is often used in counting systems and logic circuits. For example, refer to the bistable multivibrator circuit shown in figure 11-8. In this circuit, the output of a transistor amplifier is fed back through the coupling resistor R1 to the input of a second transistor amplifier. It was shown in Chapter 3 that the output voltage signal of a common emitter amplifier is 180 degrees out of phase with the input signal. In figure 11-9, there is a 180 degree phase shift in amplifier Q1 and a 180 degree phase shift in amplifier Q2. These phase shifts in Q1 and Q2 force the output signal of Q2 to be in phase with the input to Q2. As a result, because of the positive feedback, one of the amplifiers is driven to cutoff and the other is driven to saturation. The two amplifiers are held in this condition until a new input signal of opposite polarity is applied. The amplifiers then both drive to the limit of their output in the opposite direction. There are circuits using FETs or vacuum tubes which function in a similar way. Linear electronic amplifiers using positive feedback *alone* are uncommon. However, many linear amplifiers combine positive feedback with negative feedback of some kind to stabilize the overall circuit and insure that its operation is linear. The parallel LC circuit in the parametric amplifier is an example of the use of positive feedback to increase the amplifier gain.

An amplifier using positive feedback has the following features.

* The gain is higher than is possible without feedback.

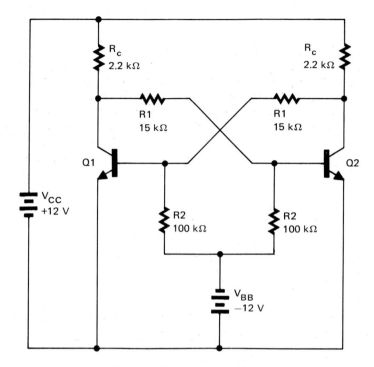

Fig. 11-8 Bistable multivibrator

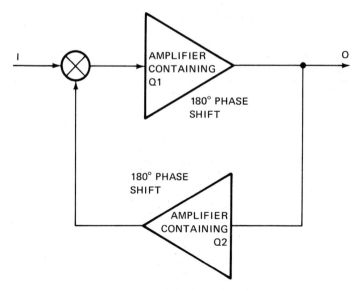

Fig. 11-9 Flip-flop circuit

- Distortion is *increased* by the factor $1/(1 - AF)$. The distortion is increased whether the positive feedback is due to changes in the signal frequency or nonlinearities of the amplifying device.

- Unstable oscillations in the output can occur with positive feedback. Such oscillations are more likely with positive feedback than with negative feedback because any change in the output is amplified even more with positive feedback. For this reason, a complex amplifying system using positive feedback usually has a negative feedback feature as well to provide for stabilization.

Stabilization is difficult, if not impossible, to achieve with positive feedback alone. The temperature effect in transistors actually is a positive feedback phenomenon. This effect must be stabilized by some form of compensating negative feedback.

Explain how the temperature instability of a transistor is a positive feedback phenomenon.   *(R11-10)*

AMPLIFIER CIRCUITS USING FEEDBACK

There are four feedback configurations for electronic amplifiers. They are:

1. voltage series feedback
2. current series feedback
3. current shunt feedback
4. voltage shunt feedback

Voltage Series Feedback — The Voltage Amplifier

Voltage series feedback is shown in figure 11-10. The input of the feedback element is connected across the output terminals of the amplifier. The output of the feedback element is connected in series with the amplifier input. This arrangement is commonly used for circuits designed to amplify voltage. Such circuits are known as voltage amplifiers because they produce an output voltage which is proportional to the input voltage and is independent of the load and source resistances. Figure 11-6 is an example of such a circuit. A similar circuit using a bipolar transistor is shown in figure 3-10. To analyze a circuit with voltage series feedback, refer to the amplifier shown in figure 11-11. An equivalent circuit for this amplifier is made by combining figure 11-11 with figures 4-7(a) and 4-9(a). The resulting circuit is shown in figure 11-12, page 456. Generally, C and h_{oe} can be neglected in the calculations. The calculations are affected by C only for signal frequencies that are below normal. As a result, figure 11-12 is simplified to the circuit of figure 11-13, page 456.

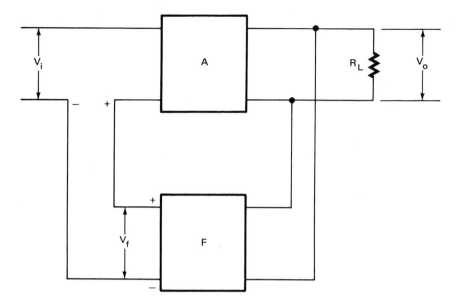

Fig. 11-10 Voltage series feedback

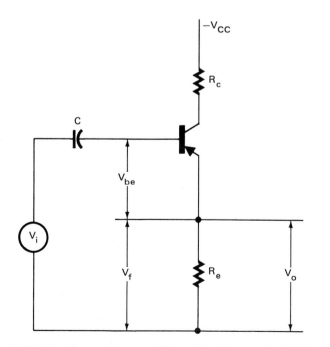

Fig. 11-11 Simple transistor amplifier with voltage series feedback

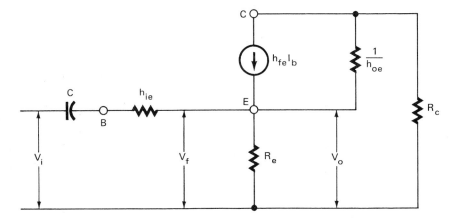

Fig. 11-12 Approximate equivalent circuit for the amplifier of figure 11-11

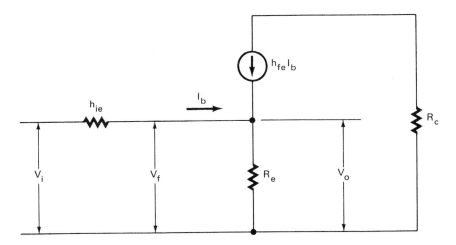

Fig. 11-13 Simplified equivalent circuit of amplifier in figure 11-11

The output voltage is equal to the current from the current source $(h_{fe}I_b)$ times the value of the emitter resistor (R_e). The feedback voltage is the voltage drop across R_e as well as the output voltage:

$$V_o = V_f = (I_b + h_{fe}I_b)R_e \qquad \text{Eq. 11.17}$$

From Equation 11.2, the gain of the feedback element is:

$$F = \frac{V_f}{V_o} = 1 \qquad \text{Eq. 11.18}$$

The transistor input voltage, V_{be}, is also the amplifier input when there is no feedback. Figure 11-14 shows the amplifier of figure 11-11 without feedback.
V_{be} is:

$$V_{be} = I_b h_{ie} \qquad \text{Eq. 11.19}$$

The gain without feedback is:

$$A = \frac{V_o}{V_{be}} \qquad \text{Eq. 11.20}$$

Substitute Equations 11.19 and 11.17 into Equation 11.20 to obtain an expression for the gain.

$$A = \frac{(I_b + h_{fe}I_b)R_e}{I_b h_{ie}} = \frac{(1 + h_{fe})R_e}{h_{ie}} \qquad \text{Eq. 11.21}$$

The gain with feedback is obtained by substituting Equations 11.21 and 11.18 into Equation 11.4.

$$A_f = \frac{(1 + h_{fe})R_e}{h_{ie} + (1 + h_{fe})R_e} \qquad \text{Eq. 11.22}$$

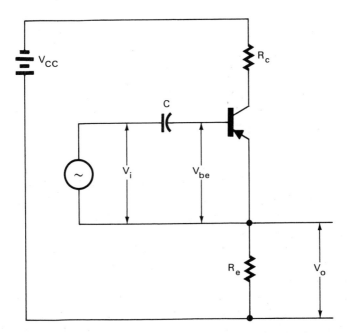

Fig. 11-14 Transistor amplifier with voltage series feedback removed

In general, $h_{fe} \gg 1$ and $h_{ie} \ll (1 + h_{fe})R_e$. Therefore,

$$A_f \approx 1$$

By using negative feedback, changes in the amplifying devices will have less effect on the overall amplifier performance. Note in the preceding analysis that changes in the transistor parameters h_{fe} and h_{ie} have little or no noticeable effect. A_f is always approximately 1.

An amplifier has a voltage gain without feedback of 4500. Voltage series feedback is added with a feedback gain (F) of 0.02. What is the voltage gain with feedback?   *(R11-11)*

An amplifier, similar to the one shown in figure 11-11, has $h_{fe} = 50$, $h_{ie} = 200$ ohms, $R_e = 200$ ohms, and $R_c = 2000$ ohms. Compute the gain with and without feedback.   *(R11-12)*

In general, voltage series feedback increases input impedance and decreases output impedance. As proof of this statement, consider a circuit such as that of figure 11-11. Determine the input and output impedances for the input signal connected from the base to the emitter ($V_{be} = V_i$), as shown in figure 11-14, and compare these values with the input and output impedances determined for the input signal applied as shown in figure 11-11.

Using figures 11-11 and 11-12 and the parameters given in problem (R11-12), compute the output impedance of the amplifier with and without voltage series feedback. Assume that $1/h_{oe}$ is large in comparison to the other resistances in the circuit and that the input signal is a pure voltage source.   *(R11-13)*

Calculate the input impedance for the amplifier described in problems (R11-12) and (R11-13), for the situation where (1) voltage series feedback is used, and (2) voltage series feedback is not used.   *(R11-14)*

Current Series Feedback — The Transconductance Amplifier

Current series feedback is shown in figure 11-15. In this arrangement, the output current is sampled by connecting the feedback element input in series with the load. The output of the feedback element is in series with the input voltage. This type of feedback connection is most useful in amplifiers

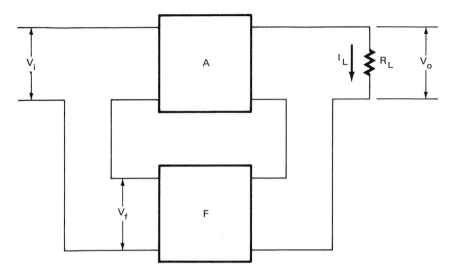

Fig. 11-15 Current series feedback

designed to produce an output current proportional to an input voltage, regardless of the load and signal source resistances. This type of amplifier is called a *transconductance amplifier.*

The FET amplifier shown in figure 11-16, page 460, is a form of transconductance amplifier. Other devices that can be used to build transconductance amplifiers include pentodes and bipolar transistors. The resistor in the source circuit provides current series feedback. Bias connections are omitted because the concern is only with the response to a time varying signal. A gate coupling resistor and coupling capacitors are omitted to simplify the discussion. Figure 11-17, page 460, shows an equivalent circuit for the transconductance amplifier. The resistance r_L is the parallel combination of the bias resistor R_D and the load resistance. The gain of this amplifier is defined by the relationship between the output current and the input voltage. Therefore, the transconductance (g_m) can be used as the amplifier gain without feedback. In Chapter 8, Equation 8.17 expressed the transconductance.

$$g_m = \frac{i_d}{v_{gs}} = \frac{I_L}{V_{gs}} = A \qquad \text{Eq. 11.23}$$

The gain of the feedback element is the ratio of the feedback voltage to the output current (I_L). Assuming that the gate current is negligible, the load current is:

$$I_L = \frac{g_m V_{gs} r_d}{R_S + r_L + r_d} \qquad \text{Eq. 11.24}$$

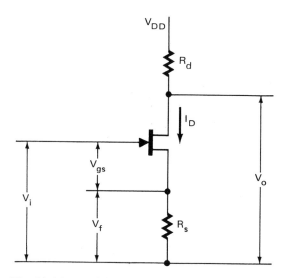

Fig. 11-16 Amplifier using current series feedback

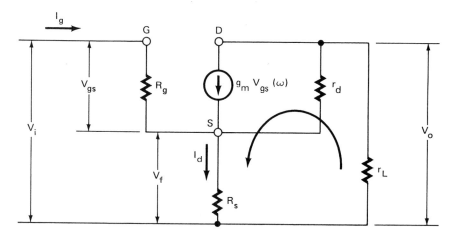

Fig. 11-17 Equivalent circuit of an amplifier using current series feedback

The feedback gain is:

$$F = \frac{V_f}{I_L} = \frac{I_L R_s}{I_L} = R_s \qquad \text{Eq. 11.25}$$

The gain of the transconductance amplifier with feedback is:

$$A_f = \frac{I_L}{V_i} = \frac{A}{1 + AF} = \frac{g_m}{1 + g_m R_s} \qquad \text{Eq. 11.26}$$

If R_s is large enough,

$$A_f \approx \frac{1}{R_s}$$

Eq. 11.27

As a result, any changes in g_m are made harmless by the current series feedback. Current series also provides stabilization for a voltage amplifier. The output voltage is given as:

$$V_o = I_L r_L = \frac{g_m V_{gs} r_L r_d}{R_s + r_L + r_d}$$

Eq. 11.28

The input voltage is:

$$V_i = V_{gs} + V_f$$

Eq. 11.29

Combine Equations 11.29, 11.28, 11.24, and 11.25 to obtain the voltage gain.

$$A_V = \frac{V_o}{V_i} = \frac{\dfrac{g_m V_{gs} r_d r_L}{r_d + R_s + r_L}}{V_{gs}\left(1 + \dfrac{g_m r_d R_s}{r_d + R_s + r_L}\right)}$$

$$= \frac{g_m r_d r_L}{r_d + R_s + r_L + g_m r_d R_s}$$

Eq. 11.30

If the $g_m R_s r_d$ term is large enough,

$$A_V \approx -\frac{r_L}{R_s}$$

Eq. 11.31

Thus, the voltage gain also is unaffected by changes in the transconductance. A_V is stable as long as r_L and R_s are constant. Unlike voltage series feedback, current series feedback increases the amplifier output impedance because R_s is in series with the output terminals of the amplifying device.

An FET has the following values: g_m = 2.4 millimhos and r_d = 63.3 kilohms. Use this device in the amplifier of figure 11-16 with R_d = 2 kilohms and R_s = 100 ohms.

a. *Compute the gain with and without current series feedback, using the circuit as a transconductance amplifier.*
b. *Compute the voltage gain of the amplifier.*
c. *Compute the output resistance of the amplifier with and without current series feedback.* (R11-15)

Voltage Shunt Feedback — The Transresistance Amplifier

Voltage shunt feedback is shown in figure 11-8. A circuit using voltage shunt feedback is shown in figure 2-16. This circuit stabilizes the operating point of a transistor, as discussed in Chapter 2. Voltage shunt feedback is a typical feature of an amplifier which supplies an output voltage proportional to an input current. Such an amplifier is known as a *transresistance amplifier*. The gain without feedback is expressed by

$$A = \frac{V_o}{I_i}$$

Eq. 11.32

The output voltage is fed back to provide a change of input current. Figure 11-19 is an ac equivalent circuit of figure 2-16. The feedback signal is the current defined by

$$I_f = \frac{V_o - V_i}{R} \approx \frac{V_o}{R}$$

Eq. 11.33

This expression is obtained by applying Kirchhoff's Voltage Law around the outer loop defined by the arrow in figure 11-19. V_i is the same voltage as V_{be} and is small enough to be neglected. The feedback gain is:

$$F = \frac{I_f}{V_o} = \frac{V_o/R}{V_o} = \frac{1}{R}$$

Eq. 11.34

Equations 11.34 and 11.32 can be combined with Equation 11.4 to obtain an expression for the gain with feedback.

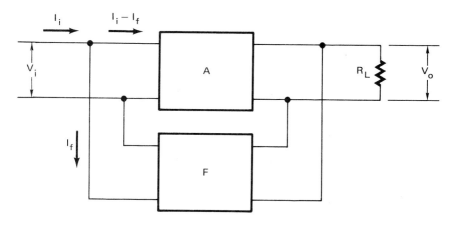

Fig. 11-18 Voltage shunt feedback

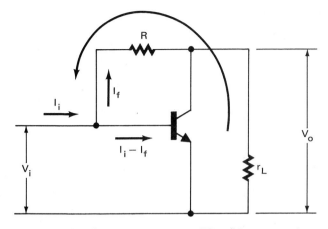

Fig. 11-19 Example of transresistance amplifier (bias connections omitted)

$$A_f = \frac{A}{1 + AF} = \frac{\dfrac{V_o}{I_i}}{1 + \dfrac{V_o}{I_i R}}$$

Eq. 11.35

This amplifier is a transresistance type if the load is light. This means that $A \gg 1$ and $A_f \approx R$. The gain of a transresistance amplifier, as defined by the ratio of the output voltage to the input current, depends only on the feedback resistor R. As a result, the gain cannot be affected by transistor nonlinearities and instabilities. One well-known amplifier using voltage shunt feedback is the *operational amplifier*, shown in figure 11-20. Ideally, such an amplifier has the

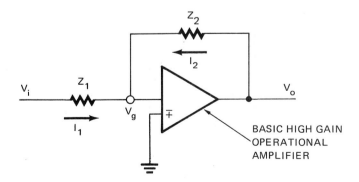

Fig. 11-20 Schematic diagram of an operational amplifier

following characteristics: (1) a voltage gain much greater than 1000, (2) a phase shift of 180 degrees between the input and output voltages, (3) infinite bandwidth for the signal frequencies, and (4) a zero output voltage when the input voltage is zero. This device is the basic building block of the analog computer, which was widely used before modern digital computers were available.

Theory and Applications of the Operational Amplifier

The operational amplifier has an input voltage that is so low with respect to the output voltage that it can be assumed to be zero. For example, if the output voltage of an amplifier having a gain of 1000, is 50 volts, then the input voltage must be $50/1000 = 0.05$ volt.

The operational amplifier also has a negligible amount of current flowing to ground since the input terminal is at ground potential as compared to the output terminal. To illustrate this condition, apply Kirchhoff's Current Law for the currents to ground at the input terminals of the operational amplifier. The input resistance of the amplifier is R_i. The currents to and out of the amplifier input terminal are determined as follows:

$$\frac{V_i - V_g}{Z_1} + \frac{V_o - V_g}{Z_2} = \frac{V_g}{R_i} \qquad \text{Eq. 11.36}$$

V_g is almost zero; therefore,

$$\frac{V_i}{Z_1} + \frac{V_o}{Z_2} = 0$$

OR

$$\frac{V_i}{Z_1} = -\frac{V_o}{Z_2} \qquad \text{Eq. 11.37}$$

As a result, the current through the feedback resistance is equal and opposite to the current through the series input resistor (R_1). Additionally, the voltage gain with feedback (A_f) can be determined from Equation 11.37.

$$A_f = \frac{V_o}{V_i} = -\frac{Z_2}{Z_1} \qquad \text{Eq. 11.38}$$

Because of its characteristics, the operational amplifier has many applications. If $Z_2 = Z_1$, the output voltage is equal to the input voltage, except for a sign reversal. In other words, the operational amplifier can be used as a phase inverter.

The operational amplifier can also be used as a scale factor changer. A scale factor changer is produced if Z_2/Z_1 is a fixed constant that is greater than or less than 1. Such a device can shift the phase of a sinusoidal signal without changing its magnitude. To do this, Z_2 or Z_1 is selected to be an impedance that differs in phase angle only.

The operational amplifier can also be used to add a number of signals. For example, consider the circuit shown in figure 11-21. Instantaneous values of voltage and current are used in the discussion that follows. The current is:

$$i_i = \frac{v_1}{R_1} + \frac{v_2}{R_2} + \frac{v_3}{R_3}$$

Eq. 11.39

Because $i_i \approx -i_f$ (from Equation 11.37),

$$v_o = -R_F i_i$$

Eq. 11.40

and

$$i_i = \frac{v_o}{R_F}$$

Eq. 11.41

Substitute Equation 11.41 into Equation 11.39 and solve for V_o.

$$V_o = -\left(\frac{R_F}{R_1}v_1 + \frac{R_F}{R_2}v_2 + \frac{R_F}{R_3}v_3\right)$$

Eq. 11.42

If $R_F = R_1 = R_2 = R_3$, then

$$V_o = -\frac{R_F}{R_1}(v_1 + v_2 + v_3) = -(v_1 + v_2 + v_3)$$

Eq. 11.43

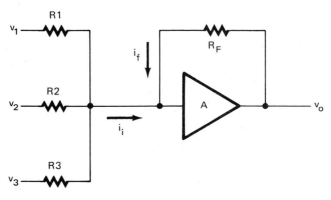

Fig. 11-21 Operational amplifier used to add three signals

In other words, the output voltage is equal to the sum of the input voltages.

Another use of the operational amplifier is to connect it to perform the mathematical operations of integration and differentiation. The operational amplifier integrates if the feedback resistor is replaced by a capacitor. The device differentiates if the input resistor is replaced with a capacitor and a resistor is connected in the feedback circuit.

Figure 11-21 shows a circuit in which an operational amplifier is connected. Let $R_F = R_1 = R_2 = R_3 = 10$ kilohms, $v_1 = 2$ volts, $v_2 = 3$ volts and $v_3 = 7$ volts. What is the output voltage, v_o?   *(R11-16)*

Using operational amplifiers, it is desired to subtract the voltage v_2 from the sum of v_1 and v_3. What changes must be made to figure 11-21 so that voltage v_o is the correct answer to the calculation?   *(R11-17)*

An operational amplifier, figure 11-20, has a voltage gain which depends only upon the externally connected resistors. Both Z_1 and Z_2 are 1-megohm resistors. What is the overall voltage gain of the amplifier?   *(R11-18)*

Figure 11-19 shows a transresistance amplifier that has the following values: $R = 40$ kilohms and $r_L = 4$ kilohms. Compute the transresistance gain with and without voltage shunt feedback. Assume the following values for the transistor: $h_{fe} = 50$, $h_{ie} = 1.1$ kilohms, and $h_{oe} = 0$.   *(R11-19)*

Current Shunt Feedback — The Current Amplifier

Current shunt feedback is shown in figure 11-22. An amplifier using current shunt feedback is shown in figure 11-23. The output current of this type of amplifier is directly proportional to the input current. This amplifier is known as a current amplifier and has a very low input resistance and a very high output resistance. It drives a low-resistance load and is driven by a high-resistance source. When Kirchhoff's Voltage Law is applied to the loop defined by the arrow in figure 11-23, the resulting expression is:

$$V_i - I_f R - V_e = 0 \qquad \text{Eq. 11.44}$$

In this case, $V_i \approx V_{be}$. This voltage is small and changes very little. Equation 11.44 can be solved for the feedback current.

$$I_f \approx \frac{-V_e}{R} \qquad \text{Eq. 11.45}$$

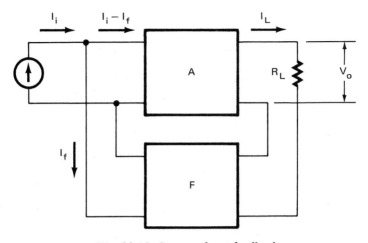

Fig. 11-22 Current shunt feedback

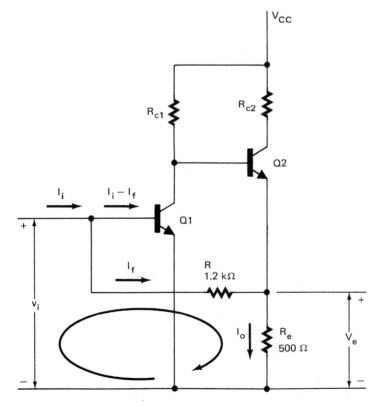

Fig. 11-23 Two-stage amplifier using current shunt feedback

I_i and the base current of Q2 are small compared to the Q2 collector current. Thus,

$$V_e \approx I_o R_e \qquad\qquad \text{Eq. 11.46}$$

F is the ratio of I_f/I_o and is negative because of the 180 degree phase shift between Q1 and Q2. Combine Equations 11.45 and 11.46 and solve for the feedback gain F.

$$F = -\frac{I_f}{I_o} = \frac{R_e}{R} \qquad\qquad \text{Eq. 11.47}$$

Now substitute Equation 11.47 into Equation 11.4. Assuming that A is very large, the gain is

$$A_f = \frac{A}{1 + AF} \approx \frac{1}{F} = \frac{R}{R_e} \qquad\qquad \text{Eq. 11.48}$$

To analyze this amplifier without feedback, first find the output current in terms of the input current with the current shunt feedback (R) disconnected. For this analysis, the appropriate ac equivalent circuit must be formed. Based on the theory presented in Chapter 4, the ac equivalent circuit is given in figure 11-24. First, observe that resistor R_e has the effect of a much higher resistance because both I_{c2} and I_{b2} flow through it. The voltage drop across R_e is $(I_{b2} + I_{c2})R_e$. But,

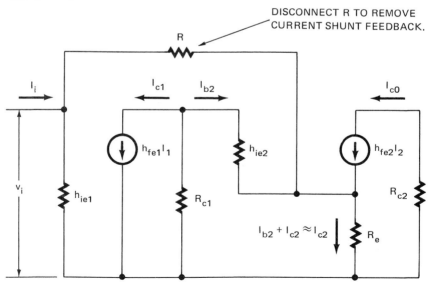

Fig. 11-24   Approximate equivalent circuit for amplifier of fig. 11-22, ac signals only

$$(I_{b2} + I_{c2})R_e = (I_{b2} + h_{fe2}I_{b2})R_e = I_{b2}(1 + h_{fe2})R_e$$

Eq. 11.49

Equation 11.49 is the voltage drop in terms of the current I_{b2}. Therefore, to the first stage, resistor R_e looks like $(1 + h_{fe2})R_e$. Thus, if R_e is 500 ohms and h_{fe2} is as low as 50, then the apparent resistance that R_e provides to the base current of Q2 is:

$$(50 + 1)500 = 25,000 \text{ ohms}$$

By the current divider theorem, I_{b2} can be defined in terms of I_{c1}.

$$I_{b2} = I_{c1}\left[\frac{R_{c1}}{R_{c1} + h_{ie} + (h_{fe2} + 1)R_e}\right]$$ Eq. 11.50

Since

$$I_{c1} = h_{fe1}I_{b1}$$ Eq. 11.51

and

$$I_{c2} = h_{fe2}I_{b2}$$ Eq. 11.52

I_{c2} can be defined in terms of I_{c1}. Equations 11.50, 11.51 and 11.52 are combined to yield:

$$I_{c2} = \frac{h_{fe2}h_{fe1}I_{b1}R_{c1}}{R_{c1} + h_{ie} + (h_{fe2} + 1)R_e}$$ Eq. 11.53

If the current I_{c2} is considered to be the load current and i_{b1} the input current, then the following expression gives the gain without current shunt feedback.

$$A_I = \frac{I_o}{I_i} = \frac{I_{c2}}{I_{b1}} = \frac{h_{fe2}h_{fe1}R_{c1}}{R_{c1} + h_{ie} + (h_{fe2} + 1)R_e}$$

Eq. 11.54

For most transistors, the value of h_{fe} is high enough to insure that the magnitude of A_I is high. Thus, the gain with current shunt feedback will behave according to the approximate form of Equation 11.48.

In current shunt feedback, as in the other feedback modes, the gain is stabilized by forcing it to depend almost entirely on fixed resistances. Current shunt feedback provides the advantages of most negative feedback situations. In addition, the input resistance is decreased and the output resistance is increased in a current amplifier using this type of feedback.

A current shunt feedback amplifier, like that of figure 11-23, has the following typical values: R_{c1} *= 3 kilohms,* R_{c2} *= 500 ohms, R = 1.2 kilohms, and* R_e *= 500 ohms. Both transistors have* h_{ie} *= 1.1 kilohms and* h_{fe} *= 50. It is assumed that* h_{oe} *is negligible. Compute the current gain of the amplifier with and without feedback.* *(R11-20)*

OSCILLATORS

Both the amplitude and phase angle of a transfer function vary with frequency. If AF becomes -1 for some level of signal frequency, then in theory, the gain of a feedback amplifier can become:

$$A_f = \frac{A}{1 - 1} = \infty$$

This is an unstable condition which can occur even with negative feedback, if at some frequencies there is a 180 degree phase shift in AF at a magnitude of one.

The simplest circuit for which instability is possible is the RLC circuit shown in figure 11-25. Recall that an inductance can store magnetic energy if current is passed through it. Also recall that a capacitance can store electrostatic energy if a voltage is applied across it. The circuit of figure 11-25 is arranged so that the energy stored in the capacitor is exchanged with the energy stored in the inductance. In other words, the energy stored in the inductance recharges the capacitor with energy, and vice versa. This energy exchange continues until the energy is expended as heat in the resistance. Assume that a sinusoidal voltage is applied to the circuit. The output voltage is determined by the total current times the parallel combination of L and C.

$$V_o = \frac{(I)(j\,\omega L)\left(-\dfrac{j}{\omega C}\right)}{j\,\omega L - \dfrac{j}{\omega C}}$$ Eq. 11.55

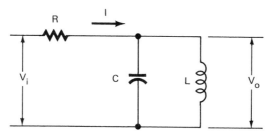

Fig. 11-25 RLC circuit

The input voltage is determined by the total current times the series parallel combination of R, L, and C.

$$V_i = I\left[R + \frac{(j\omega L)\left(-\dfrac{j}{\omega C}\right)}{j\omega L - \dfrac{j}{\omega C}}\right]$$

Eq. 11.56

The transfer function is obtained by dividing Equation 11.56 into Equation 11.55. An algebraic manipulation of the resulting expression yields:

$$\frac{V_o}{V_i} = \frac{1}{R\left(j\omega C - \dfrac{j}{\omega L}\right) + 1}$$

Eq. 11.57

A comparison of Equation 11.57 with Equations 11.4 and 11.16 shows that figure 11-25 is an example of an elementary feedback circuit. Figure 11-26 is a block diagram of this comparison. If the total equivalent resistance in the circuit is small, the frequency at which the maximum gain occurs is such that

$$\omega C = \frac{1}{\omega L}$$

Eq. 11.58

This frequency is:

$$f = \frac{1}{2\pi\sqrt{LC}} \quad \text{hertz}$$

Eq. 11.59

where

$$f = \frac{\omega}{2\pi}$$

Eq. 11.60

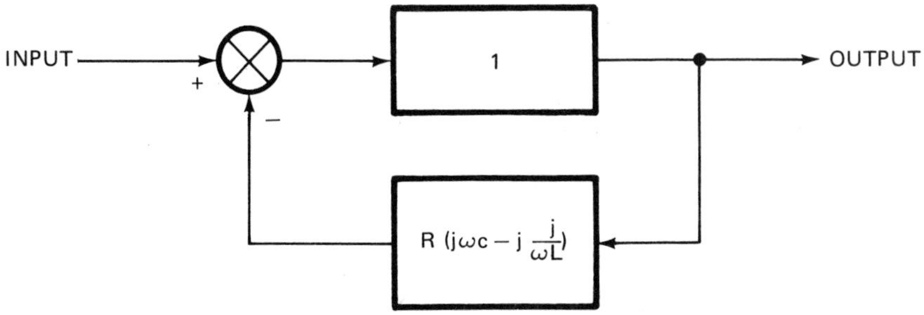

Fig. 11-26 Block diagram of feedback system represented by figure 11-25

The j-terms in Equation 11.57 indicate that the feedback term, $R(j \omega C - j/\omega L)$, contains imaginary numbers and adds vectorially to the real number, one (1). Even at resonance, V_o/V_i cannot exceed one. Figure 11-27 is a frequency response graph of the circuit. Note the peak at the resonant frequency.

The exchange of stored energy between the capacitance and the inductance is greatest when the input voltage is at the resonant frequency. Because there is some loss of energy through heating, it is physically impossible for the power gain to be greater than 1. For a power gain greater than 1, an additional power source would be required to supply the losses.

At the resonant frequency, the inductive reactance in figure 11-25 is equal to the capacitive reactance. Theoretically, what is the gain of the circuit at this frequency?   *(R11-21)*

What is the resonant frequency of the RLC circuit in figure 11-25 if C = 0.0001 microfarad, L = 0.3 millihenry and R = 1000 ohms?   *(R11-22)*

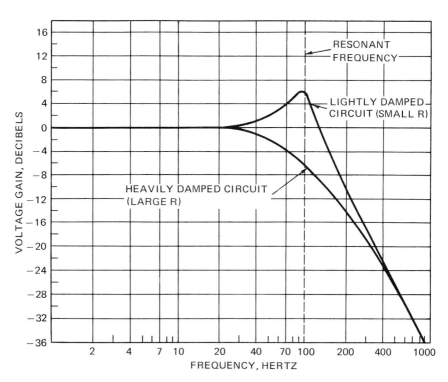

Fig. 11-27 Frequency response of parallel RLC circuit to a sinusoidal input

Another way of testing the circuit of figure 11-25 is to apply a dc voltage at the input terminals. The output is then observed as a function of time immediately after the dc voltage is applied. A typical output voltage response is shown in figure 11-28. This curve is an *exponentially damped sinusoid.* That is, it is approximately a sinusoidal waveform that decays in amplitude with a decay envelope that follows an exponential function. The frequency of the sinusoid is approximately the same as the frequency at which resonance occurs. Figure 11-28 shows that eventually there is no more transfer of energy between the inductance and the capacitance. After the dc voltage is applied for a long enough period of time, the pure inductance connected across the output terminals constitutes a shorted output and the output must be zero. It is only while electrical energy is oscillating between the inductance and the capacitance that there is an output voltage. If the energy is made to continue its oscillation, an

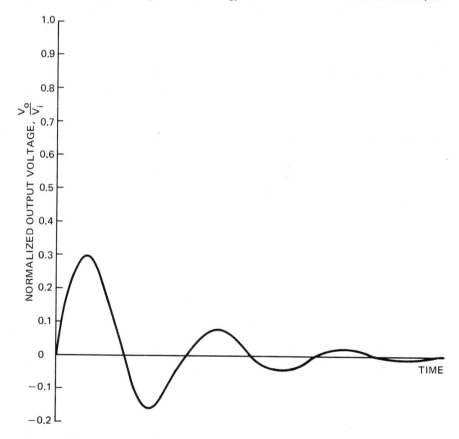

Fig. 11-28 Response of lightly damped parallel RLC circuit to dc step input voltage

alternating output voltage is produced. A circuit that behaves in this way is called an oscillator and has many applications in electronics.

Oscillator Circuits

An oscillator can be constructed by combining the circuit in figure 11-25 with a simple amplifier, such as those shown in figures 2-10, 6-21, or 7-22. The bias power for the amplifier supplies the energy to sustain the oscillation. The output of the RLC circuit is fed to the amplifier input, and the amplifier output signal is fed back to the RLC circuit to sustain the oscillation. Figure 11-29 is a typical resonant circuit oscillator. The inductance for this circuit is the combined leakage and magnetizing inductance of the coupling transformer. R_g and C_g represent a gate leak bias as described in Chapter 6. The ac voltage drop across the inductor is 180 degrees out of phase with the voltage across the transformer winding. The transformer primary winding is connected to the gate circuit of the FET. If the secondary winding is connected to introduce an additional phase shift of 180 degrees, then the total loop phase shift is $360°$ or zero. This means that positive feedback is produced. The absolute value of the gain without the transformer is $|\mu|$. The absolute value of μ is used because the sign of μ was already included in this discussion. The gain of the LC circuit is expressed by Equation 11.57. The gain of the transformer is the ratio of the secondary turns to the primary turns (n). The various gains are combined by the circuit so that:

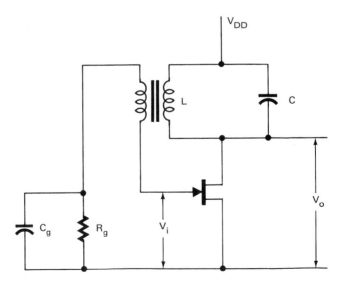

Fig. 11-29 Resonant circuit oscillator

$$V_o = nV_i[R\left(j\omega C - \frac{j}{\omega L}\right) + 1]$$

Eq. 11.61

V_o is also expressed by:

$$V_o = |\mu|V_i$$

Eq. 11.62

Equations 11.61 and 11.62 are combined and shown diagrammatically in figure 11-30. Oscillations begin in the RLC circuit, as a result of the application of a dc bias voltage. These oscillations are amplified and forced to continue by the feedback connection. Energy supplied by the bias power source (V_{DD}) makes up the losses in the RLC circuit. The R in the RLC circuit includes the dynamic drain resistance of the FET.

An oscillator can be made without an oscillating RLC circuit. The only requirement is a positive feedback situation that is limited by the nonlinearities of an amplifying device. One example of such a circuit is the phase shift oscillator in figure 11-31, page 476. This type of oscillator uses a voltage shunt feedback with three RC circuits. The three RC circuits produce sufficient phase shift to make the oscillations start and continue. An equivalent circuit for figure 11-31 is shown in figure 11-32, page 476. Each RC circuit (loops 1, 2, and 3) provides from 50 to 90 degrees of phase shift. Resistor R_i is a Thévenin equivalent resistance made up of R3 and R4. There is a phase shift of 180 degrees in any transistor voltage amplifier (see Chapter 3, pages 61 to 74). This phase shift coupled with the RC network phase shift produces a total phase shift of approximately 360 degrees. When the RC network output is fed back from B2 to B1, figure 11-32, the conditions for oscillation are satisfied. This situation is expressed by Equation 11.16:

$$A_f = \frac{A}{1 - AF}$$

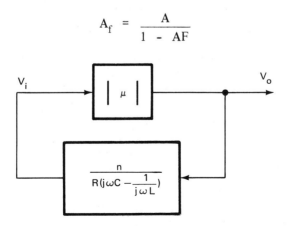

Fig. 11-30 Block diagram of resonant circuit oscillator function

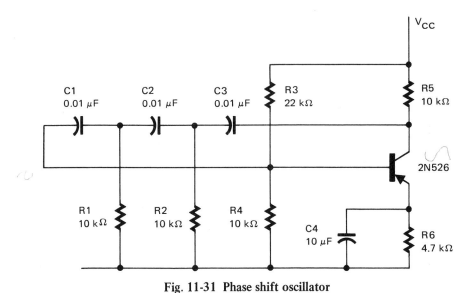

Fig. 11-31 Phase shift oscillator

Fig. 11-32 Equivalent circuit for phase shift oscillator

The value of AF is adjusted to unity by means of voltage dividers formed by resistors R_i and h_{ie}. If $A_f = \infty$ (as it is when $AF = 1$), an output voltage must exist even if there is no measurable applied signal voltage. This condition is called the *Barkhausen criterion.* For $R1 = R2 = R4$, the approximate frequency of oscillation is:

$$f = \frac{1}{2\pi R_1 C_1 \sqrt{6 + \dfrac{4R_5}{R_1}}} \qquad \text{Eq. 11.63}$$

Sustained oscillations occur when

$$h_{fe} > \frac{4R_5}{R_1} + 23 + \frac{29R_1}{R_5} \qquad \text{Eq. 11.64}$$

What is the approximate oscillation frequency of the circuit of figure 11-31? Assume that R_6 is shorted out by capacitor C4. (R11-23)

What is the minimum value of h_{fe} permitted if transistor 2N526 is to be used in the circuit of figure 11-31? Compare this value with the one listed in a transistor manual for the 2N526 device. (R11-24)

Oscillators can be constructed using vacuum triodes, pentodes, bipolar transistors, or FETs. For example, a *Hartley* oscillator is shown in figure 11-33, and a *Colpitts oscillator* is shown in figure 11-34, page 478. Both of these oscillators can be built using either tubes or transistors. Figure 11-35, page 478, shows the general form of these oscillators. This circuit is a feedback amplifier with impedances Z_1, Z_2, and Z_3 in a delta connection. Impedances Z_1, Z_2, and Z_3 are combinations of pure inductive and capacitive reactance. Oscillation occurs when the vectorial sum of Z_1, Z_2, and Z_3 is zero.

The feedback mode in figure 11-35 is (a) voltage shunt (b) voltage series (c) current shunt (d) current series (e) none of these. Choose the correct answer and explain the reason for selecting this answer. (R11-25)

The oscillators shown in figures 11-33 and 11-34 are useful in the range from low frequencies to radio frequencies. However, the analysis and design of

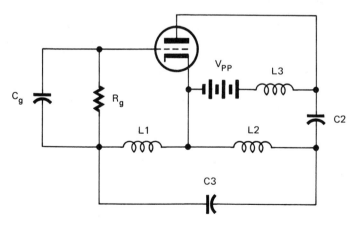

Fig. 11-33 Hartley oscillator (vacuum tube)

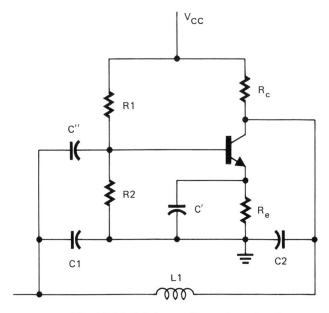

Fig. 11-34 Colpitts oscillator (transistor)

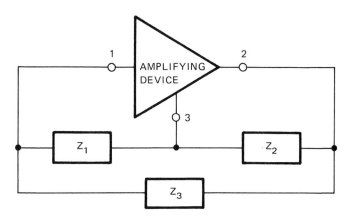

Fig. 11-35 Basic oscillator configuration

radio-frequency oscillators is complicated by inherent frequency dependent parameters in the amplifying devices and other components as well as in the wiring and component layout.

Crystal Oscillators

Oscillators designed for constant frequency applications above frequencies of several thousand hertz use piezoelectric crystals to produce the initial

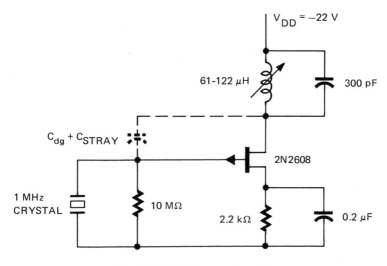

Fig. 11-36 FET crystal oscillator

oscillations. A piezoelectric crystal is a quartz crystal whose dimensions and mounting are adjusted so that if a voltage is applied across its surfaces, it begins to vibrate mechanically, like a spring. This mechanical vibration causes a change in the impedance of the crystal. As a result, the crystal acts in a manner similar to that of the RLC circuit described previously. If the crystal is placed in an amplifier circuit, such as the one in figure 11-35, the circuit acts like an oscillator. In this circuit, the crystal is connected in place of Z_1. One type of crystal oscillator is shown in figure 11-36. This circuit has excellent frequency stability, because the crystal vibrates more if the frequency changes a small amount, and forces the circuit to follow the natural resonant frequency of the crystal.

FEEDBACK AMPLIFIER ANALYSIS

A high fidelity audio amplifier may begin oscillating spontaneously if the feedback is improperly specified or adjusted. Such a condition is considered to be unstable and undesirable. The phenomena is sometimes observed in loud-speaker amplifying systems when the speakers are installed too close to the microphone. The microphone picks up the signal from the speakers and feeds it back to the amplifiers, resulting in a loud squealing noise. This is the result of improper feedback. Improper feedback can be internal to the amplifier also. Various methods must be used to insure that the system will not produce sustained oscillations. Some of these methods are discussed in the following paragraphs.

The transfer function for an amplifier is determined either by theoretical derivation or experimentation. The amplifier then is examined without the feedback connected using a steady state sinusoidal analysis. This analysis determines if the amplifier will be unstable when feedback is applied. If the analysis is to be successful, the characteristics of the amplifying devices and the resistance, inductance, and capacitance of the associated components must remain constant as the frequency and amplitude of the signal are changed. There are a number of analytical methods that can be used. Two of the methods are considered in the remaining pages of this chapter: (a) the Nyquist method and (b) the Bode method. Both of these analyses require a knowledge of the open loop transfer function.

The open loop transfer function is the transfer function of the amplifier and its feedback elements with the feedback lead disconnected. This function is AF in Equation 11.4, as shown in figure 11-37. The feedback element output is divided by the system input signal to represent the open loop gain (AF). It must be determined if the denominator of the closed loop transfer function, 1+ AF, can ever become zero. This situation will occur when AF = -1. The Nyquist and Bode methods both use this as a criterion.

What is the open loop transfer function for the circuit in figure 11-25?
(R11-26)

What is the open loop transfer function for the circuit in figure 11-16?
Hint: the loop is opened by connecting the input signal directly to the FET so that $V_{gs} = V_i$.   *(R11-27)*

The fact that an amplifier is not unstable is not enough. The amplifier must not be close to instability because slight changes in the components due to temperature or other factors can cause the situation to worsen. The stability analysis should also determine the margin of stability that exists for the amplifier.

The open loop transfer function can be obtained in two ways: (1) by laboratory testing of the amplifier or (2) by derivation.

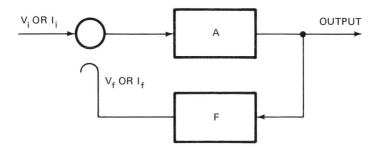

Fig. 11-37 The open loop transfer function

To test an amplifier in the laboratory, a low-amplitude sinusoidal input signal is applied and the feedback signal amplitude and phase relationship are measured. The phase relationship used is the shift in phase between the amplifier input signal and the feedback signal. The physical meaning of phase shift is that it takes time for a signal to pass from the input to the output. The output amplitude and phase are measured as the signal frequency is varied through the expected frequency range.

If equipment or the amplifier is not available, the transfer function can be derived from the known parameter values and characteristics of the amplifying devices, using their equivalent circuits. This derivation is based on a sinusoidal input and is similar to the derivations just presented for the sample circuits.

The typical open loop transfer function has the form,

$$AF = \frac{K(1 + j\omega T_a)(1 + j\omega T_b)(1 + j\omega T_c)\ldots\ldots \text{ and so forth}}{(1 + j\omega T_1)(1 + j\omega T_2)(1 + j\omega T_3)\ldots\ldots \text{ and so forth}}$$

Eq. 11.65

K is a parameter which is not proportional to frequency and is derived from the circuit constants. T_a, T_b, T_1, T_2, and so forth are time constants which are not proportional to frequency. When the frequency is at a specific measured value, AF is a complex number expressed as: (1) a + jb, or (2) $\sqrt{a^2 + b^2}$ at the phase angle, arctan b/a. (Note that a is the real part and b is the imaginary part of a complex number. To review the use of complex numbers, the reader is referred to the Appendix.) The Nyquist method of analysis uses form (1), a + jb. The Bode method uses form (2).

To use the Nyquist method, a table is made of the calculated or measured values of the open loop gain (AF) at various frequencies. These values are plotted on a set of coordinate axes known as the complex plane, as shown in figure 11-38.

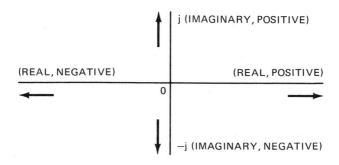

Fig. 11-38 Complex plane

Positive real values (+a's) are plotted to the right of the origin (0). Negative real values (-a's) are plotted to the left of (0). Positive imaginary values (+jb's) are plotted above the origin (0), and negative imaginary values (-jb's) are plotted below (0). The path of the curve described on this complex plane determines whether or not the amplifier will be stable when the feedback loop is connected.

If there are more terms $(1 + j\omega T_a, 1 + j\omega T_b,$ and so forth) in the numerator than there are in the denominator, the circuit is unstable. This means that a change must be made before the analysis can be continued. If there are any "negative" time constants in the denominator (so that $1 + j\omega T_1$ becomes $1 - j\omega T_1$, for example), the circuit is unstable. Again, a change must be made before any further analysis is attempted. In general, the transfer function will have more frequency dependent terms in the denominator than in the numerator. These terms will all be positive. If this is not the case, the analysis should be checked for possible errors or a change must be made in the circuit.

The following procedure is used to determine the amplifier stability using the Nyquist method.

1. Establish adequate scales on a graph where the j-axis is the ordinate and the real axis is the abscissa. Generally, the scales can be determined using linear graph paper, after calculating the value of the open loop transfer function at $\omega = 0$ and several other selected values of ω.

2. Plot the real and imaginary values of the open loop gain on the graph for increasing levels of frequency.

3. Mark the coordinates on the graph: $-1, j0$.

4. If the plotted curve encircles the $-1 + j0$ point on the graph in a clockwise direction, the system is unstable. Otherwise, the system is stable. Figure 11-39 shows various stable and unstable system configurations.

As an example of the calculations involved in the Nyquist method, some points will be plotted for

$$AF = \frac{10}{j\omega(1 + j0.25\,\omega)(1 + j0.06\,\omega)}$$

This function is sketched in figure 11-39 (c).
At $\omega = 0$:

$$AF \approx \infty$$

At $\omega = 1000$ rad/sec.:

$$AF = \frac{10}{j1000(1 + j250)(1 + j60)} \approx 0$$

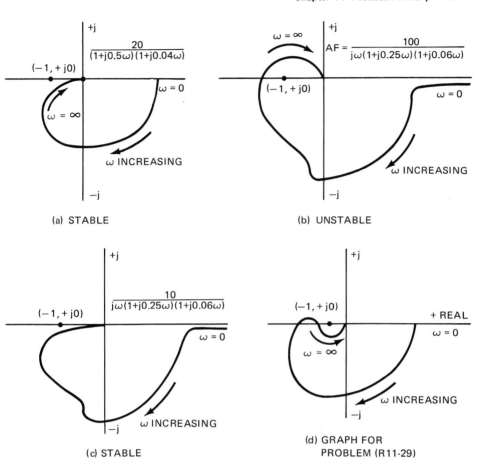

Fig. 11-39 Typical stable and unstable circuit configurations

At $\omega = 10$ rad/sec.

$$AF = \frac{10}{j10(1 + j2.5)(1 + j.6)}$$

$$= \frac{10}{10\ \underline{/90}\ \times\ 2.7\ \underline{/68.2}\ \times\ 1.165\ \underline{/31^\circ}}$$

$$= 0.318\ \underline{/-189}$$

$$= -0.314 + j0.051$$

At ω = 1 rad/sec.

$$AF = \frac{10}{j1(1 + j0.25)(1 + j0.06)}$$

$$= \frac{10}{1 \, \underline{/90} \times 1.03 \, \underline{/14} \times 1 \, \underline{/3.5}}$$

$$\approx 10 \, \underline{/-107.5} = -3 - j9.5$$

The points at ω = 0 and ω > 1000 rad/sec are indicated on figure 11-39 (c). The student is encouraged to check these calculations using the rules given in the Appendix.

Calculate the real and imaginary parts of the transfer function of figure 11-39(a) for various values of ω. Begin with ω = 0. Plot the values on linear graph paper and verify the Nyquist plot for figure 11-39(a). (Refer to the methods outlined in the Appendix for handling complex numbers.) *(R11-28)*

The graph for figure 11-39(d) indicates a (stable) (unstable) circuit configuration. *(R11-29)*

To determine the stability of an amplifying system, the Bode method relies on a mathematical relationship between the magnitude and the phase angle of the gain. The Bode method actually is the easiest and most often used procedure to determine if a feedback amplifier is stable. In fact, actual knowledge of the circuit transfer function is not necessary if a gain amplitude versus frequency curve is available. The Nyquist method has the disadvantage that both the phase shift and the amplitude must be measured at each frequency level. To use the Bode method, only the gain amplitude is needed. To measure the frequency response, a small amplitude sinusoidal signal is applied as the frequency is varied. The input and output signal amplitudes are measured simultaneously at each frequency level. Voltage or current signals may be used, depending on the type of amplifier that is being studied. The gain is expressed in decibels and is defined as:

$$\text{Decibels of Gain} = 20 \log \frac{\text{output signal}}{\text{input signal}} \qquad \text{Eq. 11.66}$$

The gain is plotted on semilog paper as a function of frequency in radians per second. The results are similar to the curve in figure 5-1. Using a linear approximation of the frequency response curve, the circuit transfer function and time constants can be determined, as well as circuit stability. Figure 11-40 shows a typical Bode plot for the transfer function of the simple circuit of figure 11-3. The dashed line shows the exact frequency response curve and the solid line is the linear or straight line approximation.

The linear approximation is made according to the following steps.

1. Each time constant causes a 20 decibel per decade slope. A decade means the frequency has changed by a factor of 10. A time constant in the numerator causes a positive slope and a time constant in the denominator causes a negative slope. Two time constants in the denominator can cause a negative slope of 40 decibels per decade when the frequency is high enough.

2. As the frequency is increased, the slope of the curve is constant until the frequency reaches $\omega = 1/T$, where T is one of the time constants in the

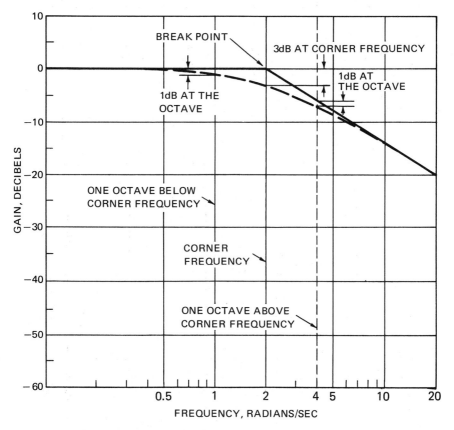

Fig. 11-40 Bode plot for a single time constant system

transfer function. At this frequency, the slope of the linear approximation changes abruptly by 20 decibels per decade because of the time constant (T). For each point at which the frequency causes $\omega T = 1$ (for each time constant in the transfer function), the slope changes by 20 decibels per decade. A point at which the slope changes is called a *breakpoint* or *corner frequency*.

3. At a breakpoint, the linear approximation is 3 decibels from the actual frequency response curve. If two time constants are approximately the same, the linear approximation may be as much as 6 decibels from the actual curve at the corner frequency.

4. At a frequency of one octave from a breakpoint, the linear approximation is one decibel from the actual curve. An *octave* is a frequency that is either one-half or twice the frequency at the breakpoint. There are two points near any breakpoint at which the linear approximation is one decibel away from the actual curve. If two time constants have values which are nearly the same, the combined effect of both of these time constants appears on the linear approximation. Therefore, for some amplifier transfer functions, there may be a point at which there is a difference of two decibels between the actual curve and the linear approximation at a frequency of one octave from the corner frequency.

Based on such criteria, a linear straight line approximation can be made of a frequency response curve. The transfer function can be derived graphically. Conversely, knowing the transfer function, a frequency response curve can be drawn by constructing a linear approximation based on the corner frequencies.

For example, consider a basic ac amplifier configuration having capacitor or transformer coupling to the load and to the signal source. There is a phase lead portion, such as at the low-frequency end of figure 5-1, where the gain is low because of the ac coupling. There is also a phase lag portion, such as at the high-frequency end of figure 5-1 or in figure 11-40, where the gain is again low. The phase lag portion is caused by various shunt capacitances in the active amplifying device together with capacitances across other components of the amplifier. In addition, a capacitor may be connected across the circuit to prevent unwanted high-frequency signals from being fed into the load. Figure 11-41 is an equivalent circuit for the high-frequency end of a Bode plot. R is a Thévenin equivalent resistance representing all resistances looking into the circuit from the output terminals. C is the total shunt capacitance across the output terminals. The triangle is an ideal amplifier with a gain, A. The transfer function is given by Equation 11.8, except that it is multiplied by A. It is assumed that A = 1, R = 2000 ohms, and C = 500 pF. The gain (in decibels) of the circuit of figure 11-41 at zero frequency is:

$$20 \log \frac{1}{1 \ + \ j0} \ = \ 20 \log 1 \ = \ 0 \text{ decibels}$$

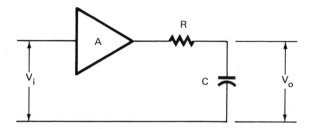

Fig. 11-41 An equivalent amplifier circuit for high-frequency performance

For the straight line approximation, the gain remains constant at 0 until the corner frequency is reached, as in figure 11-40. The corner frequency is reached when the j-term is 1 or when

$$\omega = \frac{1}{RC} \text{ rad/sec.}$$

Eq. 11.67

This value occurs at a frequency of 10^6 rad/sec. At all frequencies much greater than 10^6 rad/sec., the slope of the frequency response curve is 20 decibels per decade. Note where the curve is 3 decibels below the straight line approximation and 1 decibel below the straight line approximation.

The following three questions refer to the example just presented.

Show that the corner frequency occurs at 159 kilohertz using Equation 11.67. *(R11-30)*

At what frequencies does the straight line approximation differ by (a) 3 decibels and (b) 1 decibel? What is the relationship of these frequencies to the corner frequency? *(R11-31)*

What is the phase shift between the input and output at (a) 0 frequency, (b) infinite frequency, and (c) the corner frequency? *(R11-32)*

The amplifier example of figures 11-40 and 11-41 can never be unstable because the phase shift cannot become as much as 180 degrees. However, multistage amplifier circuits having one or more feedback loops (which may contain reactive elements), have a more complex transfer function. The Bode method can be used to examine the more complex circuits for stability.

Stability criteria using the Bode method are based on the straight line linear approximation of the frequency response curve. This curve is defined according to the following procedure.

1. Plot the straight line approximation of the open loop transfer function (AF) versus the frequency. (For an example, see the curve in figure 11-40 for the circuit of figure 11-41.)

2. In the region where the curve has a negative (downward) slope of 20 decibels per decade, the phase lag of the system approaches 90 degrees. A 40 decibel per decade slope indicates a phase lag approaching 180 degrees. Zero decibels indicates an open loop gain of unity (1). An open loop gain of unity or greater at 180 degrees of phase shift means instability. This condition indicates a positive feedback situation. Refer to Equation 11.16. Therefore, to insure amplifier stability, *the negative slope of the approximate frequency response curve must not be much greater than 20 decibels per decade at an open loop gain of 0 decibels.*

As an example, assume that a feedback amplifier has the following open loop transfer function:

$$\frac{316}{[j5(10)^{-6} \omega + 1] [j(10)^{-6} \omega + 1] [j(10)^{-7} \omega + 1]}$$

The frequency response in the high-frequency region is shown by dashed lines in figure 11-42. The straight line approximation is the solid line. Below 10^5 radians per second, the gain is constant at 50 decibels (gain ratio of 316). The energy storage elements in the circuit cause breakpoints at $2(10)^5$ radians per second, 10^6 radians per second, and 10^7 radians per second. The zero decibel level is crossed at $8(10^6)$ radians per second. According to the Bode criterion, this system is unstable. The amplifier can be made stable by (1) lowering the dc gain (lower than 316), (2) changing the design to add phase lead terms in the transfer function, or (3) modifying the time constants for the phase lag terms. These changes must insure that zero decibels is not crossed at a slope as great as 40 decibels per decade.

Verify by computation that the breakpoints in the previous example occur as specified in figure 11-42. Hint: Use Equation 11.67 for each time constant.   *(R11-33)*

Figure 11-42 shows the high-frequency response of an unstable amplifier. The slope is 40 decibels per decade at the frequency at which the zero decibel line is crossed. At the crossover frequency of $8(10)^6$ radians per second, the 5-microsecond and 1-microsecond time constants also affect the downward slope of the curve (breakpoints 1 and 2). The influence of the third time constant (0.1 microsecond) is also present at this frequency. At the corner frequency for the third time constant, the phase shift has passed 180 degrees and is approaching 270 degrees.

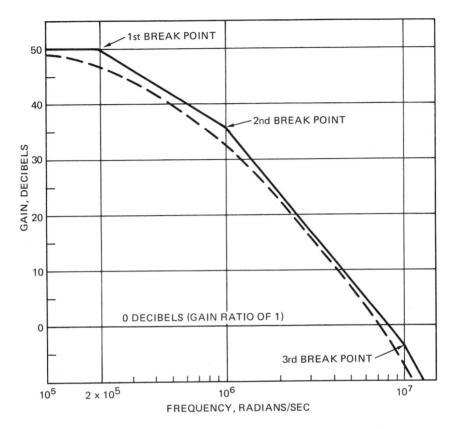

Fig. 11-42 Frequency response for a transfer function, showing linear approximation. The transfer function is

$$AF = \frac{316}{(1 + j\omega T_1)(1 + j\omega T_2)(1 + j\omega T_3)}$$

where

T_1 = 5 microseconds; T_2 = 1 microsecond; T_3 = 0.1 microsecond

Suggest a change in gain that will stabilize the circuit described previously. (Note: with the gain plotted in decibels, the curve can be moved vertically downward until it intersects the zero decibel line at a safe frequency.)(R11-34)

The Bode and Nyquist techniques are for linear amplifiers and are not directly applicable to nonlinear amplifying systems. Nonlinear analysis techniques are applied to nonlinear oscillator circuits and to amplifiers that are used near the limits of stable operation. However, if it is determined that an amplifier operates at a safe margin of stability, as determined by linear analysis, its performance under nonlinear conditions will usually be stable as well.

LABORATORY EXPERIMENTS

The following experiments were tested and found useful for illustrating the theory in this chapter.

LABORATORY EXPERIMENT 11-1

OBJECTIVE

To illustrate the difference in performance between a simple amplifier with negative feedback and the same amplifier with negative feedback removed.

EQUIPMENT AND MATERIALS REQUIRED

1	Field effect transistor, 2N4224
1	Transformer, TR-116 or TR-120
1	Resistor, about 500 kilohms, 1/4 watt, 10%
1	Resistance substitution box, Heathkit EU-28A or equivalent
1	Power supply, dc, 0-30 volts, 0.5 ampere, regulated and filtered
1	Oscilloscope, dual channel, time based, sensitivity to 5 millivolts per centimeter
1	Signal generator, capable of low-level, adjustable amplitude sinewave output between 10 hertz and 50 kilohertz
1	Capacitor, 10 microfarads, 50 volts

PROCEDURE

1. Construct a common drain amplifier as in figure 11-43, using a 2N4224 JFET. The connection diagram and the maximum allowable voltage and current values for the 2N4224 JFET are given in the description of Laboratory Experiment 6-3.

R_s serves as the drain bias resistor as well as the gate bias resistor. The value of this resistor is selected theoretically as the proper value to insure normal operation of the FET in a common drain circuit. However, it may be necessary to adjust the value for satisfactory operation in the experimental circuit.

The value of R_g is arbitrarily selected at approximately 500 kilohms. This resistance couples the bias voltage developed across R_s to the gate of the FET.

Operate the transformer selected as a step-down transformer for this circuit. The transformer coupling is used because of grounding problems which otherwise would occur in this experiment.

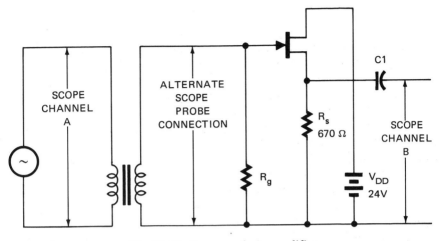

Fig. 11-43 Common drain amplifier

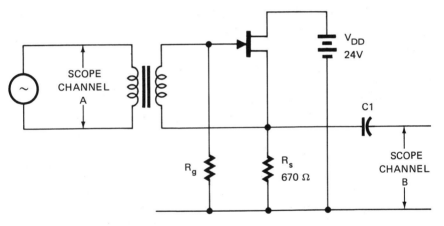

Fig. 11-44 Amplifier without negative feedback

2. Operate the amplifier for an input signal small enough to prevent distortion. Consider the transformer as a part of the amplifier and measure the input signal across the transformer primary. Measure the output voltage also and compute the voltage gain. This amplifier is operating with feedback.

3. Measure the voltage across the transformer secondary. Compute the turns ratio of the transformer. Note that there is no phase inversion between the input to the FET gate and the output.

4. Change the circuit to the form shown in figure 11-44.

5. Measure the input voltage across the transformer primary and the amplifier output voltage. Compute the voltage gain. This amplifier is operating without feedback.

6. Compare the gain obtained in step 2 with the gain obtained in step 5. Discuss the reason for the difference in gain based on feedback theory.

7. Using the turns ratio, compute the voltage gain of the amplifier from the transformer secondary to the amplifier output. Compute this gain for both circuit connections.

8. Experiment with the signal amplitudes applied to both circuits. Which circuit can tolerate the largest input signal without distortion? Comment on the reason for this behavior.

LABORATORY EXPERIMENT 11-2

OBJECTIVE

To construct and operate a phase shift oscillator using a bipolar transistor.

EQUIPMENT AND MATERIALS REQUIRED

1	Transistor, 2N404
2	Resistors, 4.7 kilohms, 1/2 watt, 10%
3	Capacitors, 0.047 microfarad, 50 volts
1	Capacitor, 10 microfarads, 50 volts
2	Resistance substitution boxes, Heathkit EU-28A or equivalent
1	Oscilloscope, dual channel, time based, sensitivity to 5 millivolts per centimeter
1	Power supply, dc, 0-30 volts, filtered and regulated, with both output terminals isolated from ground
1	Volt-ohm-milliammeter, preferably of the high-impedance, FET type

PROCEDURE

1. Construct a phase shift oscillator according to the schematic of figure 11-45.

Since transistor parameters vary so widely, it may be necessary to readjust R_c and R_b to obtain different biasing. Basically, the transistor must operate as an ordinary common emitter amplifier before it can be used as an oscillator. Thus, the bias levels must be established for the transistor. For this reason, resistance substitution boxes are used for R_c and R_b. A resistance substitution box is also used for R so that the oscillation frequency can be changed.

2. If the circuit is operating satisfactorily, it produces a sinusoidal waveform at the output terminals without requiring an input signal. The frequency of oscillation is between 250 hertz and 500 hertz. The peak-to-peak value

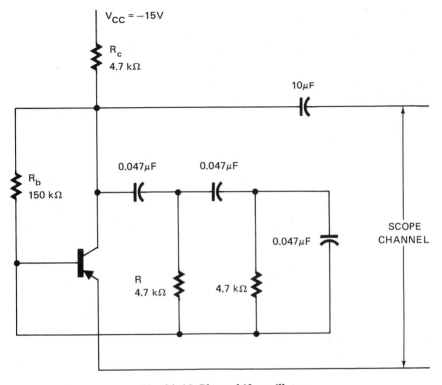

Fig. 11-45 Phase shift oscillator

extends almost to the level of V_{CC}. If the waveform is not sinusoidal, adjust R_c and/or R_b or V_{CC} until the waveform is sinusoidal. Record the new values for R_c and R_b.

3. With the oscilloscope, measure the frequency of the waveform and its peak-to-peak amplitude. Sketch the waveform shown on the scope and record the measurements.

4. The frequency of the oscillation can be changed by adjusting R. Vary R and determine and record the maximum variation in frequency.

5. At some setting of R or one of the capacitors in the network, the oscillations either cease or are not dependable. Experiment with the circuit values until this point is established. Record the changes in the values.

6. The frequency of oscillation can be verified approximately by Equation 11.63. Perform this calculation and compare the actual and theoretical frequency values.

LABORATORY EXPERIMENT 11-3
OBJECTIVE

To construct and operate a Colpitts oscillator using a bipolar transistor.

EQUIPMENT AND MATERIALS REQUIRED

1	Transistor, 2N404
1	Radio frequency coil, 10 millihenries
1	Capacitance substitution box, Heathkit IN-3147 or equivalent
1	Capacitor, 0.1 microfarad, 50 volts
3	Resistance substitution boxes, Heathkit EU-28A or equivalent
1	Power supply, 0-30 volts dc, 0.5 ampere, filtered and regulated
1	Oscilloscope, time based controls

PROCEDURE

1. Construct a Colpitts oscillator circuit using the schematic of figure 11-46. Since transistor parameters vary so widely, it may be necessary to readjust R_1, R_2, and R_e to obtain different biasing. Basically, the transistor must operate as a common collector amplifier and bias levels must be established for this connection.

2. If the circuit is operating properly, it produces a sinusoidal waveform at the output terminals without an input signal. The frequency of oscillation is 5 kilohertz and 10 kilohertz with a fairly large amplitude. If the waveform is not sinusoidal, adjust the bias resistors until it is sinusoidal. Record the new values.

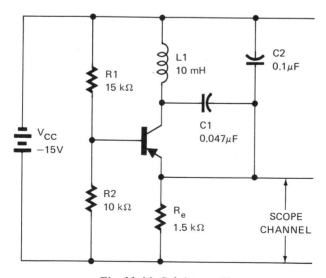

Fig. 11-46 Colpitts oscillator

3. Using the oscilloscope, measure the frequency of the waveform and its peak-to-peak amplitude. Sketch the waveform on the scope and record the measurements.

4. The frequency of the oscillation can be changed by adjusting the capacitance (C) to another value. Vary the capacitance and record the maximum variation in frequency that results.

5. At some setting of the parameters in the network, the oscillations either cease or are not dependable. Experiment with the circuit values until this behavior occurs. Record the changes that are made.

6. The frequency of oscillation can be verified approximately by Equation 11.57:

$$f = \frac{1}{2\pi\sqrt{LC}}$$

where L is the inductance and C is the capacitance in the circuit. Calculate the frequency using this equation and compare the calculated value with the frequency measured in step 3.

EXTENDED STUDY TOPICS

1. Explain the process of the transfer of electrical energy between the inductor and the capacitor in a RLC circuit where the capacitor, inductor, and resistor are in series. Assume that the input is sinusoidal and the output voltage is taken across the inductor.

2. List the characteristics of an amplifier having negative feedback, as compared to an amplifier having no feedback.

3. List the characteristics of an amplifier using positive feedback, as compared to an amplifier having no feedback.

4. What are the four feedback circuit connections that can be used with electronic amplifiers? What kinds of amplifiers use each type of feedback configuration? Sketch some circuits that are examples of these feedback configurations.

5. In figure 11-11, using the parameters specified in problem (R11-12), compute the input impedance of the amplifier with and without voltage series feedback. Neglect any effect due to the coupling capacitors and h_{oe}.

6. What is the basic difference between the Hartley and Colpitts oscillator circuits? What is the basic similarity of the two circuits?

7. Using library reference information, summarize the properties of a piezo-electric crystal that make it useful as a device for producing sustained oscillations.

8. List several ways in which the circuit represented by figure 11-39(b) can be stabilized.

9. The transfer function for an amplifying system has negative imaginary parts in the terms of the denominator. In addition, there are more terms in the numerator than in the denominator. What must be done before additional analysis is attemped? Explain the reason for this action.

10. Using the frequency response curve of figure 5-1 and its straight line approximation, determine the approximate transfer function for the circuit these curves represent.

12

Power Amplifiers
and Tuned
Amplifiers

OBJECTIVES

After studying this chapter, the student will be able to:

- calculate the power gain and evaluate the performance of power amplifiers
- list and describe the various classes of amplifiers with regard to their bias settings
- evaluate transformer coupled amplifiers
- describe amplitude modulation and frequency modulation
- describe the various types of tuned amplifiers
- evaluate the performance of tuned amplifiers

POWER AMPLIFICATION

The power gain of an amplifier is the ratio of the power delivered to the load to the power delivered to the amplifier input. An amplifier designed specifically for power amplification is known as a *power amplifier*. The power amplifier is usually the last stage of amplification in a servomechanism and is used to drive a servomotor. The power amplifier is the last stage of amplification in a sound system and is used to drive the voice coil.

A power amplifier may use either capacitors or transformers to provide for ac coupling. Transformers are used more often because they block dc quiescent currents and also allow a fixed load impedance to be matched to that of the amplifying device. The final load of a power amplification system, such as a motor, voice coil, or solenoid, is a very low-impedance device. This device must be carefully matched to the amplifying device so that it will operate properly.

The impedance matching properties of a transformer can be seen by studying the simple ideal transformer circuit of figure 12-1. V_1, I_1, and N_1 are the voltage, current, and turns of the primary winding, and I_2, V_2, and N_2 are the current, voltage, and turns of the secondary winding. The magnetomotive force of the source must be equal to the magnetomotive force of the load.

$$N_1 I_1 = N_2 I_2 \qquad \text{Eq. 12.1}$$

The rapidly changing magnetomotive force produces a changing flux in the magnetic core. By Lenz's law, a changing flux induces a voltage in both the primary and secondary windings. The flux change in the primary windings is nearly the same as the change in the secondary windings since the windings share a common magnetic core. Thus,

$$\frac{\Delta \phi_1}{\Delta t} \cong \frac{\Delta \phi_2}{\Delta t} \qquad \text{Eq. 12.2}$$

where $\phi_1 = \phi_2$ = flux in the magnetic core. From Lenz's law,

$$V_1 = N_1 \frac{\Delta \phi_1}{\Delta t}$$

$$\text{and} \quad V_2 = N_2 \frac{\Delta \phi_2}{\Delta t} \qquad \text{Eq. 12.3}$$

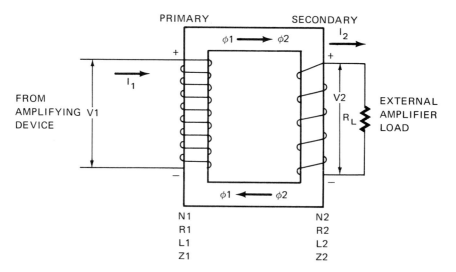

Fig. 12-1 Amplifier output transformer

Combine Equations 12.2 and 12.3 to obtain:

$$\frac{V_1}{N_1} = \frac{V_2}{N_2}$$

Eq. 12.4

For an ideal transformer, the volt-ampere input is the volt-ampere output. In equation form,

$$V_1 I_1 = V_2 I_2$$

Eq. 12.5

According to Ohm's law, Equation 12.5 can be changed to

$$(I_2)^2 Z_2 = (I_1)^2 Z_1$$

Eq. 12.6

where Z_1 and Z_2 are the primary and secondary impedances. Substitute Equation 12.1 into 12.6 to obtain an expression for the secondary impedance:

$$Z_2 = \left(\frac{N_2}{N_1}\right)^2 Z_1 = n^2 Z_1$$

Eq. 12.7

where n = the transformer turns ratio.

For power amplifiers, the load resistance generally is much smaller than the dynamic resistance of the amplifying device. If the dynamic resistance of the amplifying device is much larger than the load resistance, the output *voltage* is very low. As a result, there is a limitation on the power that can be delivered to the load. If the load resistance is much higher than the dynamic resistance of the amplifying device, the output *current* is very low. Again, the power that can be delivered to the load is limited. The load must receive adequate voltage *and* current to deliver power. The maximum power that can be delivered to the load occurs when the two resistances are equal. For this reason, a transformer primary winding used as a load for a power amplifier must be selected so that it presents a load impedance that is nearly equal to the dynamic impedance of the amplifier. The transformer must also have a turns ratio that allows the secondary winding impedance to be similarly matched to the load impedance connected to the secondary (reference Equation 12.7). The resistive component of the impedance (rather than the reactive component) is needed to transfer power. This means that the maximum power is delivered when the load *resistance* connected across the transformer secondary is equal to the dynamic *resistance* of the amplifying device and connected bias resistors.

Using Equation 12.7, the transformer turns ratio can be specified so that maximum power is transferred to the load. This occurs when

$$n = \sqrt{\frac{R_L}{R_L{}'}}$$

Eq. 12.8

where R_L is the actual load resistance and R_L' is the resistance needed to load the amplifying device so that maximum power is transferred.

A single-stage transistor amplifier having h_{oe} = ∞ and a collector bias resistance = 700 ohms, is used to supply a load of 50 ohms through transformer coupling. Select a transformer turns ratio to obtain maximum power transfer.
(R12-1)

A transformer is rated at 117 volts for the primary winding and 24 volts, 1 ampere for the secondary winding. What is the turns ratio of the transformer? If 85 volts are applied to the primary, what is the secondary voltage? What is the current in the primary winding when the transformer secondary is operated at the rated voltage and current? *(R12-2)*

The rate of change of magnetic flux determines the voltage across the secondary winding of a transformer. Explain why the voltage transformation ratio remains about the same regardless of frequency. *(R12-3)*

Equations 12.1 through 12.8 are based on a transformer having negligible power losses. However, the winding currents do result in I^2R losses. At low frequencies, the winding reactance decreases. A lower reactance causes the winding currents to be too high if the same voltage is applied at the lower frequency. Therefore, lower voltage ratings must be observed when lower signal frequencies are used. Not only is more heat generated, but Equations 12.1 through 12.8 no longer correctly describe the transformer performance because I^2R losses cannot be neglected.

Any of the amplifiers described in this text can be considered as power amplifiers. Signals applied to these amplifiers cause them to operate over the full range of their active region, from cutoff to saturation. Considerable power amplification results with some noticeable distortion. To examine the power effects in large signal operation, consider the power transistor shown in figure 12-2. The load line in figure 12-2 is selected for a V_{CC} of 15 volts and a load impedance of $V_{CC}/I_C = 15/3 = 5$ ohms. If the power amplifier is transformer coupled, as in figure 2-12, the dc load is much greater than the ac load. The transistor must dissipate more heat when there is no ac signal than when there is a signal input. This condition results because the dc impedance (resistance) of a transformer primary is so much smaller than its ac impedance. Frequently, the only collector bias resistance is the transformer primary coil.

A transistor having the characteristic curves shown in figure 12-2 is used in a power amplifier, figure 2-12. The Q-point is at I_B = 30 milliamperes and

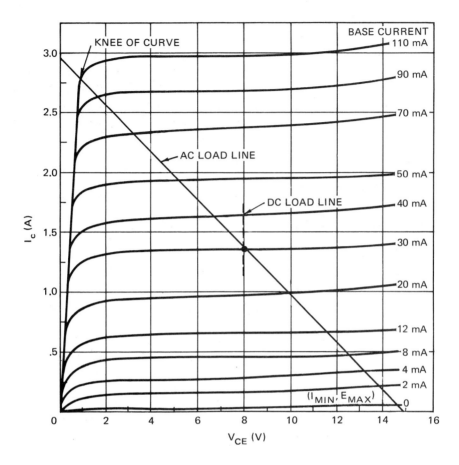

Fig. 12-2 Volt-ampere characteristic for a power transistor. Load lines are drawn for problems (R12-4) through (R12-9)

$V_{CE} = 8$ volts. The dc resistance of the transformer winding is so small that it can be considered zero. (In actual practice, this assumption is common.) Then, $V_{CC} = V_{CE}$, and the dc load line is vertical. What is the quiescent collector current? (R12-4)

Note: Review questions (R12-5) through (R12-9) refer to the amplifier described in review question (R12-4).

What is the average power input from the dc power supply ($R_L = \infty$)? (R12-5)

Since the transformer is not supplying dc power to the load, and there are negligible dc losses in it, the entire power input from the dc power supply is being dissipated in the transistor.

What is the power dissipated in the transistor when there is no ac input to the amplifier?   *(R12-6)*

An ac signal of 40 milliamperes peak-to-peak is applied to the transistor base at a frequency that is well within the frequency response of the transformer. The response of the transistor is along the ac load line shown in figure 12-2. A linear amplifying action is assumed. What is the peak-to-peak change in the collector current? What is the peak-to-peak change in V_{CE}?   *(R12-7)*

There is no ac voltage across the dc power supply terminals. In addition, the collector current is also the transformer primary current. Assuming a sinusoidal ac signal and a perfect transformer, what power is transmitted to the load?   *(R12-8)*

Power is transmitted to the load when the ac signal is applied. The dc power supply is still delivering the same average power. The total average power consumed in the circuit evidently is being shared between the transformer with its attached load and the transistor. What is the power now being dissipated in the transistor? Compare this value with the power computed in review question (R12-6).   *(R12-9)*

The assumption is made that the load line in figure 12-2 is the ac load line. For Class A amplifier operation, the quiescent point is located near the center of the load line. (The classes of amplifiers are described in the next section.) Power output with the least amount of signal distortion is achieved when the ac signal along the load line is as small as possible. As the signal amplitude is increased, the power output increases, but so does the distortion. The signal amplitude for a tolerable amount of distortion must be determined. This value fixes the limits of amplifier operation along the load line. For example, in figure 12-2, the minimum output signal is fixed by zero base current with I_C = 0.05 ampere and V_{CE} = 14.8 volts. The maximum level of output power may be fixed at 110 milliamperes of base current where I_C = 2.82 amperes and V_{CE} = 0.9 volt. In practice, these values are fixed by the maximum distortion allowed in the amplifier specifications. It must be pointed out that the frequency of the time varying signal must be within the operating frequency range of the output transformer. Otherwise, the operation is along a different and possibly steeper load line.

Maximum Power Capabilities of Power Amplifiers

A graphical analysis can be used to establish the fact that the small signal value of the collector resistance is equal to the load resistance when the ac load line is drawn through the center of the knee of the characteristic curve, as in figure 12-2. This "dynamic" collector resistance is equal to the load resistance for the maximum power delivered. The maximum output power of this circuit (and of any circuit) occurs when its load resistance is equal to its internal resistance at the load terminals. For figure 12-2, the maximum power output occurs at 5 ohms.

When making connections to transistor power amplifiers, the following criteria must be observed.

1. The maximum value of V_{CC} must be less than the maximum safe V_{CE} established by the manufacturer.

2. The maximum instantaneous collector current reached must be less than the maximum safe saturation collector current.

3. The maximum power dissipation must be less than the allowable values stated by the manufacturer.

Note that the maximum power dissipation occurs not for the maximum power output signal, but at the maximum expected quiescent power level plus the peak of the power output signal. This value can be quite a bit higher than the maximum power output.

Chapter 2 described the criterion that the Q-point must be located below $V_{CC}/2$. However, it may not be possible to meet this criterion for transformer coupled amplifiers because of the low dc resistance of the primary winding. Precautions must be taken to insure that any expected (or unexpected) power supply variations do not drive the quiescent point up until the maximum allowable power dissipation is exceeded for the device. (The dc load line for a transformer coupled amplifier can be almost vertical.) In addition, the signal frequencies must be high enough so that they are in the range of normal transformer operation. For sinusoidal signals, one rule of thumb is to rate the transistor at twice the rating of the power amplifier. That is, if the amplifier is to deliver 5 watts of a sinusoidal signal, then the transistor is rated at 10 watts of maximum power dissipation.

The maximum ratings of the transistor whose curves are shown in figure 12-2 are:

Collector-to-emitter voltage = 15 volts
Collector current = 3 amperes
Collector power dissipation = 10 watts

Is the operation of the amplifier discussed in the solutions of problems (R12-4) through (R12-9) within the safety limits of this transistor (figure 12-2)?
(R12-10)

The operating path of the transistor along the load line is selected so that it is tangent to the maximum power dissipation curve. An operating path above the maximum power dissipation curve results in the destruction of the transistor. An operating path below the curve results in a lower maximum power output.

The preceding criteria are also used for other active power amplifying devices having characteristic curves similar to those of the transistor. Such devices include pentodes, beam power tubes, and the various forms of field effect transistors that were described in previous chapters. The operation of transformer coupled amplifiers is similar regardless of the amplifying device used. An advantage of vacuum tubes is that they have a greater heat capacity. Thus, these devices do not need the same degree of protection against transient power surges as that required by semiconductors. In fact, to achieve economy and the greatest operating efficiency, the maximum power dissipation in vacuum tube circuits is somtimes allowed to exceed the maximum allowable levels momentarily during each cycle, as long as the *average* power dissipation does not exceed the maximum allowable levels.

The vacuum triode whose curves are shown in figure 7-9 is to be used in a transformer coupled amplifier that is similar to the one described in figure 2-12. The dc load line in figure 7-9 is assumed to be the ac load line of the triode for the transformer coupled circuit. What is the maximum value of the average power dissipation for the circuit? Is the tube being used safely in this amplifier?
(R12-11)

CLASSES OF AMPLIFIERS

Vacuum tube and semiconductor amplifiers are classified in terms of their bias settings and signal amplitudes. The bias setting determines the location of the Q-point on the characteristic curves of the amplifying device. It is assumed that a symmetrical sine wave is applied to the amplifier. The classifications are then explained in terms of the transmission of this sine wave.

The classes of amplifiers are defined as follows:

Class A. The Q-point is selected to be in the center of the active region of the characteristic curves. The load current flows throughout the full cycle of the signal. Distortion is a minimum. The linear amplifiers described in this text are largely examples of Class A amplifiers. Class A operation for a transistor amplifier is shown in figure 3-1 and the accompanying explanation. Figure 6-3 and the explanation of figure 6-21 show Class A operation for an FET or vacuum tube amplifier.

Class A_1. Same as Class A. However, for a vacuum tube or a JFET, the grid or gate current is not allowed to flow. A transistor is not allowed to saturate.

Class A₂. Same as Class A, but the grid or gate current is allowed to flow for a portion of the cycle. A definite division between A_1 and A_2 operation does not exist for the bipolar transistor because the signal does not change in polarity as saturation is approached.

Class B. The Q-point is set at cutoff. Load current flows only when a signal drives the amplifying device into its active region. The load current waveform looks like that of a half-wave rectifier. Figure 3-4 shows a transistor being operated as a Class B amplifier.

Class B₁. Same as Class B. Operation is not allowed in the saturation region.

Class B₂. Same as Class B, except that the amplifying device *is* allowed to operate in the saturation region. For a vacuum tube or JFET, the grid or gate current is allowed to flow.

Class AB. The Q-point and the maximum input signal are set so that for a part of the cycle (less than one-half cycle), the device is operating in the cutoff region. The Q-point is in the active region but just above cutoff. Figure 8-5 is an example of Class AB operation for an FET amplifier.

Class AB₁. Same as Class AB. Operation is not allowed in the saturation region.

Class AB₂. Same as Class AB, except that the amplifying device is allowed to operate in the saturation region for a portion of the cycle. In the case of the JFET or vacuum tube, the gate or grid current is allowed to flow for a portion of the cycle.

Class C. The Q-point is selected well into the cutoff region. As a result, the load current flows for less than one-half cycle. Figure 12-3, page 506, shows the input and output voltage waveforms for the FET of figure 8-8, with the Q-point adjusted to $V_{GS} = -10$ volts. The operation is Class C in figure 12-3 since the FET is sensitive to less than 1/2 cycle of the input wave.

Class C₁. Same as Class C. The amplifying device is not allowed to saturate.

Class C₂. Same as Class C, except that the input signal amplitude is such that the amplifying device operates in the saturation region for a small portion of the cycle.

Sketch the approximate shape of the output current waveform for a Class B FET amplifier.   *(R12-12)*

Sketch the approximate shape of the output voltage waveform for a Class AB FET amplifier.   *(R12-13)*

Figure 12-3 shows the operation for one-half cycle of the input signal during Class C operation. Sketch the approximate output waveform for the second half cycle. (R12-14)

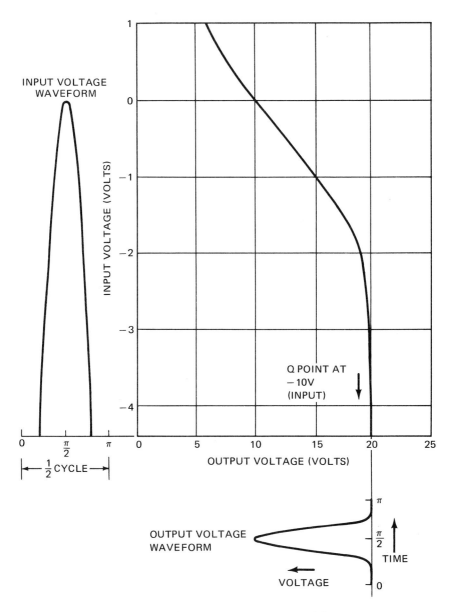

Fig. 12-3 Class C operation, FET amplifier

POWER AMPLIFIER SPECIFICATIONS

Two factors that are not so important for other types of amplifiers, are used to describe the performance of power amplifiers: (1) power conversion efficiency, and (2) power gain or power output.

Power gain is the output power divided by the input power. The power that is measured or calculated is the average ac signal power, not the dc power. In many cases, the output power is the only criteria specified because the input signal power is such a small amount compared to the total power transferred by the amplifier. Most of the power transferred comes from the dc bias supply. For this reason, power conversion efficiency is a more useful index of performance than the power gain.

The power conversion efficiency measures the ability of an amplifying device to convert the dc power of the bias supply into the ac power delivered to an external load. The conversion efficiency is defined as:

$$\text{Power conversion Efficiency} = \frac{\text{average ac signal power delivered to the load}}{\text{dc power supplied to the amplifier}} \qquad \text{Eq. 12.9}$$

The average power delivered to a load is always half the power received by the source if the circuit has been adjusted for the maximum average power delivered. Refer to the schematic shown in figure 12-4. In this circuit, R and V are the Thévenin equivalent resistance and voltage respectively of a power source. To deliver maximum power, $R = R_L$. This condition means that half the power is dissipated in the load and half is dissipated in the source. This is true for dc power and for ac power using rms values of sinusoidal voltage and current. For a nonsinusoidal signal, however, the situation is different because the average value is defined differently. The maximum possible conversion efficiency for a class A power amplifier is 50%, if it is designed for the maximum delivered power.

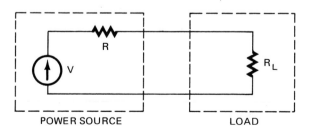

POWER SOURCE   LOAD

Fig. 12-4 Typical loaded power source

An ideal triode is operating along a load line as described in figure 7-9. The grid voltage is sinusoidal and varies between 0 and –10 volts. R_g is assumed to be infinity. (a) In figure 7-10, what is the power dissipated in resistor R_p? (b) What is the power gain?   *(R12-15)*

Assume that the vacuum tube and bias resistor load line described in figure 7-9 applies to the circuit of figure 7-21. What external load resistance is required to draw maximum power from the circuit? What is the approximate power conversion efficiency?   *(R12-16)*

PUSH-PULL AMPLIFIERS, CLASSES B AND AB

The class B and class AB amplifiers are particularly useful for power amplification. Examine the push-pull arrangement shown in figure 3-21 where the amplifying devices are biased for class B operation. The amplifying devices are biased at cutoff so that no current flows in their output circuits (either plate or collector) when there is no input signal. Conduction can take place only when the polarity of the input signal is such that the device turns on. In the case of the circuit shown in figure 3-21, there is no base bias ($I_B = 0$). For this condition, the transistor is at cutoff until it is turned on by the input signal. For a sinusoidal input signal, the collector current of either transistor has the form shown in figure 12-5. This form is the same as that of a half-wave rectifier. The average current in a half-wave rectifier is:

$$I = \frac{I_M}{\pi}$$

       Eq. 12.10

where I_M is the peak current value. The power supplied to the amplifier is expressed by voltage times current, or

$$P = \frac{2I_M V_{CC}}{\pi}$$

       Eq. 12.11

Equation 12.11 multiplies the average power supplied to one transistor by two because there are two transistors in the push-pull circuit. The output current

Fig. 12-5 Collector current waveform for one transistor in a push-pull amplifier with a sinusoidal input signal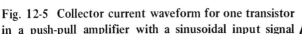

in a push-pull amplifier consists of the half-wave currents in each transistor combining to make a sine wave in the output. The power in a sinusoidally excited circuit is

$$P_o = IV \qquad \text{Eq. 12.12}$$

where I and V are the rms values of the current and voltage. When the rms values are converted to peak values, the power becomes:

$$P_o = \left(\frac{I_M}{\sqrt{2}}\right)\left(\frac{V_M}{\sqrt{2}}\right)$$

$$= \frac{I_M V_M}{2} \qquad \text{Eq. 12.13}$$

V_M is defined as follows:

$$V_M = V_{CC} - V_{Min} \qquad \text{Eq. 12.14}$$

V_{Min} is the lowest level that the output voltage reaches. For efficient use of the power supply, $V_{Min} \approx 0$. Thus, Equation 12.13 becomes:

$$P_o = \frac{I_M V_{CC}}{2} \qquad \text{Eq. 12.15}$$

Equations 12.15 and 12.11 represent the power values that are used to find the conversion efficiency as expressed by Equation 12.9. Substituting these equations into Equation 12.9, the conversion efficiency becomes:

$$\text{Conversion Efficiency} = \frac{P_o 100\%}{P}$$

$$= \frac{100\% \dfrac{I_M V_{CC}}{2}}{\dfrac{2 I_M V_{CC}}{\pi}} = 78\%$$

This conversion efficiency is for an ideal Class B push-pull amplifier. Note that this value is a decided improvement over the Class A power amplifier which has a conversion efficiency of only 50%.

With the collector bias set at cutoff, there is no power dissipation in the amplifier when there is no input signal. For battery operated amplifiers that have intermittent uses, this condition is an advantage because there is no drain on the battery when there is no signal. When the impedances are selected for the maximum transfer of power, the transistors in a Class B push-pull amplifier can deliver four times as much output power as one of the same transistors in a Class A amplifier, without exceeding the maximum dissipation in either transistor. The reason for this situation is that each transistor dissipates power only every other half cycle, and the total output power to the load is *twice* that received for each half cycle.

The output of a push-pull amplifier is sinusoidal for a sinusoidal input signal, because it is reconstructed from a pair of half cycles. Unfortunately, before a transistor can conduct, it requires a small forward bias voltage across the base-to-emitter junction. This requirement, combined with the natural variations between two transistors of the same type number, causes a form of distortion in the output signal. This distortion is known as *crossover distortion* and typically takes the form shown in figure 12-6. Note that the distortion is most evident at signal levels near zero, indicating that the greatest effect occurs near the cut-in point of each transistor. The distortion can be minimized by inserting a small positive bias in the base circuit of each transistor. As a result of this bias, a small standby current flows and keeps each transistor away from the cutoff region at small input signals. This is an example of the use of Class AB operation. The presence of the standby current flow in the transistors results in a loss of efficiency and a waste of power. This is the price paid for lower distortion.

To further minimize the distortion, pairs of matched transistors especially designed for Class B push-pull amplifiers are used. Satisfactory results are obtained if the transistor parameters are matched to within 10%.

Assume a push-pull amplifier, such as the one shown in figure 3-21, is operated with Class B biasing. The peak collector current is 1 ampere. What is the average collector current, based on a sinusoidal input signal?   (R12-17)

If the dc supply voltage of the amplifier of review question (R12-17) is 25 volts, what is the average power supplied to the amplifier?   (R12-18)

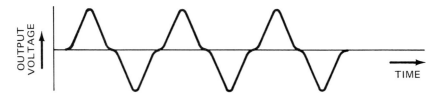

Fig. 12-6 Crossover distortion

The amplifier described in questions (R12-17) and (R12-18) is used. The ac output voltage at its lowest level is 0 volts, indicating the most efficient use of the dc power supply. What is the power output of the amplifier at this level?

(R12-19)

Stabilization against changes in transistor parameters due to temperature can be achieved by using an emitter resistor for each transistor. This arrangement provides negative feedback and is an example of current series feedback, as discussed in Chapter 11. The action of the circuit is that of the emitter self-bias circuit described in Chapter 2. Figure 12-7 shows how the circuit is connected. The resistors (R_e) usually have small values (< 10 ohms) because they lower the gain and conversion efficiency. The resistors also reduce the distortion in the output signal because of the effect of negative feedback (see Chapter 11). A capacitor cannot be used to bypass each emitter resistor as in other emitter self-bias circuits (figure 5-11, for example), because the direct voltages developed across the capacitors act to keep the transistors cut off in spite of the input signals. Temperature stabilization is further enhanced if the emitter resistors are of a temperature sensitive type.

Although the previous discussion was based on the use of transistors, any high-impedance amplifying devices (such as vacuum tubes or FETs) may

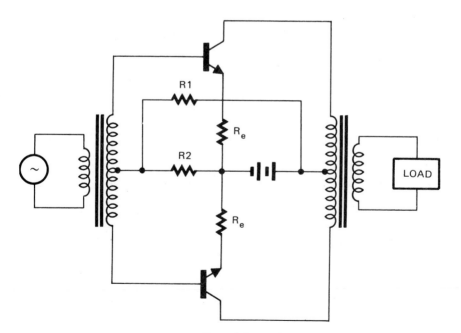

Fig. 12-7 **Push-pull amplifier with emitter resistors**

be used in push-pull amplifiers. The operation of the push-pull amplifier is the same regardless of the device used, including the necessity for eliminating the crossover distortion.

Other types of Class B push-pull circuits are shown in figures 12-8 and 12-9. The circuit of figure 12-8 may result in a lighter and less expensive unit because it does not require an output transformer. The circuit in figure 12-9 has no transformers and allows capacitive input coupling. The disadvantage

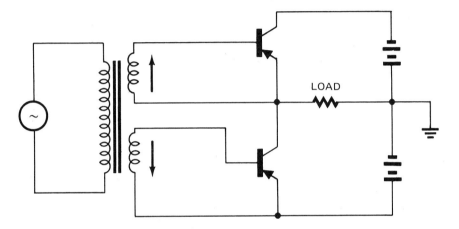

Fig. 12-8 Class B push-pull amplifier which does not need an output transformer

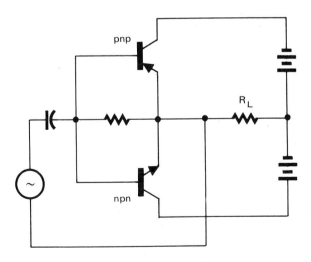

Fig. 12-9 Push-pull amplifier requiring no transformers; this amplifier allows capacitive input coupling

of the circuit in figure 12-9 is that matched complementary (pnp and npn) transistors are required. Pnp and npn transistors having the same basic behavior are available, but they must have matched characteristics to produce a high quality amplifier. Another disadvantage of this circuit is that two separate bias power supplies are required. However, it is a convenient and satisfactory arrangement for integrated circuit push-pull amplifiers.

Push-pull amplifiers can also be used with Class A operation. However, since both amplifying devices are always operating in the active region, the analysis of these circuits is difficult. In addition, the performance of such amplifiers usually does not justify the amount of analysis required.

CLASS C POWER AMPLIFIERS

When the signal frequency is constant, there is an advantage to Class C operation with regard to power amplification. To convert a Class A amplifier to a Class C amplifier, the only change to be made is the polarity of the input circuit bias. For an npn transistor, the base is negative. For a vacuum tube, the grid is made more negative until operation is entirely in the cutoff region, except for extreme positive values of the input signal. The output waveform of the Class C amplifier is less than one-half cycle and thus is decidedly nonsinusoidal. However, the amplifier output can be made very nearly sinusoidal if the plate or collector load resistance is replaced by a parallel RLC circuit which is resonant at the signal frequency. Such a circuit is shown in figure 12-10. The RLC circuit is known as a *tank circuit* because of its ability to store electrical energy temporarily. C2, L, and the equivalent resistance across the LC circuit (including $1/h_{oe}$ and the coil resistance) represent the tuned load. Capacitor C3 is added to shunt the ac around the power supply to reduce the losses due to the ac voltage drop across the power supply. C1 is the load coupling capacitor. As in any parallel RLC circuit, when $X_C = X_L$, the impedance across

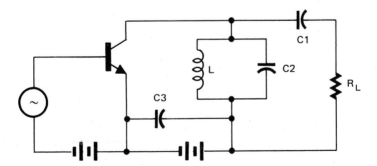

Fig. 12-10 Class C tuned amplifier (equivalent resistances of the circuit elements are omitted for simplicity)

the circuit has the greatest value and the largest output voltage is achieved. X_C = X_L when the signal frequency is

$$f = \frac{1}{2\pi\sqrt{LC}}$$ Eq. 12.16

where C is the equivalent shunt capacitance across the collector. At other frequencies, more of the total power supply voltage is dropped across the transistor collector-to-emitter terminal, and the output voltage is lower. As the frequency of the signal moves further from the resonant frequency, the voltage output decreases and less alternating current is transferred to the load. At a frequency of zero, the drop across L is zero. At infinite frequency, the drop across C_2 is zero. The collector current waveform for the transistor is similar to the output voltage waveform for the FET in figure 12-3. This waveform only recharges the RLC circuit and keeps it oscillating. The harmonics generated by this nonlinear amplification are greatly suppressed and very few of these frequencies are transferred to the load. The signal transferred to the load is nearly a perfect sinusoid. Imperfections in the signal exist because the harmonic components of the distorted wave are not completely eliminated. The distortion can be very severe, depending upon the amount of negative base bias present. Almost all of the power stored in the tank circuit during each cycle is delivered to the load (the equivalent resistance is kept low to minimize losses). At the same time, since the quiescent current is zero and current is drawn from the bias supply for only a short portion of the cycle, the conversion efficiency can exceed 90% under the most favorable conditions. For an ideal amplifying device with no losses, the collector dissipation is zero. For this condition, the period of time during which the collector current flows can approach zero. If the collector current is zero, there can be no losses in the amplifying device and the conversion efficiency is 100% for the ideal amplifier.

The preceding discussion of Class C transistor amplifiers is applicable directly to FET or vacuum tube amplifiers, using the appropriate analogous terms. In actual practice, vacuum tubes are often used in Class C power amplifiers and oscillators because of their advantages for high-power and high-frequency applications.

POWER MEASUREMENT AND CALCULATIONS

The measurement of sinusoidal voltages and currents can be made so as to produce an automatic measurement of the combined quantity, power. The same statement can be made for dc power as well. The device used to take the measurements is a wattmeter. However, to measure nonsinusoidal or non-dc power, it may be necessary to apply graphical and mathematical methods using the actual waveform measurements of voltage and current. It must be recognized that most meter movements, such as the common D'Arsonval meter, that are used for ac measurements are calibrated and designed for sinusoidal voltages and currents only. For any other waveforms, the meter must be recalibrated for the specific waveform being measured, or a special meter must be used that can measure true rms values for any waveform.

If the waveform deviates from a sinusoid by too great an amount, accurate power measurements and calculations cannot be made using the instrumentation that is available. For this situation, a choice must be made between two possible approaches.

1. The waveform can be broken down by Fourier analysis into its dc, fundamental, and harmonic components. Then the standard equations for computing dc and sinusoidal ac power can be used. The ac calculations are based on the rms values of the sinusoidal components that make up the Fourier series.

2. Instantaneous values of voltage and current throughout the waveform are measured or calculated using the amplifier input-output characteristics and the instantaneous values of power calculated at each point. The average power is then determined by integrating these values over one complete period of the waveform.

The first method is based on the theory that any periodic, continuous wave can be expressed in terms of sinewave components of different frequencies. The resulting expression is called a *Fourier series*. If the amplifier is linear, then the sinewave components that come out of the amplifier are the same as the ones that go into it. (This assumes that all frequencies are within the frequency range of the amplifier.) If the amplifier is nonlinear for the frequency and magnitude of the applied signal, the signal frequencies that exit from the circuit are different from those applied. If the input is a pure sinewave, the sinewave components of the output can be written in the form of a Fourier series. Then the power, voltages, and currents can be analyzed on the basis of an ordinary steady-state sinusoidal circuit analysis of each sinewave component. The results for each component are combined to describe the complete signal. If the input is nonsinusoidal, it must also be broken down into its components. The distortion is considered for each component as it passes through the amplifier. This type of analysis becomes exceedingly difficult and is not covered in detail in this text. If the input signal is sinusoidal and the nonlinear distortion is less than 10%, the output power can be computed with less than 1% error by the following procedure:

1. Determine by test or calculation a transfer curve expressing the relationship between the output current and the input signal, figure 12-11, page 516. The input signal, X, can be either voltage or current, but it must be a sinusoidal function of time.

$$X = X_M \cos \omega t \qquad\qquad \text{Eq. 12.17}$$

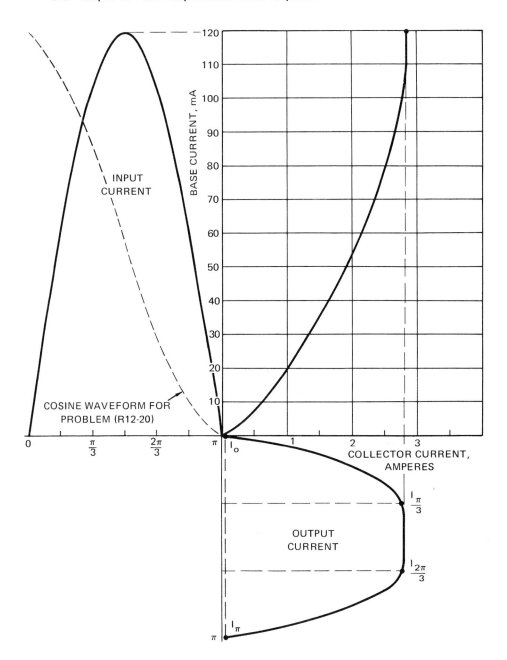

Fig. 12-11 Transfer curve expressing the relationship between the input and output current

where X_M is the maximum value.

2. The output power can be expressed approximately by:

$$P_o = \frac{B_1{}^2 R_L}{2}$$
Eq. 12.18

where R_L is the load resistance and B_1 is expressed as:

$$B_1 = \frac{1}{3}\left(I_o + I_{\pi/3} - I_{2\pi/3} - I_{\pi}\right)$$
Eq. 12.19

I_o, $I_{\pi/3}$, $I_{2\pi/3}$, and I_{π} are the values of the output current at $\omega t = 0$, $\pi/3$, $2\pi/3$, and π respectively. These values are read from straight lines intersecting with the transfer curve.

Using figure 12-11, find the values of the current at 0, $\pi/3$, $2\pi/3$, and π. Assuming R_L = 5.4 ohms, compute the approximate output power. (R12-20)

Graphical Method of Calculation

The second method of determining the power is graphical. It is based on determining instantaneous values of the waveform and graphically combining these values to yield the desired information. The method is applicable regardless of the distortion and for any function of the input signal with respect to time. The following procedure is used.

1. Determine the characteristic curves for the device or group of devices (such as an integrated circuit array or a pair of transistors connected in a push-pull arrangement). If the amplifier is a common emitter type with a single transistor, the published curves for that transistor can be used. If the curves are not available, or are thought to be inaccurate, they may be determined experimentally. A set of curves can be generated for a complete amplifier circuit, if necessary, using the amplifier as a three-terminal generalized amplifying device (refer to Chapter 1). The curves for the complete amplifier may be more helpful than those of a single device if the amplifier is a push-pull amplifier or one of the other multitransistor amplifiers described in this text. As an example, consider a simple common emitter amplifier which uses a power transistor having the characteristic curves shown in figure 12-2. The amplifier itself has a schematic like figure 2-12.

2. To define the operating path for the amplifier draw the ac load line on the curves (as in figure 12-2).

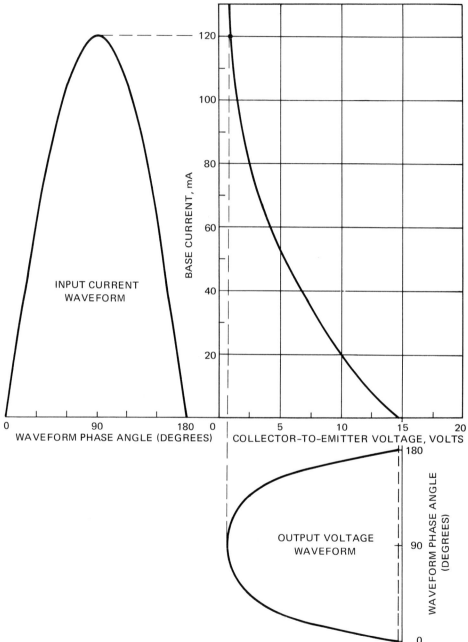

Fig. 12-12 Transfer curve relating the input current to the output voltage

3. Using the characteristic curves, draw a transfer curve relating the input signal to the output voltage. Then draw a curve relating the input signal to the output current. These curves are shown in figures 12-11 and 12-12 for the power transistor whose characteristics are shown in figure 12-2. The input signal is assumed to be a half sinusoid of the base current (Class B_2 operation). Note that this results in a collector current and collector-to-emitter voltage which are both nonsinusoidal due to saturation. This example is for an input signal that is like the output of an unfiltered and unregulated power supply with half-wave rectification. The example under discussion illustrates two points regarding this graphical method of calculating the power values.

    a. The input signal does not have to be sinusoidal for the method to work.

    b. When a large sinusoidally varying signal is applied (only one-half cycle, in this case), both the output current *and* voltage show distortion due to saturation.

This means that the ac output may not be measured or calculated accurately by the use of steady-state ac circuit analysis techniques. The point will be illustrated in the numerical calculations that follow.

4. All data is now available to calculate the output power of the amplifier. Instantaneous values of output current and output voltage are read from the output current and voltage waveforms. The power is calculated at those points in time. The waveform of the instantaneous power output can be drawn for the assumed input signal. Figure 12-13(a), page 520, shows the output current and voltage waves of figures 12-11 and 12-12 recopied so that the student can check the contours of the resultant power wave in figure 12-13(b), page 520.

5. From the graph paper, determine the scale factors for degrees per division and watts per division. Plot the power curve, as in figure 12-13(b). In figure 12-13(b) there are two degrees per vertical division and 0.2 watt per horizontal division.

6. Count the number of squares under the curve. In figure 12-13(b) there are 2723 small squares. (Each square is bordered by two vertical and two horizontal divisions.)

7. Calculate the average power using the following equation:

$$P_{ave} = \frac{\text{(No. of squares)(degrees per division)(watts per division)}}{\text{Total number of degrees over which calculations are made}}$$

Eq. 12.20

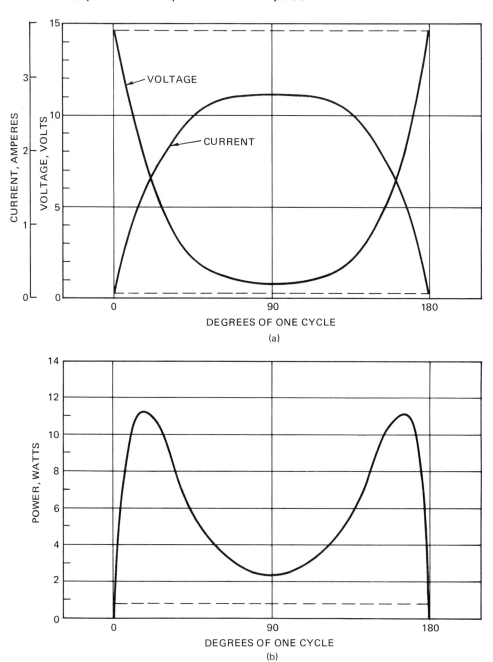

Fig. 12-13   Calculation of power for a nonsinusoidal voltage and current waveform

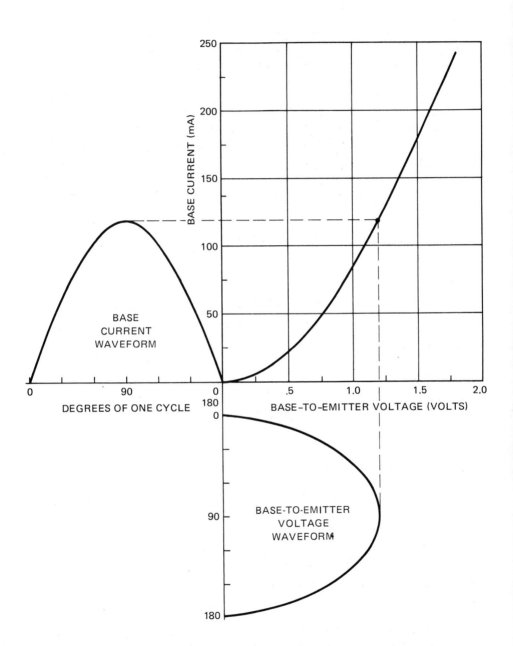

Fig. 12-14 Transfer curve for the input circuit of a transistor amplifier

As an example of the use of Equation 12.20, refer to figure 12-13(b). The average power over one-half cycle is:

$$\frac{(2723)(2)(0.2)}{180} = 6.05 \text{ watts}$$

It is difficult to compare this approach with another method of calculating the values for the following reasons:

1. a nonsinusoidal waveform is applied to the amplifier (the waveform of figure 12-12);

2. the distortion in the amplifier (as seen by figures 12-11 and 12-12); and

3. the 180 degree phase displacement between the output current and the output voltage.

For small signal Class A operation, using sinusoidal signals, it is quite easy to measure the power using rms values of the voltage and current on standard meter scales.

If the input current and voltage waveforms are examined to determine the input power, two methods of calculation can be compared. For the transistor shown by figure 12-2, the relationship between the base current and the base voltage is nonlinear, as it is for all transistors. This relationship is shown in figure 12-14. Figure 12-14 also shows the waveform for the base voltage that results when the transistor is driven with a sinusoidal base current. A graphical method of calculation is used to find the average input power that is similar to the method described for the output power. At simultaneous values of base current and base voltage, the input power is calculated. A waveform for the input power is then plotted as in figure 12-15.

Verify the curve in figure 12-15, for at least eight points along the curve. *(R12-21)*

The average input power can be determined from figure 12-15 where the scale factors are 0.002 watt per horizontal division and three degrees per vertical division. The number of squares under the curve are counted and Equation 12.20 is used. The average power over one-half cycle is:

$$\frac{(2363)(3)(0.002)}{180} = 0.079 \text{ watt}$$

Assume that the input characteristic curve for the transistor is linear. As a result, the base voltage waveform is the same as the base current waveform. The average power in this case is calculated in the same way as for a half-wave

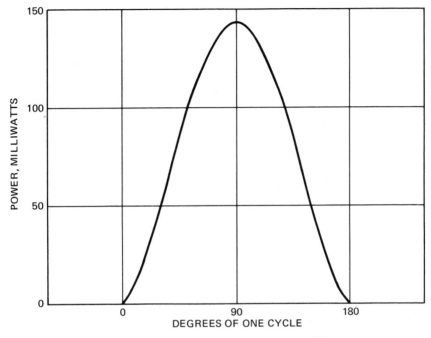

Fig. 12-15 Input power to transistor amplifier

rectifier.   The power output of an unfiltered, unregulated full-wave rectifier
is:

$$P_{ave} = \frac{I_M V_M}{2}$$ Eq. 12.21

Using Equation 12.21, the average power is:

$$\frac{(0.120)(1.2)}{2} = 0.072 \text{ watt}$$

The error that arises because a linear input characteristic is assumed is:

$$\frac{0.079 - 0.072}{0.079} = 0.0886 = 8.86\%$$

*Refer to figure 12-16, page 524.   Calculate the maximum power output
for an amplifier where these curves show its instantaneous output voltage and
current variations.*   (R12-22)

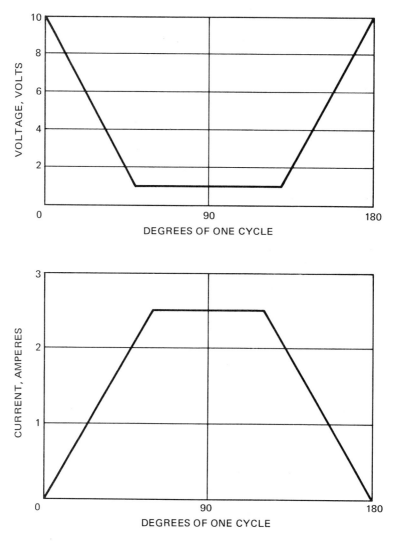

Fig. 12-16 Output voltage and current of a power amplifier over one-half cycle, for (R12-22)

TUNED AMPLIFIERS

Tuned amplifiers are amplifiers which are sensitive to only one frequency or a narrow band of frequencies. In other words, such amplifiers are "tuned" to a specific frequency or a specific range of frequencies. Tuned amplifiers are used to amplify signals used in high-frequency communication circuits. In the radio-frequency region, most amplifiers are sensitive to frequency because of the lead inductances and shunt capacitances in the amplifying devices and other components in the circuits. This disadvantage can be overcome by connecting additional inductances and capacitances to achieve a resonant condition at a certain signal frequency. The circuit then becomes almost entirely resistive at that frequency. The gain potential of most amplifying devices is severely reduced at these high frequencies. However, the use of these additional inductances and capacitances in the circuit means that greater amplification can be obtained than is otherwise possible.

A tuned amplifier takes a high-frequency input signal that serves as the *carrier* of the desired signal, and amplifies it, while attenuating signals of all other frequencies. Since the tuned amplifier is sensitive to only one frequency, it can discriminate and separate this one frequency from a group of frequencies. A high-frequency carrier signal is used because high-frequency electrical signals can be transmitted efficiently over great distances without the use of wires. These signals are received and amplified through high-frequency tuned amplifiers, which are important parts of radio, radar, and television communication circuits. The high-frequency signal is received through an antenna which is tuned to resonate at the carrier frequency. The antenna is a part of the input circuit of the first stage of a tuned amplifier.

There are three methods by which the radio frequency carrier signal can be modulated:

1. amplitude modulation
2. frequency modulation
3. phase modulation

Amplitude modulation, commonly abbreviated AM (as in "AM radio"), has a carrier frequency which is at a fixed level. The signal is transmitted over long distances and its amplitude is varied to carry any human intelligence, such as speech or music. The frequency of the amplitude variation is known as the *audio frequency.* In an AM radio, the signal is sent through a detector or demodulating circuit to remove the carrier frequency used for transmission. The remaining audio frequency is used to drive a speaker to produce audible sound. The audio frequency is much lower than the carrier frequency. Carrier frequencies for standard commercial AM broadcasting range between 550 kilohertz and 1600 kilohertz. Audio frequencies range from 30 hertz to 20 kilohertz. A plot of an amplitude modulated signal is shown in figure 12-17, page 526.

Frequency modulation, common abbreviated FM (as in "FM radio"), is accomplished by varying the frequency of the carrier wave about a specified reference level. The change in the frequency is very small when compared to the average frequency level. Tuned amplifiers are designed to amplify all frequencies within a bandwidth that includes the greatest expected variation in frequency. The reference frequency may be at the level of about 100 megahertz.

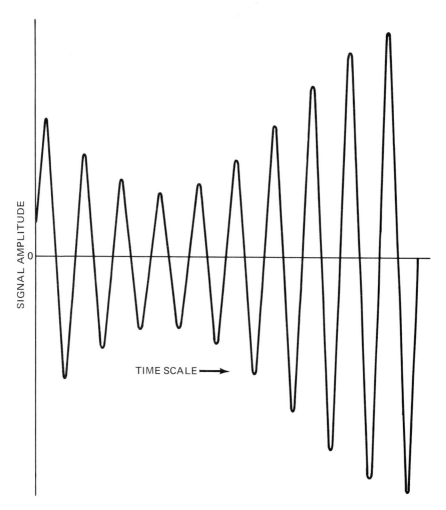

Fig. 12-17 Amplitude modulated wave

The high level of carrier or reference frequency allows the bandwidth for audio frequencies to be very small by comparison. The amplitude remains constant in frequency modulation. Any change in amplitude can be inhibited by the circuit design to reduce the effect of extraneous signals (noise or static) in the output. For this reason, FM is preferred over AM for obtaining pure, clean audio signals. A plot of a frequency modulated signal is shown in figure 12-18.

Phase modulation is a way to modulate a carrier signal used in the generation of frequency modulated waves. In this method of modulation, the phase of the carrier signal is changed or modulated. The modulation frequency is difficult to produce in FM circuits because it is so much lower than the carrier

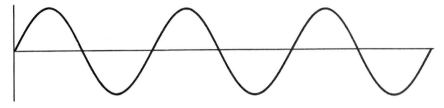

(a) WAVEFORM OF MODULATING SIGNAL

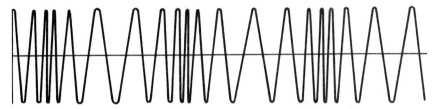

(b) THE CARRIER FREQUENCY, DECREASED AND INCREASED ACCORDING TO THE AMPLITUDE AND POLARITY OF MODULATING SIGNAL, PRODUCING FREQUENCY MODULATION.

Fig. 12-18 Frequency modulation

frequency. That is, the modulation frequency is so low that it falls within the normal variation in carrier frequency occurring as a result of changes in temperature and humidity. Phase modulation techniques allow these natural changes in carrier frequency to take place without affecting the reception or transmission of the audio frequency. The voltage-time function of an unmodulated carrier signal is:

$$v = V_M \cos \omega_c t \qquad \text{Eq. 12.22}$$

where V_M is the peak value, ω_c is the carrier frequency in radians per second, and t is the time in seconds. A phase modulated carrier is described by:

$$v = V_M \cos(\omega_c t + \theta_M \sin \omega_M t) \qquad \text{Eq. 12.23}$$

where θ_M is the maximum value of the phase angle and ω_M is the frequency at which the phase is modulated. Phase modulation is produced by first modulating the amplitude of the carrier frequency by the incoming audio signal. Then this amplitude modulated signal is added to an unmodulated carrier frequency signal which has been previously shifted in phase by a constant 90 degrees. The sum of these two signals is then applied to the tuned amplifier input. The output of the tuned amplifier is a phase modulated signal that is expressed by Equation 12.23. Recall that ω_M is the modulation or audio frequency. Equation

12.23 also expresses a frequency modulated signal, if the frequency modulating signal is sinusoidal.

The effect of modulation is to produce new frequencies which are known as side frequencies or *sidebands.* The tuned amplifier responds to the carrier frequency and to the sidebands produced by the modulating signal. For an amplitude modulated carrier, the voltage time function can be expressed as:

$$v = V_M \cos \omega_c t + V_{AM} \cos(\omega_c + \omega_M)t + V_{AM} \cos(\omega_c - \omega_M)t$$

Eq. 12.24

The first term of Equation 12.24 is the unmodulated waveform expressed by Equation 12.22. For an amplitude modulated waveform, there are two sidebands: the upper sideband as expressed by the second term of Equation 12.24 and the lower sideband as expressed by the third term.

If a carrier signal of an amplitude modulated wave is 100 kilohertz, and the modulating frequency is 500 hertz, what are the upper and lower sideband frequencies?   *(R12-23)*

Sideband frequencies also exist in the frequency modulated waveform. However, these sidebands cannot be easily defined because, theoretically, there are an infinite number. The equation describing a frequency modulated waveform is the same as that for the phase modulated waveform. This is due to the fact that phase modulation is used to generate frequency modulation. Equation 12.23 can be rewritten to express a frequency modulated waveform.

$$v = V_M \cos(\omega_c t + M_f \sin \omega_M t) \qquad \text{Eq. 12.25}$$

where M_f is still the maximum value of the phase angle. For this case, however, M_f is known as the *modulation index* of the frequency modulated wave. The smaller the value of M_f, the fewer sideband frequencies there are in the FM wave. If M_f is approximately one-half or less, there is only one pair of significant sideband frequencies.

Frequency Response of Tuned Amplifiers

In the untuned RC coupled amplifier, the frequency response is similar to the curve in figure 5-1. The indicated dropoff at the low and high frequencies is undesirable. The bandwidth for this type of amplifier should be as wide as possible. A transmitted AM radio signal is "demodulated" after it leaves the tuned amplifier. This process eliminates the carrier frequency and leaves only the modulating waveform. This waveform is the desired audio frequency signal. The demodulating circuit is similar to that of a half-wave rectifier with a filter.

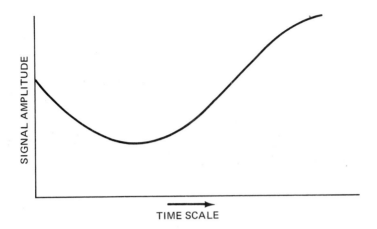

Fig. 12-19 Amplitude modulated waveform of figure 12-17 after detection (drawn to the same time scale)

The circuit is sometimes known as a *detector* since it "detects" the audio signal. The amplitude modulated signal of figure 12-17 is shown in figure 12-19 after it is demodulated. All frequencies contained in the demodulated waveform should be amplified with good fidelity. Therefore, an untuned amplifier follows the detector.

For selective transmission, reception, and amplification of radio signals, only the carrier and its sidebands must be amplified by the tuned amplifier. The sidebands must be included, but signals having a wider variation in frequencies are not allowed. The desired result is a frequency response similar to that of figure 5-1, but at radio frequencies and with a bandwidth that is a much smaller portion of the total frequency spectrum. The required bandwidth for an amplitude modulated wave is two times the highest audio frequency. The bandwidth required for an FM amplifier depends upon the value of the modulation index. The smaller the modulation index, the fewer sidebands there are and the narrower the bandwidth. Since phase modulation is essentially a form of frequency modulation, the amplifier in a phase modulation system has similar bandwidth requirements.

In radio broadcasting, explain why FM has less static or unwanted noise than AM.   *(R12-24)*

It is much more difficult to obtain a flat frequency response within the bandwidth for a tuned amplifier, as compared to an untuned amplifier. The frequency response for a simple single-tuned amplifier circuit takes the form

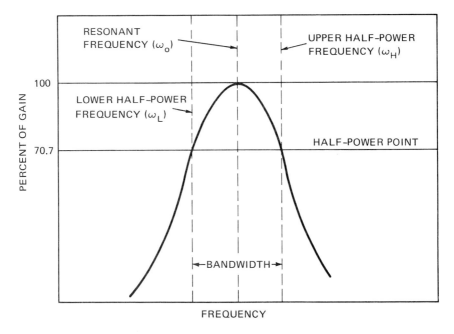

Fig. 12-20 Typical frequency response curve for a single-tuned amplifier

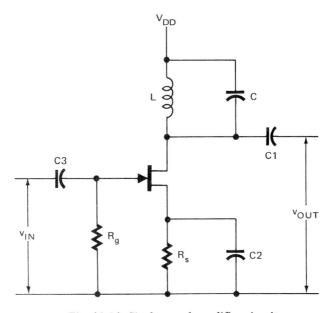

Fig. 12-21 Single-tuned amplifier circuit

shown in figure 12-20. This form is typical of the frequency response for the circuit of figure 12-21. This amplifier includes an RLC circuit that is resonant at the desired center frequency ω_0. The resistance R of the circuit is an equivalent resistance composed of all of the resistances in the circuit, including the coil resistance and the internal shunt resistances of the capacitors. The ac equivalent circuit for this tuned amplifier is shown in figure 12-22. R_C is the series resistance of the coil. C1, C2, and C3 are large capacitances that serve as shorts at the signal frequency. The output side of this equivalent circuit in a Norton equivalent with many parallel branches. The most convenient method of analyzing the circuit is to change R_c to its Thévenin equivalent, R_p. Thus, the ac equivalent circuit is changed to the circuit shown in figure 12-23.

Gain Calculations for Tuned Amplifiers

A typical inductance coil specification includes the inductance L and a value for Q which for this analysis is called Q_0. The equivalent parallel resistance of the coil (R_p) is computed from:

$$R_p = Q_0 \omega_0 L \qquad \text{Eq. 12.26}$$

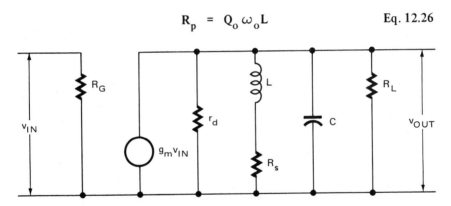

Fig. 12-22 Ac equivalent circuit for single-tuned amplifier of figure 12-21

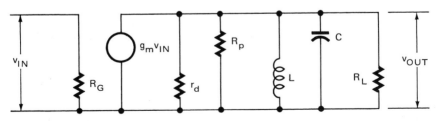

Fig. 12-23 Revised ac equivalent circuit for the tuned amplifier in figure 12-21

where ω_0 is the resonant frequency. The ac equivalent circuit can be simplified to the circuit shown in figure 12-24, where

$$R = \frac{1}{1/r_d + 1/R_p + 1/R_L}$$ Eq. 12.27

An expression can be written for the voltage gain of the amplifier by defining v_0, using an analysis of the circuit of figure 12-24, and dividing v_i into v_0.

$$v_0 = g_m v_i \left(\frac{1}{1/R + 1/j\omega L + j\omega c}\right)$$ Eq. 12.28

$$A = \frac{v_0}{v_i} = \frac{g_m}{1/R + j(\omega C - 1/\omega L)}$$ Eq. 12.29

In Equation 12.29, the denominator has a form that is characteristic of a single-tuned amplifier. There are two such terms in the denominator of the gain equation for the double-tuned amplifier.

The maximum gain occurs when resonance is reached, or when

$$\omega_0 C = \frac{1}{\omega_0 L}$$

Thus,

$$\omega_0 = \frac{1}{\sqrt{CL}}$$ Eq. 12.30

A plot of the gain A as a function of frequency yields a curve having the shape of figure 12-20.

A circuit Q (different from the coil Q, Q_0) can be defined.

$$Q = \frac{R}{\omega_0 L}$$ Eq. 12.31

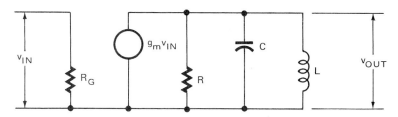

Fig. 12-24 Ac equivalent circuit of figure 12-21, revised and simplified

This Q can be used to define the bandwidth (B) in radians per second.

$$B = \frac{\omega_o}{Q}$$

Eq. 12.32

The bandwidth lies between the half-power points ω_L and ω_H. (See figure 12-20.) The amplifier gain at ω_o is:

$$A = g_m R$$

Eq. 12.33

At the half-power frequencies ω_L and ω_H, the voltage gain is:

$$A_v = \frac{g_m R}{\sqrt{2}}$$

Eq. 12.34

What is the percent reduction in the voltage gain of a tuned amplifier at the half-power points, from the voltage gain at the resonant frequency? (R12-25)

A single-tuned amplifier, similar to the one in figure 12-21, is to be used. This amplifier has a g_m of 2000 micomhos, an equivalent load resistance of 100 kilohms, a total equivalent capacitance of 60 picofarads, and an inductance of 420 microhenries. Compute the frequency at resonance and the voltage gain at that frequency. *(R12-26)*

Compute the circuit Q and the bandwidth of the amplifier described in question (R12-26). *(R12-27)*

There are a number of other configurations for single-tuned amplifiers. Figure 12-25, page 534, shows a common configuration that uses a bipolar transistor. The primary of the transformer (L_1) serves as the inductance coil L. The secondary inductance (L_2) is selected so that it is negligibly small in this transformer and the behavior is that of a single-tuned amplifier. The secondary can be coupled directly to the load or to the base of the next transistor stage. The advantage of transformer coupling is that the output impedance of the amplifier can be matched to the load or to the input of the following stage.

A simpler circuit can be obtained by tapping a single coil midway in its length and using it as an autotransformer (a single-winding transformer). This alternative is shown in the inset for figure 12-25. Using this method, the impedance transformation can then be accomplished at less cost.

The analysis of the tuned amplifier is the same as that for the untuned amplifier, if all signal frequencies are assumed to be within the bandwidth. This assumption is called the *narrow-band approximation.*

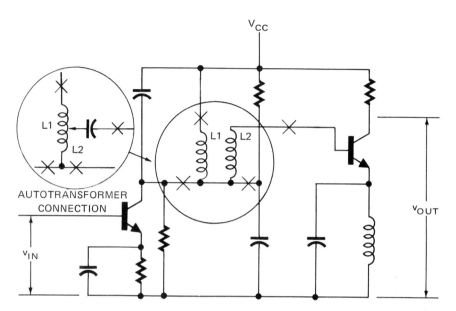

Fig. 12-25   Single-tuned amplifier using transformer coupling. Inset shows an alternate autotransformer connection

Under this condition, the circuit is at or near resonance and there is no significant net reactance in the circuit. The circuit is reduced to one which contains only resistances. Therefore, the circuit can be analyzed using the techniques presented in previous chapters of this text. If frequencies other than the resonant frequency are to be considered, the circuit can be studied by plotting the gain and phase shift versus frequency on the complex plane or Bode plot. The exact form of the gain equation is used in this case (Equation 12.29). It may be necessary to change the ac equivalent circuit depending upon the signal frequencies being considered. At low frequencies, coupling and bypass capacitors must be included. To develop equations to complete the analysis, the equivalent circuit used is the one that best represents the circuit at the expected signal frequency. (Use the information presented in Chapter 5.)

The circuit of figure 12-25 is also a two-stage amplifier. The overall gain is the gain of the first stage (adjusted to include the effect of the tuned circuit) times the gain of the second stage. The second stage is not affected by the tuning.

A double-(or higher order) tuned circuit is obtained when the secondary inductance of the transformer is too large to be considered negligible.

Under what conditions can the inductances and capacitances in a tuned amplifier be assumed negligible so that the analysis can be based entirely on the resistance in the circuit?   *(R12-28)*

Amplifiers with More Than One Tuned Circuit

An amplifier may have more than one tuned circuit. The resonant frequency (ω_o) of each tuned circuit may be the same or they may be different, depending upon the frequency response. The amplifier may consist of:

1. one stage of amplification, with two or more tuned circuits included in the stage; or

2. more than one stage of amplification with each stage having one or more tuned circuits and each circuit having its own resonant frequency.

For example, consider a second tuned circuit for the amplifier of figure 12-25. If the secondary inductance cannot be neglected, then the secondary must also be tuned by adding a capacitor across the secondary winding. Figure 12-26 shows a typical circuit configuration in which the tuned transformer couples the signal to the next stage. This circuit has tapped transformer windings so that it is easier to choose the series resistance in each of the tuned circuits.

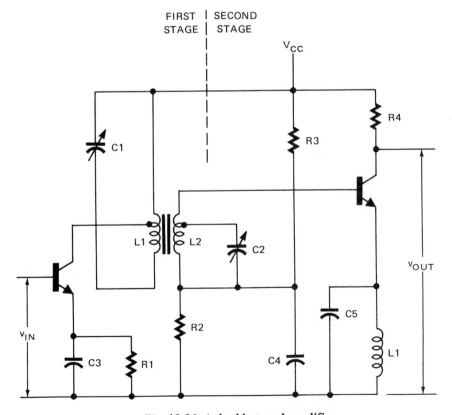

Fig. 12-26 A double-tuned amplifier

This circuit is called a double-tuned circuit. A single-stage version of the circuit of figure 12-26 has the transformer secondary feeding directly to a load, rather than to a second transistor. Since the load of the second stage may be frequency sensitive, it may also require tuning to a resonant frequency, resulting in *triple tuning.*

Analysis of More Complex Tuned Amplifiers

To examine more complex amplifier circuits and relate them to the single-tuned amplifier, Equation 12.29 is revised, using the definition of the j-operator.

$$A = \frac{g_m}{1/R + 1/j\omega L + j\omega C} \qquad \text{Eq. 12.35}$$

Multiply the numerator and denominator by $j\omega$.

$$A = \frac{g_m j\omega}{j\omega/R + 1/L - \omega^2 C} \qquad \text{Eq. 12.36}$$

If there are two stages of amplification (as in figure 12-26), the overall voltage gain is the gain of the first stage times the gain of the second stage. Using the subscripts 1 and 2 for the parameters of the first stage and second stage, the following expression for the voltage gain can be written.

$$A_V = \left[\frac{g_{m1}(j\omega)}{j\omega/R_1 + 1/L_1 - \omega^2 C_1}\right]\left[\frac{g_{m2}(j\omega)}{j\omega/R_2 + 1/L_2 - \omega^2 C_2}\right]$$

$$= \frac{g_{m1}g_{m2}(j\omega)^2}{(j\omega/R_2 + 1/L_2 - \omega^2 C_2)(j\omega/R_1 + 1/L_1 - \omega^2 C_1)}$$

$$\text{Eq. 12.37}$$

Equation 12.37 is the same whether the second tuned circuit follows or precedes the second-stage amplifying device. If both tank circuits are tuned to the same frequency, the terms "$1/L_2 - \omega^2 C_2$" and "$1/L_1 - \omega^2 C_1$" are zero at the same resonant frequency (ω_0). This condition is known as *synchronous tuning.* When each circuit has a different resonant frequency (usually only slightly different), the amplifier is said to have *staggered tuning.* Figure 12-27 shows the frequency response of an amplifier having two synchronously tuned tank circuits, with various degrees of stagger tuning. The half-power bandwidth is indicated on each curve by the intersections that each curve makes with the half-power gain level. The half-power gain level marks the gain level that is 3 dB down from the gain at 1000 radians per second. Synchronous tuning does not represent the ideal bandpass characteristic. By staggering the tuning of the

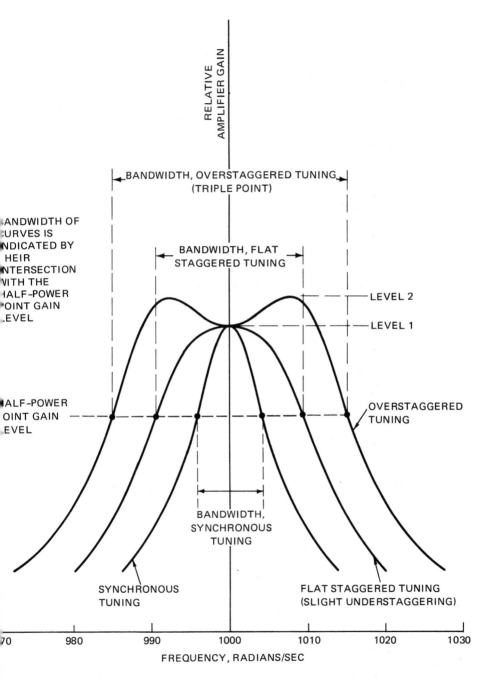

Fig. 12-27 **Frequency response characteristics of double-tuned amplifiers**

stages a wider and flatter response characteristic can be obtained in the region of the frequency of the desired signal. The ideal characteristics are flat staggered stages or slightly understaggered stages. By adding more stages and by flat staggering or slightly overstaggering some stages and understaggering other stages, amplitude characteristics such as those in figure 12-28 can be obtained. The tuning and circuit configuration that is known as the *Butterworth configuration* has a characteristic that is most nearly flat throughout the bandwidth and then drops off the most at the edges of the band.

Amplifier specifications include values for the center frequency ω_0 and the bandwidth B. If the tuning is overstaggered, or is to be overstaggered by the circuit designer, the specifications may also include the *peak-to-valley ratio*. This is the ratio of the gain of the amplifier at the resonant peaks to its minimum gain within its bandwidth. In figure 12-27, this ratio is level 2 over level

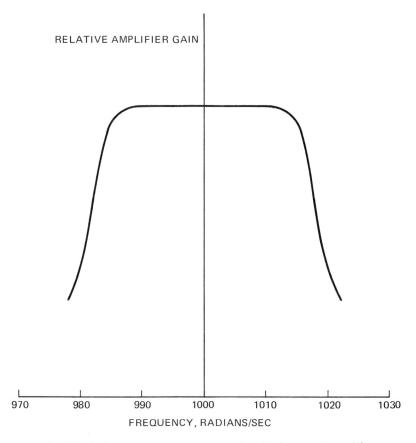

Fig. 12-28 Frequency response characteristic for tuned amplifier

1. There may be several half-power bandwidths for the overstaggered, double-tuned, or multistage amplifier. In this case, the triple-point bandwidth may be obtained. The *triple-point bandwidth* is the band of frequencies between the outside half-power points, as identified in figure 12-27.

Compute the half-power bandwidths for the three frequency response curves of figure 12-27.   *(R12-29)*

The above criteria may be applied either to the double-tuned, single-stage amplifier or an amplifier consisting of a number of single-tuned stages.

Figure 12-26 is a typical double-tuned amplifier. The high frequency ac equivalent circuit of the first stage of this amplifier is shown in figure 12-29. R_0 is the equivalent output resistance of the amplifying device and bias circuits. The numerical subscripts 1 and 2 (for example, 1 in R_1 and 2 in C_2) represent the capacitances, inductances, and resistances of the tuned circuits 1 and 2. Using the techniques developed in Chapters 4 and 5, the voltage gain can be derived. According to the discussion in Chapter 4, the no load voltage gain of two stages is the gain of one stage multiplied by the gain of the other stage. The resulting gain is expressed by Equation 12.37. An analysis of figure 12-29 shows that the voltage gain equation of this circuit has the same form as Equation 12.37. That is, there will be two quadratic terms in the denominator and a $(j\omega)^2$ term in the numerator. Similar results are obtained for capacitively coupled double-tuned amplifiers. The stagger tuning concept is the same for both types of circuits.

It is customary for both the primary and secondary tuned circuits to have the same inductive and capacitive reactances in these amplifiers. This condition results in ease of adjustment and analysis. Tapped coil and tapped capacitor arrangements are frequently used to match impedances.

In general, double-tuned circuits are more desirable than single-tuned circuits because the frequency response characteristics have steeper sides and it is easier to obtain a flat response within the bandwidth.

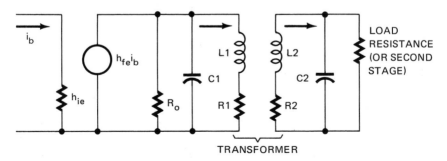

Fig. 12-29 Ac equivalent circuit of a double-tuned, single-stage amplifier

Controlling Stability in Tuned Amplifiers

Tuned amplifiers are particularly susceptible to the types of instability described in Chapter 11. Instability results because these amplifiers have relatively high gain produced in part by the resonant condition in the tank circuits, the coupling between stages, and inadvertent coupling due to the circuit design and construction. In addition, the signal frequencies are usually in the high or radio frequency region. Undesirable feedback is introduced by one or more of the following causes:

1. Improper layout or inadequate shielding. Unwanted capacitive and inductive coupling can be produced by leads that are unnecessarily long or in close proximity to sensitive signals. Coupling also occurs between components that are mounted too close to one another. Components must be mounted so that they will provide the least coupling with one another. Some components carry signals that contain undesirable harmonics of the signal frequency or signal frequencies unlike the frequency being amplified. Such components must be located for minimum coupling and must be shielded to prevent the transmission of undesirable signals. It is possible for the unwanted signals to be fed back into the input side of the amplifier where they are amplified to the extent that the amplifier becomes unstable.

2. Conductive coupling can occur. This is caused by improper grounding practices and wires that are connected so that they share power supply currents as well as signal currents.

3. Capacitive coupling within the amplifying devices. This form of coupling is due to the junction capacitance in semiconductor amplifying devices and interelectrode capacitance in vacuum tube amplifiers.

Instability in tuned amplifiers is treated the same way as it is in untuned amplifiers. The only difference is in the complexity of the circuits involved. Tuned amplifiers have more time constants and more complicated transfer functions. The technique called *neutralization* frequently is used to stabilize tuned amplifiers. With this technique, feedback signals are deliberately introduced to compensate and cancel out the feedback signals causing the instability. Such techniques are required for high-frequency amplifiers because the junction or interelectrode capacitances in the amplifying devices cannot be eliminated by careful design and construction precautions. The only way to compensate for these capacitances is to make circuit changes.

Is a two-stage tuned amplifier automatically a double-tuned circuit? Explain.   *(R12-30)*

List several special construction features that may be included in radio frequency amplifiers, to prevent the amplification or generation of undesirable feedback signals, or extraneous signals close to the signal frequency.   (R12-31)

LABORATORY EXPERIMENTS

The following experiments were tested and found useful for illustrating the theory in this chapter.

LABORATORY EXPERIMENT 12-1

OBJECTIVE

To build and operate a transformer coupled amplifier and to take measurements to determine its performance.

EQUIPMENT AND MATERIALS REQUIRED

1	Transistor, 2N1541
1	Resistor, 65 ohms, 10%, 7-watts minimum capacity
1	Transformer, AR504 or AR503
2	Capacitors, 10 microfarads
1	Resistance substitution box, Heathkit EU-28A or equivalent
1	Signal generator, sinusoidal waveform up to 20 kilohertz, 0 to 10 volts peak-to-peak
1	Power supply, dc, 0-40 volts, capable of producing 0.5 ampere of filtered, regulated current
1	Oscilloscope, dual channel, time-based, medium sensitivity requirements
1	Volt-ohm-milliammeter, high-resistance FET type, with 10-millivolt lowest scale, battery operated

PROCEDURE

1. Connect a transformer coupled amplifier, as in figure 12-30, page 542. The connection diagram for the transistor is given in figure 12-31, page 542.

2. Apply a dc bias voltage of about –20 volts and a sinusoidal input signal of about 2 kilohertz. Adjust R_b and the input signal magnitude to achieve a maximum voltage gain without distortion across the secondary winding of the transformer.

3. Use the oscilloscope to measure the voltage input (signal generator output) and voltage output (voltage across the full transformer secondary).

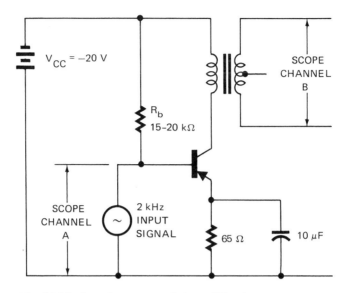

Fig. 12-30 Transformer coupled amplifier (Experiment 12-1)

LOOKING AT TERMINALS ON CASE

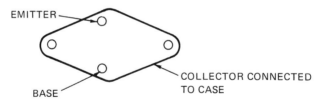

Fig. 12-31 Connections for 2N1541 transistor

4. Compute the voltage gain of the amplifier.

5. Measure the transformer primary and secondary voltages, using a volt-ohm-milliammeter. Measure the voltage across the full transformer secondary winding. Compute the voltage transformation ratio:

$$\text{Transformation Ratio} = \frac{\text{Secondary Voltage}}{\text{Primary Voltage}} \qquad \text{Eq. 12.38}$$

6. The circuit in this experiment is a transformer coupled amplifier. Transformer coupling frequently is used for the last stage of power amplifiers. The transformer also provides complete isolation of the input signal from the output signal. The transformer can also provide phase inversion. When used as a phase inverter, the center tap of the transformer is tied to

common and two voltages can be produced. These voltages are $180°$ out of phase with one another. Tie the center tap to common and measure the phase and magnitude of the two voltages as part of this experiment.

LABORATORY EXPERIMENT 12-2

OBJECTIVE

To construct and operate a transformer coupled push-pull amplifier

EQUIPMENT AND MATERIALS REQUIRED

2   Transistors, 2N1541
1   Transformer, AR504
1   Transformer, AR503
2   Power supplies, dc, 0-20 volts, capable of producing 0.5 ampere of filtered, regulated dc power
1   Signal generator, ac, capable of producing a sinusoidal waveform up to 20 kilohertz, at 1 volt peak-to-peak
1   Resistor, 65 ohms, 10%, rated at 7 to 10 watts
2   Resistance substitution boxes, Heathkit EU-28A or equivalent

PROCEDURE

1. Connect a circuit as in figure 12-32.

2. Apply a dc bias voltage of about −15 volts to both transistor collector terminals, as shown in figure 12-32.

3. Use a sinusoidal input signal of about 2 kilohertz. The magnitude is to be low enough that neither transistor is saturated, and yet high enough that crossover distortion does not affect the amplifier performance.

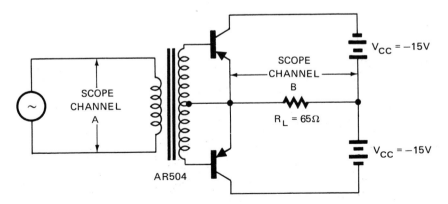

Fig. 12-32 Transformer coupled amplifier (Experiment 12-2)

4. Use the common emitter connection as ground and measure (a) the output voltage across the 65-ohm load resistance and (b) the input voltage across the primary winding of the transformer.

5. Calculate the voltage gain of this amplifier.

6. Sketch the output waveform. What are the reasons for any distortion present?

7. Change the circuit to the one shown in figure 12-33.

8. Using the volt-ohm-milliammeter, measure the voltage across the 10-ohm resistor in the transformer primary circuit and the voltage across the 65-ohm load resistance.

9. Compute the input current, the output current, and the current gain of the circuit.

10. Calculate the power gain of the circuit.

11. Change the circuit to that of figure 12-34.

12. Use one dc power supply voltage of -15 volts and apply a sinusoidal input signal of about 2 kilohertz. Adjust the magnitude of the input signal to provide a limited amount of distortion in the output signal.

13. Adjust R1 to minimize crossover distortion in the output. Does the adjustment of R2 have any effect?

14. Measure the voltage across the secondary winding of the AR503 transformer and the input voltage across the primary winding of the AR504 transformer.

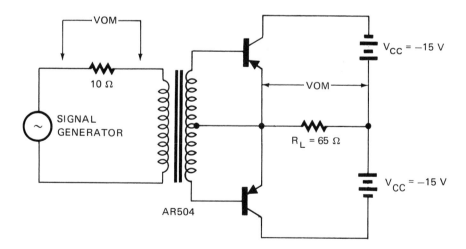

Fig. 12-33 Transformer coupled amplifier (Experiment 12-2)

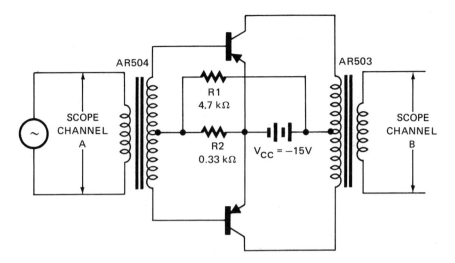

Fig. 12-34 Transformer coupled push-pull amplifier (Experiment 12-2)

15. Calculate the voltage gain of the amplifier.

16. Sketch the input and output voltage waveforms.

EXTENDED STUDY TOPICS

1. Write an equation for the power gain of an amplifier. Use the input and output voltage and current as known parameters. Define the terms used.

2. Equation 12.37 is the voltage gain equation for a two-stage, double-tuned amplifier. What change is made to the equation for a single-stage, double-tuned amplifier?

3. If a power amplifier is to deliver a 5-watt sinusoidal signal, what is the minimum collector power dissipation rating of the transistor in the final stage of amplification?

4. Compare the power conversion efficiency of a Class A amplifier with that of a Class B push-pull amplifier.

5. In what part of a stabilized push-pull amplifier is the use of capacitors undesirable? Why?

6. Compare the advantages and disadvantages between the types of push-pull amplifiers described in figures 3-21, 12-8, and 12-9.

7. How is the principle of negative feedback used to stabilize push-pull power amplifiers against temperature variation?

8. How is a Class A amplifier changed to a Class C amplifier?

9. Explain how a Class C amplifier with a tuned RLC tank circuit produces a sinewave output even though the actual collector current waveform is nonsinusoidal.

10. What are the number of sideband frequencies in a frequency modulated wave? What is the number if the modulation index is one-half or less?

11. How is a demodulating circuit similar to a rectifier with a filter?

12. Draw a block diagram of an AM radio communication circuit, using an antenna, an audio frequency amplifier, a radio frequency amplifier, a demodulator, and a bias power supply as the blocks.

13. What are some causes for undesirable feedback in tuned, radio frequency amplifiers?

Appendix

OBJECTIVES

This Appendix is provided as a reference that can be used as a review of the basic information given in more detail in elementary texts on electricity. This section is not meant to be a substitute for a good basic background in dc and ac fundamentals, or in the basic laws of science as taught in high schools. The Appendix is meant to serve only as a reminder of the principles and terminology that were studied previously and mastered.

LAWS FOR CIRCUIT ANALYSIS

The theory of electricity is often explained by analogy to the flow of water through a pipe where the flow is controlled by a valve or faucet. The water pressure from a municipal water system will force water out of the faucet into a wash basin if the faucet is opened. Closing the faucet causes resistance to the flow of water and less water is delivered. The control of electric current is achieved in an analogous way. The force provided by the system water pressure is analogous to the voltage in an electrical system. The water flow is analogous to the electric current. The faucet is like a variable resistor in an electric circuit. If resistance is added to an electric circuit, the voltage cannot drive as much current through the circuit. The relationship between voltage, current, and resistance is expressed by Ohm's Law:

$$R = \frac{V}{I} \qquad OR \qquad V = IR \qquad OR \qquad I = \frac{V}{R}$$

Eq. A.1

When any changes in the electric current are slow, then Ohm's law and the rules of algebra can be used to analyze any electric circuit. Trigonometry, calculus, and additional rules of physics are not required for the analysis. Any equations or analytical techniques that are used are the result of the manipulation of algebraic symbols or numbers. The actual derivations of the basic equations and proofs of the basic network theorems are covered in texts on basic electricity. The following sections give the necessary equations for network analysis.

Series Circuits

Resistances in series, figure A-1, add together as follows:

$$R_T = R1 + R2 + R3 \qquad \text{Eq. A.2}$$

where R_T is the total resistance of the circuit.

Parallel Circuits

Resistances in parallel, figure A-2, are combined according to the following expression:

$$\frac{1}{R_T} = \frac{1}{R1} + \frac{1}{R2} + \frac{1}{R3} \qquad \text{Eq. A.3}$$

where R_T is the total equivalent resistance in the circuit. For two resistances, Equation A.3 reduces to:

$$R_T = \frac{(R1)(R2)}{R1 + R2} \qquad \text{Eq. A.4}$$

R1   R2   R3

Fig. A-1 Resistances in series

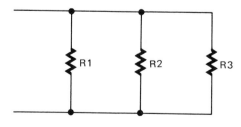

Fig. A-2 Resistances in parallel

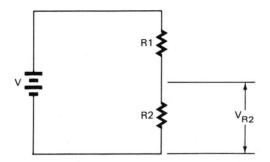

Fig. A-3 Circuit illustrating the voltage divider principle

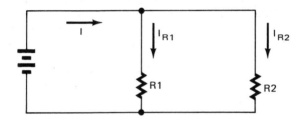

Fig. A-4 Circuit illustrating the current divider principle

Voltage Divider Principle

A shortcut method of computing the voltage across one resistor in a circuit is defined by the voltage divider principle. In figure A-3, the voltage across resistor R2 is:

$$V_{R2} = V\left(\frac{R2}{R1 + R2}\right) \qquad \text{Eq. A.5}$$

where R1 is one resistor or an equivalent resistance representing the other resistances in the circuit.

Current Divider Principle

A shortcut method of computing the current through a branch in a circuit is known as the current divider principle. In figure A-4, the current through resistor R2 is:

$$I_{R2} = I\left(\frac{R1}{R1 + R2}\right) \qquad \text{Eq. A.6}$$

where R1 is one resistor or an equivalent resistance representing the other resistances in the circuit.

Kirchhoff's Laws

The currents and voltages as determined by the current and voltage divider principles can also be determined using Equations A.1, A.2, and A.3. However, the use of the current and voltage divider principles saves time. In addition, these principles can easily be used to develop an equation describing a more complex circuit. This statement is also true for other circuit analysis techniques. As the circuit becomes more complex, the network analysis theorems become more useful. Kirchhoff's voltage and current laws are especially valuable. A review of these laws is appropriate at this point.

Kirchhoff's Voltage Law. Kirchhoff's Voltage Law states that the algebraic sum of the voltages around a circuit equals zero. For the circuit in figure A-5, Kirchhoff's Voltage Law yields:

$$V_1 + V_2 + V_3 - V_4 - V_5 = 0 \qquad \text{Eq. A.7}$$

The rule may be extended to circuits with more than one loop. For example, there are three loops in figure A-6. Kirchhoff's Voltage Law can be written for

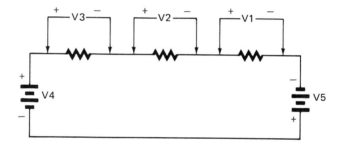

Fig. A-5 Circuit illustrating the use of Kirchhoff's Voltage Laws

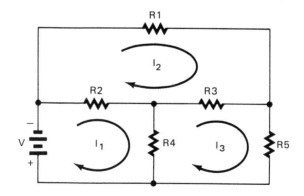

Fig. A-6 Circuit with three loops

each loop or for all of the loops. The directions of the currents in the circuit are chosen arbitrarily. It does not matter what current directions are assumed. However, it is important that the equations be consistent in using the assumed directions. The values of the current will be correct, regardless of the directions that are assumed. If the assumed directions are wrong, the calculated values have a negative sign. The presence of the negative sign then indicates what the true direction is. Write Kirchhoff's Voltage Law for the three loops in figure A-6.

$$-V = (R2 + R4)I_1 - (R2)I_2 - (R4)I_3 \qquad \text{Eq. A.8}$$

$$0 = -(R2)I_1 + (R1 + R2 + R3)I_2 - (R3)I_3 \qquad \text{Eq. A.9}$$

$$0 = -(R4)I_1 - (R3)I_2 + (R3 + R4 + R5)I_3 \qquad \text{Eq. A.10}$$

To solve for one current, only one equation is needed. To solve for unknown currents, Equations A.8, A.9, and A.10 may be solved simultaneously using well-known algebraic techniques. There must be an equation available for each unknown current. For example, in figure A-6, three unknown currents can be determined because there are three equations.

Kirchhoff's Current Law. Kirchhoff's Current Law states that the algebraic sum of the currents entering and leaving a junction must equal zero. Refer to figure A-4. According to Kirchhoff's Current Law, for either the upper or lower junction,

$$I_{R1} + I_{R2} - I = 0 \qquad \text{Eq. A.11}$$

The rule may be extended to circuits with more than two junctions. Kirchhoff's Current Law can be written for any of the junctions. To solve for only one unknown voltage, only one current equation is needed.

Thévenin's Theorem

Thévenin's Theorem states that any circuit composed of any number of resistances, voltage sources, and current sources can be represented by a simple series circuit consisting of a single voltage source and a single resistance. Figure A-7, page 552, is an example of a Thévenin equivalent circuit. The current flowing through a load will be the same whether the load is connected across terminals A and B in the complex circuit or it is connected across terminals A and B of the Thévenin equivalent circuit.

The Thévenin equivalent voltage is determined by measuring or calculating the voltage across terminals A and B when there is no load connected across terminals A and B.

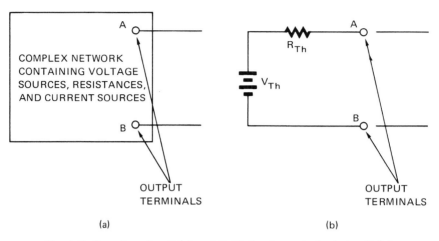

Fig. A-7 Complex circuit (a) and its Thévenin equivalent circuit (b)

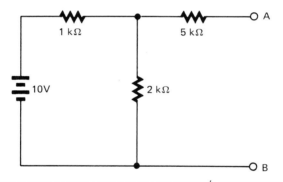

(A) CIRCUIT ILLUSTRATING THE USE OF THÉVENIN'S THEOREM

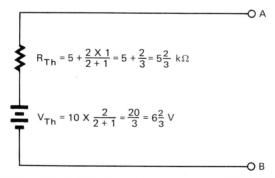

(B) THÉVENIN EQUIVALENT CIRCUIT FOR THE CIRCUIT DESCRIBED IN (A)

Fig. A-8 Thévenin's Theorem

The Thévenin equivalent resistance is determined by removing all voltage and current sources and replacing them by their internal resistances. Thus, the Thévenin equivalent resistance is the resistance looking into terminals A and B with no load connected across terminals A and B.

An example illustrating the use of Thévenin's theorem is given in figure A-8.

Norton's Theorem

Norton's theorem states that any circuit composed of any number of voltage sources, current sources, and resistances may be represented by a simple parallel circuit consisting of a single current source and a single resistance. Figure A-9 is an example of a Norton equivalent circuit. The voltage across a load resistance is the same whether the load is connected between terminals A and B in the Norton equivalent circuit or is connected across terminals A and B in the original complex circuit.

The Norton equivalent current is the current through a conductor (having zero resistance) connected between terminals A and B in the original circuit. The Norton equivalent resistance is obtained in the same way as the Thévenin equivalent resistance.

Using figure A-8(a) as an example, the resulting Norton equivalent circuit is shown in figure A-9(b). The Norton equivalent current is:

$$I_N = \left(\frac{10}{1 + \frac{5 \times 2}{5 + 2}} \right) \left(\frac{2}{5 + 2} \right) = 1.176 \text{ milliamperes}$$

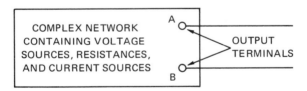

(A) COMPLEX CIRCUIT

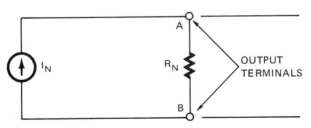

(B) NORTON EQUIVALENT CIRCUIT OF (A)

Fig. A-9 Use of Norton's Theorem

The Norton resistance is:

$$R_N \ = \ R_{Th} \ = \ 5.67 \text{ kilohms}$$

Superposition Theorem

The superposition theorem states that any circuit containing more than one power source can be analyzed by assuming that only one power source is being applied at any given time. The other power sources are replaced by their internal resistances for the analysis. The circuit is analyzed for each power source contained in the circuit. The analysis is then completed for a branch of the circuit by summing up the contributions of each power source to obtain the total currents or voltages in that branch. For example, assume that it is required to determine the total current through the 2-kilohm resistor in figure A-10. It is assumed first that only the 10-volt source produces current. The 5-volt source is replaced by its internal resistance, which is zero ohms. The contribution to the current through the 2-kilohm resistor from the 10-volt source is:

$$I_{10V} \ = \ \frac{10\left[\dfrac{(2)(5)}{2 \ + \ 5}\right]}{1 \ + \ \dfrac{(2)(5)}{2 \ + \ 5}} = \ 2.94 \text{ milliamperes}$$

The 10-volt source is then replaced by its internal resistance which is also zero ohms. The contribution to the current through the 2-kilohm resistor from the 5-volt source is:

$$I_{5V} \ = \ \frac{5\left[\dfrac{(2)(1)}{2 \ + \ 1}\right]}{\dfrac{(2)(1)}{2 \ + \ 1} \ + \ 5} = \ 0.294 \text{ milliamperes}$$

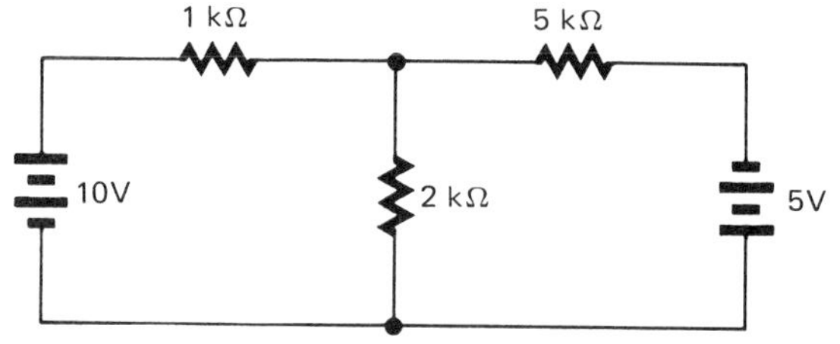

Fig. A-10 Circuit for illustrating the use of superposition

The total current is:

$$I_{2K} = I_{10V} - I_{5V} = 2.94 - 0.294 = 2.646 \text{ milliamperes}$$

The two currents subtract because the voltage sources are opposite in polarity. Frequently the actual direction of flow of the total current is unknown. The current direction is found by determining the direction for each contributing current individually. The total current is the algebraic sum of all of the contributing currents in that branch. It is important that the direction of each individual contributing current be retained through the entire calculation, so that the proper direction can be determined for the total current. The total voltage across the 2-kilohm resistor can be calculated by multiplying the total current by 2 kilohms.

Network analysis techniques are available for more complex circuits. These techniques are not covered here because they are not required for the analysis of any of the circuits in this text. The more sophisticated techniques may require that computer programs be used in the solution. Regardless of the complexity of the solutions, they all can be reduced to algebraic manipulation and Ohm's law. References that present circuit analysis techniques more completely should be consulted, as well as programming information on the type of computers available to the student.

EQUATIONS FOR ELECTRIC POWER

Current passing through an electrical device causes the device to heat. This indicates that power is being taken from the power source and is being absorbed in the device. Power is transferred to any device that has resistance. The power is expressed by:

$$P = I^2R \qquad \text{Eq. A.12}$$

or

$$P = IV \qquad \text{Eq. A.13}$$

or

$$P = \frac{V^2}{R} \qquad \text{Eq. A.14}$$

where V is the voltage across the device, I is the current through it, and R is its resistance.

BEHAVIOR OF CAPACITORS AND INDUCTORS IN A CIRCUIT

Capacitors

A capacitor is a device that temporarily stores electrical charges. The water analogy that was used in the discussion on resistance can also be applied to capacitance.

Consider a faucet that is discharging water into a pail. The pail has a small leak in it. However, the pail can be filled with water and remain full for a period of time. The capacity of the pail is a measure of how much water it holds and may be expressed in gallons or liters.

The capacitor can be filled with electrical charges by applying a voltage across it, just as a pail is filled with water by applying water pressure from the faucet. The capacitor will leak slightly and the charges will gradually drain off, but it can remain fully charged for a relatively long period of time. The measure of how many charges a capacitor holds is known as *capacitance* and is expressed in farads. A more popular unit of measure is the microfarad (μF). The equation describing the behavior of a capacitor is:

$$C = \frac{Q}{V}$$

Eq. A.15

where C is the capacitance in farads, Q is the electrical charge in coulombs, and V is the voltage in volts.

In its simplest form, a capacitor consists of two conductors separated by an insulator. The charges are stored at the junctions between the insulator and the conductors. The behavior of a capacitor in a circuit is explained by the equation:

$$i = C\frac{\Delta v}{\Delta t}$$

Eq. A.16

where i is the current in amperes, v is the voltage across the capacitor in volts, and t is the time in seconds.

Inductors

While a capacitor is a device that stores charges, an inductor is a device that stores a magnetic field. In other words, this device has an ability to retain a magnetic field for a certain length of time after it is once established. In its simplest form, an inductor consists of one or more coils wrapped around a core of magnetic (or even nonmagnetic) material. The ability of an inductor to store a magnetic field temporarily can be expressed as the *inductance* of the device. Once the magnetic field is established, the inductor contains stored magnetic energy which can be used to perform work or produce a useful voltage potential, for a short period of time.

An analogy of the behavior of an inductor is not as readily made as for the capacitor. However, one can be made. Consider a long, coiled garden hose connected at one end to a faucet. The faucet is turned on and the water pressure from the municipal water system is applied to one end of the hose. Very little water comes out of the other end of the hose initially. Eventually, more water flows out of the hose. What happens if the faucet is closed suddenly? Most faucets are closed slowly under normal conditions and there is enough air in the water system to prevent violent water pressure changes. However, if the faucet is closed suddenly and there is not enough air in the water system, the flowing water will strike the closed faucet and other parts of the plumbing. The water strikes with such force that there is a loud noise and a vibration of the pipes and the faucet. A plumber calls this condition *water hammer*. This condition can occur when there are electrically operated valves on humidifiers and automatic washers and they are improperly installed.

The inductor is usually in the form of a coil. When a voltage is applied across the coil, initially there is only the voltage (pressure) and very little current (movement of charges). As the magnetic inertia is overcome, the current increases (more movement of charges). If the switch that applies the voltage is suddenly opened so that current flow is prevented, the coil tends to continue the flow of current. As a result, the stored energy of the coil continues to discharge if there is a path available to complete the circuit. Often this current tries to make its own path by arcing across the switch contacts. The water flow and water hammer situation in the water system is analogous to the current and the arcing in the electric system.

The behavior of the coil in a circuit can be explained by the use of the equation

$$e = L\frac{\Delta i}{\Delta t}$$

Eq. A.17

where e is the voltage across the coil, L is the inductance in henries, i is the current in amperes, and t is the time in seconds.

Time Constants and Rise Time

The behavior of capacitors and inductors can be examined by connecting them in simple circuits such as that of figure A-11, page 558. The transient voltages and currents in the circuit are measured using a memory-type oscilloscope or oscillograph. As shown in figure A-12, page 558, an ordinary oscilloscope can be substituted if a square wave generator is used to supply the transient. The response of the circuit to a sudden surge of applied electrical energy can be measured. A typical response is shown by the graph of figure A-13, page 559, for the voltage across the capacitor, figure A-11(a), plotted as a function of time.

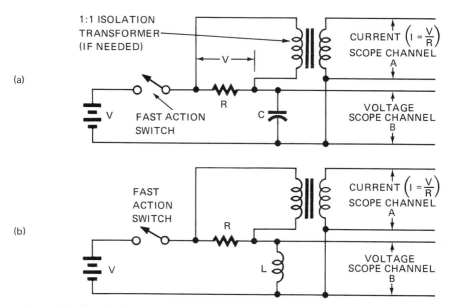

Fig. A-11 Circuits for demonstrating the behavior of inductors and capacitors

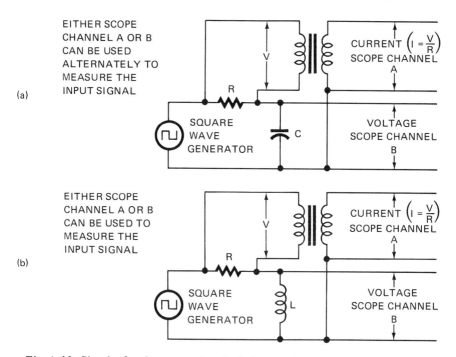

Fig. A-12 Circuits for demonstrating the behavior of inductors and capacitors

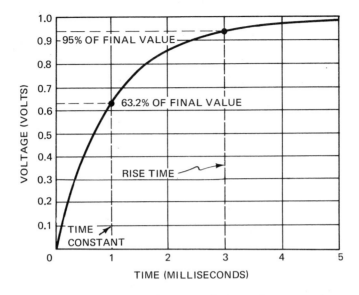

Fig. A-13   Capacitor voltage as a function of time, after a step input to the circuit of figure A-11(a)

Two criteria for examining this graph are the rise time and the time constant. *Rise time* is the time required for the response to reach 95% of its final value. The *time constant* is the time at which the response reaches 63.2% of its final value. The time constant is calculated using the equation

$$T = RC \qquad\qquad \text{Eq. A.18}$$

for the circuit in figure A-11(a). The time constant is the more useful of the two criteria. Frequently, the response is expressed in terms of the level reached after one time contant, two time constants, three time constants, or so on. The time constant concept is very useful in the analysis of the effects of inductance and capacitance in a circuit (Refer to Chapter 11).

The time constant concept is also used where the response drops to a value near zero as a function of time. This type of response is shown by plotting the voltage across the inductance in figure A-11(b) and the current through the capacitor in figure A-11(a), as a function of time. This response is shown in figure A-14, page 560. In this case, the time constant is the time required for the voltage level to reach 36.8% of the initial value. The time constant is expressed by the equation

$$T = \frac{L}{R} \qquad\qquad \text{Eq. A.19}$$

for the circuit of figure A-11(b).

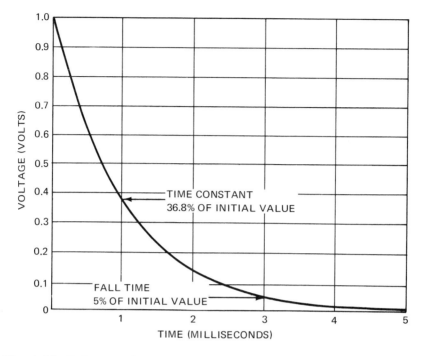

Fig. A-14   Inductor voltage as a function of time, after a step input, figure A-11(b)

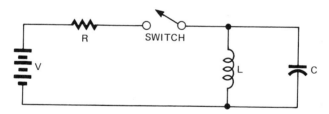

Fig. A-15 Capacitor and inductor together in one circuit

Resonance

A circuit can have more than one time constant. The number of time constants depends upon how many capacitors and/or inductors are present in the circuit and how these devices are connected. In the circuit in figure A-15 for example, there is a possibility of two time constants: one referring to an increasing signal (due to the capacitor) and one referring to a decreasing signal (due to the inductor). A condition known as *resonance* may result. The following paragraph explains this condition.

When the circuit is excited by some external voltage source, energy is absorbed from this source to store electrical field energy in the capacitor and magnetic field energy in the inductor. According to Equation A.16, the capacitors can produce current according to the change in voltage across them. Similarly, Equation A.17 shows that inductors generate voltage according to the change in current through them. If the external voltage source is removed, there is a continual transfer of energy back and forth between the capacitor and the inductor. An induced voltage across the inductor will cause a charging or discharging of the capacitor. The current produced by the charging and discharging of the capacitor causes a magnetic field in the inductor to build up or collapse. The two devices will transfer energy between each other until the energy is dissipated by heating through the inherent resistance of the devices and the connections between them. If the resistance is low, the current and the voltage in the circuit dissipate as exponentially damped sinewaves. The frequency of the sinewave is approximately:

$$f = \frac{1}{2\pi\sqrt{LC}}$$
Eq. A.20

This ability of capacitors and inductors to resonate in such a manner is a basic feature in the construction of oscillators, bandpass filters, and tuned amplifiers.

STEADY-STATE SINUSOIDAL ANALYSIS OF AC CIRCUITS

Capacitors and inductors perform according to their basic behavioral characteristics regardless of the wave shape of the signal applied to the circuits in which they are used. In power circuits and many communication circuits, the basic exciting waveform is sinusoidal. Signals composed of sinusoidal waveforms have the ability to pass through circuits containing capacitors and/or inductors without being changed in shape. If the signals are sinusoidal entering the circuit, they are sinusoidal leaving the circuit. As a result, the behavior of these circuits can be described by a sinusoidal input signal. It is convenient to describe this behavior based on a sinusoidal waveform of constant amplitude and frequency which is applied for a very long period of time. Basically, a steady-state analysis of the operation is produced. Some basic definitions and mathematical background are required for steady-state analysis. The definitions are as follows:

Capacitive reactance

$$X_C = \frac{1}{2\pi fC} \quad \text{or} \quad \frac{1}{\omega C} \text{ ohms}$$
Eq. A.21

Inductive reactance

$$X_L = 2\pi fL \quad \text{or} \quad \omega L \text{ ohms}$$
Eq. A.22

where f is the frequency in hertz. The reactances are combined vectorially with resistances in the circuit to form an impedance, Z. This impedance can be written using the j-operator to denote that it involves a vectorial addition or subtraction rather than an arithmetic process. The relationship for impedance with inductive reactance is:

$$Z = R + jX_L \qquad\qquad \text{Eq. A.23}$$

Impedance with capacitive reactance is expressed as:

$$Z = R - jX_C \qquad\qquad \text{Eq. A.24}$$

Impedance with both inductive and capacitive reactance is written:

$$Z = R + j(X_L - X_C) \qquad\qquad \text{Eq. A.25}$$

The j-operator in these expressions is merely a mathematical symbol used to remind the analyst that the reactive term is $90°$ out of phase with the resistance term. A vector representation of the impedance is shown in figure A-16 for the impedance with capacitive reactance and in figure A-17 for the impedance with inductive reactance. The phase relationship is due to the inherent characteristics of a capacitor and an inductor. A voltage applied across a capacitor results initially in a short across the capacitor. Then, as the capacitor charges, the voltage across it builds up. Assume a circuit like the one in figure A-18. The voltage across the capacitor lags *behind* the voltage developed across the resistance. This is the reason for the minus sign in Equation A.24 which shows that the capacitive reactance lags behind the resistance. The waveforms of the voltage across the resistor and the voltage across the capacitor are shown in figure A-19.

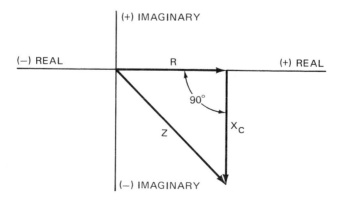

Fig. A-16 Vector representation of impedance, capacitive reactance predominating

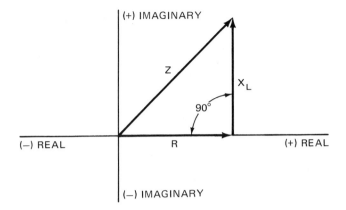

Fig. A-17 Vector representation of impedance, inductive reactance predominating

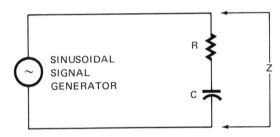

Fig. A-18 Series RC circuit with sinusoidal signal applied

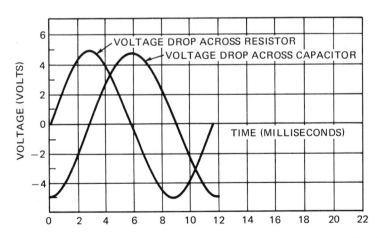

Fig. A-19 Waveforms of voltages across the resistor and capacitor in the circuit of figure A-18

A similar circuit for the inductor is shown in figure A-20. The *current* through the inductor lags behind the voltage applied to it. Voltage does not develop across the resistor until the current begins to flow in the inductor. It can be said that the voltage across the inductor *leads* the voltage across the resistor. The impedance is now expressed by Equation A.23. It is apparent that the inductive reactance appears 90° ahead of the resistor, resulting in a positive sign for the j-term in the impedance. The waveforms of the voltage across the resistor and the voltage across the inductor are shown in figure A-21.

Impedance is a vector quantity. An impedance in an ac circuit affects not only the magnitude of the currents and voltages, but also the phase or

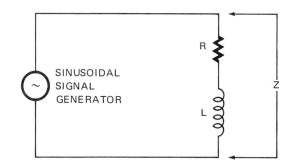

Fig. A-20 Series RL circuit with sinusoidal signal applied

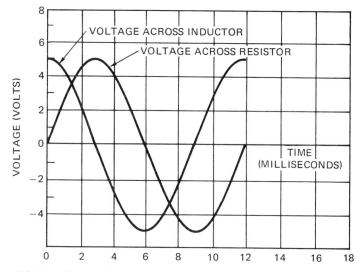

Fig. A-21 Waveforms of voltages across the resistor and inductor in the circuit of figure A-20

time relationships between them. When analyzing ac circuits these phase relationships must be considered throughout the calculations. The resulting numbers are known as complex numbers. These numbers have phase (or time) relationships as well as magnitudes. The proper calculation of the equivalent parallel and series impedances and the currents and voltages for ac circuits requires the analyst to be familiar with the manipulation of complex numbers.

ALGEBRA FOR AC CIRCUIT ANALYSIS; THE USE OF COMPLEX NUMBERS

Complex numbers can be expressed in either one of two forms. The first form, mentioned previously, consists of a *real* value and an *imaginary* value. Equations A.23 and A.24 show rectangular forms of complex numbers. The resistance is the real value and the j-term or reactance is the imaginary value. A complex number can also be expressed in a *polar* form. That is, after the vectorial mathematics are completed, the result is a number and an angle. The number is derived from the expression:

$$|Z| = \sqrt{R^2 + X^2} \qquad \text{Eq. A.26}$$

where $|Z|$ is the magnitude only of the impedance. The reactance symbol (X) is used without the C or L subscript because the equation holds true for both inductive and capacitive reactance. The angle is the phase relationship between the resistance and the reactance. The phase is expressed as an angle between $0°$ and $360°$ and is given by:

$$\text{Phase Angle} = \arctan \frac{X}{R} \qquad \text{Eq. A.27}$$

An example of the impedance expressed in the polar form is the case where R = 3 kilohms and X = 4 kilohms inductive. The impedance is expressed in rectangular form as

$$Z = 3 + j4 \text{ kilohms}$$

In the polar form, the impedance is:

$$Z = \left(\sqrt{3^2 + 4^2}\right)\left(\arctan \frac{4}{3}\right) = 5\underline{/53.1°} \text{ kilohms}$$

The value is read as "5 kilohms at an angle of 53.1 degrees."

If R = 3 kilohms and X is 1/1000 of the original assumed value (perhaps due to a frequency which is 1/1000 of the original frequency), then the impedance is:

$$Z = 3 + j0.004 \approx 3 + j0$$

This value is 3 kilohms of almost pure resistance. In polar form, the impedance is:

$$Z \approx 3\underline{/0^\circ} \text{ kilohms}$$

If X = 4 kilohms and R is reduced to one-thousandth of the original value, the impedance is:

$$Z = 0.003 + j4 \approx j4$$

which is 4 kilohms of almost pure reactance. The polar form of the impedance is:

$$Z \approx 4\underline{/90^\circ}$$

Both the rectangular and polar forms of the impedance are used in calculations, depending upon which form is the most convenient at the time.

Rules for Calculations Using Complex Numbers— Rectangular Form Calculations

A number of rules must be followed when performing computations with complex numbers expressed in the rectangular form.

1. The product of two imaginary numbers is a negative real number because $j = \sqrt{-1}$ and $(\sqrt{-1})(\sqrt{-1}) = -1$. For example, multiply j4 by j3:

$$(j4)(j3) = -12$$

2. The product of a real number and an imaginary number is an imaginary number. For example, multiply j3 by 4:

$$(j3)(4) = j12$$

3. The division of a complex number by a real number is a complex number. For example, divide j12 by 3:

$$\frac{j12}{3} = j4$$

4. To add complex numbers, the real parts and the imaginary parts are added separately. For example, add the two complex numbers 1 + j2 and 3 + j4:

$$(1 + j2) + (3 + j4) = 4 + j6$$

5. To subtract complex numbers, the real numbers and the imaginary numbers are subtracted separately. For example, subtract $1 + j2$ from $3 + j4$:

$$(3 + j4) - (1 + j2) = (3 - 1) + (j4 - j2) = 2 + j2$$

6. The product of two complex numbers is treated like the product of two binomials. For example, multiply $1 + j2$ by $3 + j4$:

$$(1 + j2)(3 + j4) = (1)(3) + (3)(j2) + (1)(j4) + (j2)(j4)$$

$$= 3 + j6 + j4 - 8 = -5 + j10$$

A minus sign in front of the real number means that the vector is in the second quadrant of the complex plane. In figure A-22, the complex plane is shown with the location of the complex number $-5 + j10$. This means that the angle is 180 minus the actual angle obtained by taking the arctangent of 10/5. The polar form of the product of $1 + j2$ and $3 + j4$ is:

$$11.18\underline{/180 - 63.4} = 11.18\underline{/116.6°}$$

7. To divide two complex numbers, the numerator and denominator are multiplied by a complex number that eliminates the imaginary number in the denominator. For example, divide $3 + j4$ by $1 + j2$. To do this, multiply and divide by the complex *conjugate* of one of the two terms. This term is either $3 - j4$ or $1 - j2$. The term is selected arbitrarily. It is selected in this example to be $1 - j2$. By multiplying and dividing by $1 - j2$, the resulting expression is:

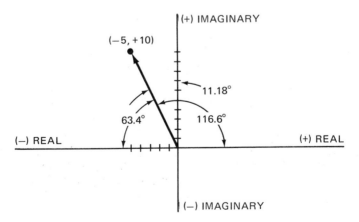

Fig. A-22 Plot of the complex number $-5 + j10$ on the complex plane

$$\frac{3 + j4}{1 + j2} = \left(\frac{3 + j4}{1 + j2}\right)\left(\frac{1 - j2}{1 - j2}\right) = 0.6 - j2$$

Polar Form Calculations

Frequently, it is more convenient to perform computations when the complex numbers are expressed in the polar form. When performing computations using the polar form, the following rules are observed:

1. The product of two complex numbers is obtained by multiplying the magnitudes and adding the angles algebraically. For example, multiply $1 + j2$ and $3 + j4$, using the polar form. In polar form, $1 + j2$ is $2.24\underline{/63.5°}$ and $3 + j4$ is $5\underline{/53.1°}$.

$$(2.24\underline{/63.5°})(5\underline{/53.1°}) = 11.2\underline{/116.5°}$$

This answer is the same as the answer obtained by multiplying the rectangular forms together. If one of the angles is negative, then the product is the difference of the two angles.

2. To divide two complex numbers, divide one magnitude by the other and subtract one angle from the other. For example, $5\underline{/53.1°}$ divided by $2.24\underline{/63.4°}$ is:

$$\frac{5\underline{/53.1°}}{2.24\underline{/63.4°}} = 2.23\underline{/-10.3°}$$

The rectangular form of $2.23\underline{/-10.3°}$ is $0.6 - j2$. This is the same number as that obtained by dividing the rectangular forms of the two complex numbers.

It may be necessary on occasion to change from the polar form to the rectangular form. In this case, the real number is the magnitude in the polar form times the cosine of the angle. The imaginary number is the magnitude of the polar form times the sine of the angle. For example, assume that the impedance $5\underline{/53.1°}$ is to be changed to the rectangular form.

Real number = $(5)(\cos 53.1°) = 3$
Imaginary number = $(5)(\sin 53.1°) = 4$
Therefore $5\underline{/53.1°} = 3 + j4$

If the angle is greater than $90°$, some rules of trigonometry are simplified if the complex plane of figure A-20 is sketched and the angle is measured counterclockwise from the horizontal line to the right of the origin. For example:

$3 + j4$ is in the first quadrant, and the polar form is $5\underline{/53.1°}$;

$-3 + j4$ is in the second quadrant and the polar form is $5\underline{/180° -53.1°} = 5\underline{/126.9°}$;

$-3 - j4$ is in the third quadrant and the polar form is $5\underline{/180° + 53.1°} = 5\underline{/233.1°}$;

$3 - j4$ is in the fourth quadrant and the polar form is $5\underline{/-53°}$, or $5\underline{/307°}$.

POWER IN AC CIRCUITS

For circuits which are subjected to sinusoidal signals, the power equations must account for the phase relationship between the current and voltage. The standard equation for ac power is:

$$P = VI \cos \phi \qquad \text{Eq. A.28}$$

where ϕ is the phase angle between the current and the voltage. Generally, this angle is also the phase angle between the resistance and the reactance of a circuit, as described by Equation A.27.

Equation A.12 can also be used for ac circuits:

$$P = I^2 R$$

If the instantaneous power is to be calculated, it can also be used for nonsinusoidally excited circuits. In this case, the actual function of current with respect to time is used. The resistance must be considered constant regardless of the signal amplitude. The average power for a symmetrical waveform over a period of time can be determined by adding the instantaneous powers calculated using Equation A.12.

SYSTEMS OF UNITS

This text uses the meter-kilogram-second (MKS) system throughout. The MKS system is popular and the units of voltage, current, resistance, magnetic flux, capacitance, inductance and other quantities used in the text are based on this system of units. However, it is wise to memorize several key conversion factors that can be applied whenever a different system of units (such as the English system) is used. Common conversion factors are:

$$2.54 \text{ centimeters} = 1 \text{ inch}$$

$$\text{milli-} \quad = \frac{1}{1000} \text{ of a unit}$$

$$\text{meg-} \quad = 1{,}000{,}000 \text{ units}$$

$$\text{kilo-} \quad = 1000 \text{ units}$$

$$\text{micro-} \quad = \frac{1}{1{,}000{,}000} \text{ of a unit}$$

Examples:

$$1 \text{ millisecond} = \frac{1}{1000} \text{ second} = 0.001 \text{ second}$$

$$1 \text{ megohm} \quad = 1{,}000{,}000 \text{ ohms}$$

It is recommended that the student concentrate on the use of milliamperes, kilohms, and volts in calculations, rather than amperes, ohms, and volts.

MATHEMATICS SYMBOLS

A number of commonly used mathematics symbols may not be completely familiar to the user of this text. The following list gives these symbols and the words for which they are substituted. In the column headed "meaning," the words enclosed by quotation marks can be substituted directly for the symbol where it appears in the text.

Symbol	Meaning
$=$	"Equals," "is equal to"
$\approx$	"Approximately equals"
Δ	"A change in"
$\|\ \|$	Vertical bars on either side of a number indicate only the absolute value or magnitude of the number.
α	Greek letter *alpha*
β	Greek letter *beta*
μ	Greek letter *mu*
ϕ	Greek letter *phi*
π	Greek letter *pi*
ρ	Greek letter *rho*
τ	Greek letter *tau*
$<$	"Is less than" or "less than"
$>$	"Is more than" or "more than"
$\ll$	"Is very much less than"
$\gg$	"Is very much greater than"
j	The imaginary number, defined algebraically as $\sqrt{-1}$
$\neq$	"Is not equal to"
∞	Infinity, or so large as to be beyond the bounds of ordinary numbers
$-$	"minus"
$+$	"plus"
$\%$	"percent," consisting of a fraction multiplied by 100.
$\&$	"and"
$@$	"at"
$f(\ )$	"a function of" (the terms in the parentheses)

SYMBOLS BY WHICH VARIOUS PARAMETERS, VOLTAGES, AND CURRENTS ARE REPRESENTED

The following is a list of common symbols.

$$r, R \;=\; \text{resistance}$$
$$v, V \;=\; \text{voltage}$$
$$i, I \;=\; \text{current}$$
$$Z \;=\; \text{impedance}$$
$$X \;=\; \text{reactance}$$
$$Y \;=\; \text{admittance}$$
$$b \;=\; \text{susceptance}$$
$$g \;=\; \text{conductance}$$

In general, lower case letters represent instantaneous quantities and capital leters represent dc or average quantities.

Subscripts are used to designate locations in the circuit for the quantities that are being described. The subscripts may be letters, if the quantities can be referenced to a particular named terminal in the circuit. Alternatively, the subscripts may be numbered, using any number between 1 and 10. The schematic must be consulted to specify the exact location of the various quantities in the circuit.

As a rule in this text, once a symbol is defined, the definition is used throughout the text. The common symbols used in circuit analysis are shown in the following lists. Note that the voltages have two subscripts. The first subscript in a voltage symbol indicates a terminal which is positive with respect to the terminal indicated by the second subscript. For example, in the symbol v_{be}, the voltage at the base is positive with respect to the voltage at the emitter. This means that $v_{be} = -v_{eb}$.

Instantaneous values of the ac components of the following quantities	are expressed by the following symbols:
Input current signal	i_{in}
Collector current	i_c
Base current	i_b
Emitter current	i_e
Base-to-emitter voltage	v_{be}
Collector-to-emitter voltage	v_{ce}
Collector-to-base voltage	v_{cb}
X to Z voltage	v_{xz}
X to Y voltage	v_{xy}

Instantaneous values of the ac components of the following quantities	are expressed by the following symbols:
Y to Z voltage	v_{yz}
Input voltage, referenced to ground	v_i or v_{in}
Y current	i_y
Z current	i_z
X current	i_x
Signal generator output voltage, unloaded	v_s
Output voltage	v_o or v_{out}
Drain current	i_d
Source current	i_s
Gate current	i_g
Grid current	i_g
Plate current	i_p
Cathode current	i_k
Current through load	i_L or i_{out}
Drain-to-source voltage	v_{ds}
Gate-to-source voltage	v_{gs}
Drain-to-gate voltage	v_{dg}
Plate-to-cathode voltage	v_{pk}
Grid-to-cathode voltage	v_{gk}
Plate-to-grid voltage	v_{pg}

Instantaneous values of the total of the ac and dc components of the following quantities	are expressed by the following symbols:
Collector current	i_C
Base current	i_B
Emitter current	i_E
Base-to-emitter voltage	v_{BE}
Collector-to-base voltage	v_{CB}
Collector-to-emitter voltage	v_{CE}
X to Z voltage	v_{XZ}

Instantaneous values of the total of the ac and dc components of the following quantities	are expressed by the following symbols:
X to Y voltage	v_{XY}
Y to Z voltage	v_{YZ}
Y current	i_Y
Z current	i_Z
X current	i_X
Drain current	i_D
Source current	i_S
Gate current	i_G
Grid current	i_G
Plate current	i_P
Cathode current	i_K
Drain-to-source voltage	v_{DS}
Gate-to-source voltage	v_{GS}
Drain-to-gate voltage	v_{DG}
Plate-to-cathode voltage	v_{PK}
Grid-to-cathode voltage	v_{GK}
Plate-to-grid voltage	v_{PG}

The dc component of the following quantities	is expressed by the following symbols:
Collector current	I_C
Base current	I_B
Emitter current	I_E
Base-to-emitter voltage	V_{BE}
Collector-to-base voltage	V_{CB}
Collector-to-emitter voltage	V_{CE}
X to Z voltage	V_{XZ}
X to Y voltage	V_{XY}
Y to Z voltage	V_{YZ}
Y current	I_Y
Z current	I_Z

The dc component of the following quantities	is expressed by the following symbols:
X current	I_X
Drain current	I_D
Source current	I_S
Gate current	I_G
Grid current	I_G
Plate current	I_P
Cathode current	I_K
Drain-to-source voltage	V_{DS}
Gate-to-source voltage	V_{GS}
Drain-to-gate voltage	V_{DG}
Plate-to-cathode voltage	V_{PK}
Grid-to-cathode voltage	V_{GK}
Plate-to-grid voltage	V_{PG}

Root-mean-square (rms) value of the ac components of the following quantities	is expressed by the following symbols:
Collector current	I_c
Base current	I_b
Emitter current	I_e
Base-to-emitter voltage	V_{be}
Collector-to-emitter voltage	V_{ce}
Collector-to-base voltage	V_{cb}
Input voltage, referenced to ground	V_{in}
Signal generator output voltage, unloaded	V_s
Output voltage	V_o or V_{out}
Drain current	I_d
Source current	I_s
Gate current	I_g
Grid current	I_g
Plate current	I_p

Root-mean-square (rms) value of the ac components of the following quantities	is expressed by the following symbols:
Cathode current	I_k
Current through load	I_{out}
Drain-to-gate voltage	V_{dg}
Drain-to-source voltage	V_{ds}
Gate-to-source voltage	V_{gs}
Plate-to-cathode voltage	V_{pk}
Grid-to-cathode voltage	V_{gk}
Plate-to-grid voltage	V_{pg}

Dc values of the following quantities at the Q (quiescent)-point	are expressed by the following symbols:
Collector-to-emitter voltage	V_{CEQ}
Drain-to-source voltage	V_{DSQ}
Gate-to-source voltage	V_{GSQ}
Drain current	I_{DQ}
Plate-to-cathode voltage	V_{PKQ}
Grid-to-cathode voltage	V_{GKQ}
Plate current	I_{PQ}
Base-to-emitter voltage	V_{BEQ}
Collector-to-base voltage	V_{CBQ}
Base current	I_{BQ}
Collector current	I_{CQ}

Table 4-1 lists the subscripts to be used for the h-parameters for specific transistor circuits. The first subscript is defined as follows:

i (as in h_{ie}) means input impedance.

o (as in h_{oe}) implies output impedance, to which it is inversely related.

f (as in h_{fe}) stands for "forward" gain. This is the forward current gain of the circuit.

r (as in h_{re}) stands for "reverse" gain. This is the reverse voltage gain of the circuit.

The second subscript is defined as follows:

e (as in h_{fe}) for the transistor circuit where the emitter is common.

c (as in h_{fc}) for the transistor circuit where the collector is common.

b (as in h_{fb}) for the transistor circuit where the base is common.

These subscripts are used in the same manner for the y-parameters. For example, y_{fs} means the forward current gain for the circuit having the source terminal common to both the input and output. A table similar to Table 4-1 can be made for the y-parameters, as follows:

Y-Parameter		Common Gate	Common Source	Common Drain
y_{11}	input admittance	y_{ig}	y_{is}	y_{id}
y_{12}	reverse transfer admittance	y_{rg}	y_{rs}	y_{rd}
y_{21}	forward transfer admittance	y_{fg}	y_{fs}	y_{fd}
y_{22}	output admittance	y_{og}	y_{os}	y_{od}

Reference Designations

Over the years, government contracts with electronics companies have given rise to a system of Reference Designations to identify components on electronic schematics. For example, the reference designation for a resistor on a schematic may be R1 or R2. "R" designates the type of component used. Each resistor in the circuit will have a different number. In this way, a "name" is assigned to every component in the circuit. Common reference designation letters are as follows:

R	resistor or potentiometer
C	capacitor
L	inductor or coil
Q	transistor
T	transformer
K	relay
CR	diode
DS	light, or audible signaling device
M	meter
S	switch
F	fuse
J	connector, on panel
P	connector, on end of wire or cable
W	wire or other transmission path, cable
V	electron tube

COMMON GRAPHIC SYMBOLS USED IN SCHEMATICS

This text observes the use of standard symbols and schematic layout conventions. For each schematic, the input is on the left side and the output is on the right side. Power is supplied from the left and it is delivered on the right. Positive voltage is at the top and the most negative (or least positive) voltage is at the bottom. Common symbols which are used in the text but not previously defined are as follows:

Ammeter

Voltmeter

Generalized amplifying device

Coil with magnetic core

Switch, single pole, single throw

Diode, semiconductor type; this symbol is also sometimes used for a vacuum diode

Diode, vacuum type

Battery; positive terminal at top, negative terminal at bottom of the symbol

Resistor

Variable resistor

Iron core transformer

Coil

Variable coil

Variable capacitor

Junction point, four leads connected

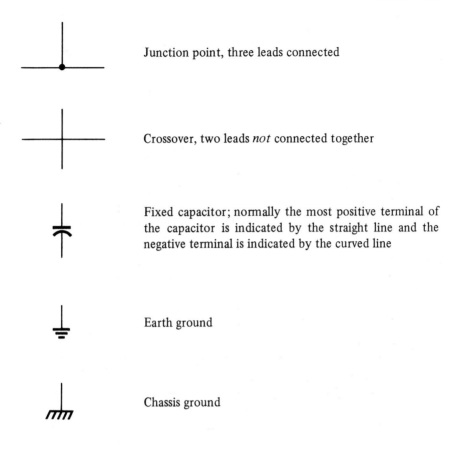

Junction point, three leads connected

Crossover, two leads *not* connected together

Fixed capacitor; normally the most positive terminal of the capacitor is indicated by the straight line and the negative terminal is indicated by the curved line

Earth ground

Chassis ground

GLOSSARY OF TERMS

Note: these terms are not defined elsewhere in the text

Abscissa. The horizontal scale on a graph, usually a plot of the independent variable.

Admittance. The reciprocal of impedance, or the impedance divided into the number one (1).

$$\text{Admittance} = \frac{1}{\text{Impedance}}$$ Eq. A.29

Amplification. A gain which is greater than one.

Attenuation. A gain which is less than one.

Bipolar. Having two poles (a north pole and a south pole, or + side and - side, etc.)

Cascading. When several amplifiers or amplifying stages are connected together to achieve a higher overall gain. The output of one amplifier has the input of another amplifier connected to it as a load.

Conductance. The reciprocal of resistance.

$$\text{Conductance} = \frac{1}{\text{Resistance}}$$ Eq. A.30

Complex plane. A pair of coordinate axes at right angles to one another, usually named X and Y(or j), arranged as in figure A-23. Values to the right along the X-axis are positive and values to the left along the X-axis are negative. Values ascending vertically on the Y-axis are positive and those descending vertically on the Y-axis are negative.

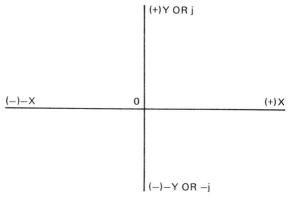

Fig. A-23 Complex plane

Decibel. A unit that expresses gain on a logarithmic scale. The actual definition is based upon whether the gain is to be in voltage decibels, power decibels, or current decibels.

$$\text{Power Decibels} = 10 \log \frac{\text{Power Out}}{\text{Power In}} \qquad \text{Eq. A.31}$$

$$\text{Voltage Decibels} = 20 \log \frac{\text{Voltage Out}}{\text{Voltage In}} \qquad \text{Eq. A.32}$$

$$\text{Current Decibels} = 20 \log \frac{\text{Current Out}}{\text{Current In}} \qquad \text{Eq. A.33}$$

Digital. As in "digital logic" or "digital system." The term refers to a type of circuit that uses a set of coded 0 and 1 or "on" and "off" modes to convey information or to control processes.

Gain. The ratio of the output of a device to its input. The ratio may be less than one or more than one.

Ground. The point in an active circuit at which there is zero voltage to a metal stake driven into the earth. *Chassis ground* refers to the use of the metal case of a device as the "earth."

Harmonic. In electrical technology, this is a high-frequency sinusoidal component of a nonsinusoidal waveform. The lower, more dominant frequency of the waveform is referred to as the *fundamental.*

Intercept. Meet with, cross over, or pass by.

Interelectrode. Between electrodes.

Linear. A relationship in which two variables are related by a constant factor. In a linear amplifier, output over input is a constant factor.

Microwave. A high-frequency radio signal having a very short wavelength, generally less than one meter.

Mobile. Having movement, not staying in one place.

Noise. In electrical usage, any unwanted signal at the output of an amplifier or signal generator.

Nonlinear. Two variables are related to several interrelated factors, or a single common factor that is changing.

Optimum. Conditions under which the best results can be achieved.

Ordinate. The vertical scale on a graph, usually used for the dependent variable of a function.

Quiescent. At rest, inactive.

Relay. A switch which is opened and closed by electromagnetic action. Voltage is applied across a coil which is wrapped around a magnetic core. The resulting magnetic force then moves one or more sets of contacts with respect to one another.

Resistance. A property of an electrical device that restricts current through it. Resistance is defined as:

$$R = \frac{\rho L}{A}$$

Eq. A.34

where ρ = resistivity, a property of the materials of which the device is made, L = the electrical length of the device between its terminals, and A = cross-sectional area of the device perpendicular to the flow of current.

Resistivity. The property of a material that tends to prevent or restrict the flow of electric current. Resistivity is known as specific resistance.

Root-mean-square or rms. The form of signal monitored by ac meters. Mathematically, the rms number is derived by squaring a group of readings, adding the squared numbers together, dividing by the total number of readings, and taking the square root of the result. In sinusoidal circuit analysis, the rms value is converted to the peak value of the waveform by multiplying by $\sqrt{2}$.

Shunt. A conductor or small resistance joining two points in a circuit. The shunt tends to divert part of the current.

Slope. In this text, the slope is the inclination of a line on a graph. It is a change in values along the ordinate resulting from a corresponding change in values along the abscissa. The slope is also the tangent of the angle formed by a vector along the complex plane, measured counterclockwise from the +X axis.

Stage. Section of an amplifier utilizing a single amplifying device.

Substrate. Base of a semiconductor crystal on which an integrated circuit or special transistor is constructed.

Susceptance. The reciprocal of reactance.

$$\text{Susceptance} = \frac{1}{\text{Reactance}}$$

Eq. A.35

Switch. Device which opens or closes an electrical circuit.

Symmetrical. In reference to an electrical signal, this term means that it has the same wave shape whether it is negative or positive with respect to ground; that is, each half-cycle has the same wave shape.

Temperature. Condition of hot or cold expressed as

Degrees Kelvin, where K = -273°C

Degrees Celsius, where 0°C equals the freezing point of water and 100°C equals the boiling point of water.

The Celsius reading is related to degrees Fahrenheit by:

$$°C = \frac{5}{9} (°F - 32)$$

Degrees Fahrenheit, where 32°F equals the freezing point of water and 212°F equals the boiling point of water. The Fahrenheit reading is related to degrees Celsius by the expression:

$$°F = \frac{9}{5} (°C) + 32$$

Acknowledgments

Source Editor: Marjorie A. Bruce
Consulting Editor: Richard L. Castellucis
Southern Technical Institute
Marietta, Georgia
Technical Reviewer: J. M. Jacob
Florence-Darlington Technical College
Florence, South Carolina
Technical Editor: Jeffrey B. Duncan
Editorial Assistant: Mary V. Miller

The author would like to express his appreciation to the electrical technology students at Milwaukee Area Technical College who have used the basic content and experiments of this text and have assisted the author in verifying its effectiveness in presenting the basic concepts of amplifiers.

Index